Probability and Mathematical Statistics (Continued)

WILLIAMS • Diffusions, Markov Processes, and Martingales, Volume I: Foundations
ZACKS • Theory of Statistical Inference

Applied Probability and Statistics

ANDERSON, AUQUIER, HAUCK, OAKES, VANDAELE, and WEISBERG • Statistical Methods for Comparative Studies
ARTHANARI and DODGE • Mathematical Programming in Statistics
BAILEY • The Elements of Stochastic Processes with Applications to the Natural Sciences
BAILEY • Mathematics, Statistics and Systems for Health
BARNETT and LEWIS • Outliers in Statistical Data
BARTHOLOMEW • Stochastic Models for Social Processes, *Second Edition*
BARTHOLOMEW and FORBES • Statistical Techniques for Manpower Planning
BECK and ARNOLD • Parameter Estimation in Engineering and Science
BELSLEY, KUH, and WELSCH • Regression Diagnostics: Identifying Influential Data and Sources of Collinearity
BENNETT and FRANKLIN • Statistical Analysis in Chemistry and the Chemical Industry
BHAT • Elements of Applied Stochastic Processes
BLOOMFIELD • Fourier Analysis of Time Series: An Introduction
BOX • R. A. Fisher, The Life of a Scientist
BOX and DRAPER • Evolutionary Operation: A Statistical Method for Process Improvement
BOX, HUNTER, and HUNTER • Statistics for Experimenters: An Introduction to Design, Data Analysis, and Model Building
BROWN and HOLLANDER • Statistics: A Biomedical Introduction
BROWNLEE • Statistical Theory and Methodology in Science and Engineering, *Second Edition*
BURY • Statistical Models in Applied Science
CHAMBERS • Computational Methods for Data Analysis
CHATTERJEE and PRICE • Regression Analysis by Example
CHERNOFF and MOSES • Elementary Decision Theory
CHOW • Analysis and Control of Dynamic Economic Systems
CLELLAND, BROWN, and deCANI • Basic Statistics with Business Applications, *Second Edition*
COCHRAN • Sampling Techniques, *Third Edition*
COCHRAN and COX • Experimental Designs, *Second Edition*
CONOVER • Practical Nonparametric Statistics, *Second Edition*
CORNELL • Experiments with Mixtures: Designs, Models and The Analysis of Mixture Data
COX • Planning of Experiments
DANIEL • Biostatistics: A Foundation for Analysis in the Health Sciences, *Second Edition*
DANIEL • Applications of Statistics to Industrial Experimentation
DANIEL and WOOD • Fitting Equations to Data: Computer Analysis of Multifactor Data, *Second Edition*
DAVID • Order Statistics, *Second Edition*
DEMING • Sample Design in Business Research
DODGE and ROMIG • Sampling Inspection Tables, *Second Edition*
DRAPER and SMITH • Applied Regression Analysis, *Second Edition*
DUNN • Basic Statistics: A Primer for the Biomedical Sciences, *Second Edition*
DUNN and CLARK • Applied Statistics: Analysis of Variance and Regression
ELANDT-JOHNSON • Probability Models and Statistical Methods in Genetics
ELANDT-JOHNSON and JOHNSON • Survival Models and Data Analysis

continued on back

Experiments With Mixtures

DESIGNS, MODELS, AND THE ANALYSIS OF MIXTURE DATA

John A. Cornell
University of Florida

JOHN WILEY & SONS
New York • Chichester • Brisbane • Toronto • Singapore

Library of Congress Cataloging in Publication Data:

Cornell, John A 1941–
Experiments with mixtures.

(Wiley series in probability and mathematical statistics)
Includes bibliographies and index.
1. Experimental design. 2. Mixtures—Statistical methods. I. Title.

QA279.C66 519.5′352 80-22153
ISBN 0-471-07916-2

Printed in the United States of America

10 9 8 7 6 5 4 3 2

To Natalie, Johnny, and Ken,
and in memory of my parents

Preface

The primary purpose of this book is to present the fundamental concepts in the design and analysis of experiments with mixtures. The book focuses on the most frequently used statistical techniques and methods for designing, modeling, and analyzing mixture data, as claimed in the literature, and includes appropriate computing formulas and completely worked out examples of each method. Most of the numerical examples were taken from real research situations.

The book is written for anyone who is engaged in planning or performing experiments with mixtures. In particular, research scientists and technicians in the chemical industries, whether or not trained in statistical methods, should benefit from the many examples that are chemical in nature. Several examples have been taken from research activities conducted in areas of food technology, while some examples were provided by research entomologists. Persons who are engaged in applied research in universities, principally from such departments as chemical engineering, chemistry, and statistics, as well as scientists in areas of agriculture such as food science, entomology, and nematology, should find the methods that are presented to be relevant and useful in their research. As a textbook on the subject of mixture experiments, the contents could serve quite nicely as a one-semester course in most applied curricula, or perhaps could supplement the coverage of a two-semester sequence of regression and response surface methodology.

Since this is the first edition, it has been necessary to exercise considerable selectivity in the choice of topics covered. Hence no claim is made that the coverage is exhaustive in either scope or depth. However, it is my feeling that the reader who works through the numerical examples in the middle five chapters (Chapters 2–6) and answers the questions listed at the end of these chapters will achieve a high level working knowledge of the tools that are used by most of the practitioners today who are involved in solving mixture problems.

The mathematical prerequisites have been kept to a minimum. Summation notation is used throughout and some background knowledge of the use of matrices is helpful. A review of matrix algebra is presented in Chapter 7 along with a discussion of the method of least squares for obtaining the parameter estimates in polynomial models. Chapter 7 could also serve as a refresher to readers who wish to review some of the fundamental ideas on the use of matrices in regression analysis. The matrix material has been placed at the end of the book so that the reader with an adequate knowledge of matrices may begin with the subject of mixture experiments in Chapter 1. Almost all of the computations throughout the book were performed on the APL system 360.

The first chapter introduces the subject of mixture experiments with several examples. Some general remarks on response surface methodology are made. An historical perspective of the relevant literature which presented most of the statistical research on mixture experiments is listed. Chapter 2 introduces the original mixture problem where the Scheffé lattice designs and associated polynomial models are applicable. Several numerical examples are provided that help to illustrate the fitting of the polynomial models to samples of mixture data that were collected at the points of the simplex-lattice and simplex-centroid designs.

In Chapter 3 a transformation is made from the system of the dependent mixture components to a system of independent variables. With the independent variables, standard regression procedures are suggested not only for the designing of the experimental runs but for the fitting of model forms as well. The idea of isolating the experimentation to a subregion of interest inside the simplex space where the region may be ellipsoidal or cuboidal in shape is also considered. Process variables, such as cooking time and cooking temperature in the preparation of fish patties, are introduced. Different types of model forms used to measure the influence the process variables could have on the blending characteristics of the components in mixture experiments are presented and discussed.

How the placing of additional constraints on the component proportions can affect the design configuration and the usual interpretation of the model parameters is considered in Chapter 4. Experimental design configurations for use in covering the restricted region of the simplex are mentioned, as are several types of polynomial model forms used for depicting the surface characteristics. Pseudocomponents are introduced, and the use of pseudocomponents rather than the original components is seen to simplify the steps in the design construction and the fitting of models when lower bound constraints are placed on the original component proportions. Some discussion on the design strategy when some or all of the component proportions are subjected to both upper and lower bound constraints is presented along with some suggested modifications that need to be made to interpret the model coefficients in these highly constrained problems. Grouping the components by categories is also studied.

Chapter 5 presents many techniques that are used in the analysis of mixture data. Testing the form of the fitted model, model reduction procedures, and the screening of unimportant components are just some of the topics covered. Investigating the shape characteristics of the surface by the measuring of the slopes of the surface along the component axes is discussed. Combining lattice designs in the mixture components with factorial arrangements of process variables is illustrated with data from a fish patty experiment in which patties that were made from three species of fish are prepared and processed by the three cooking and processing factors.

Alterations made to some of the terms of the Scheffé polynomials as well as the suggestion to use nonpolynomial models to model certain types of phenomena is the theme of Chapter six. Models that are homogeneous of degree one are shown to model additive component effects better than the polynomials. The use of ratios of the components as terms in the model is suggested particularly when relationships between the component proportions are more meaningful than the actual fraction of the mixture each component represents. Standard orthogonal designs, such as standard factorial arrangements, that can be used with independent variables are shown to be useful when working with ratios. Cox's polynomial, which is used for measuring the components effects, is compared with the several forms of Scheffé's polynomials. Two classes of octane-blending models are presented at the end of the chapter, and data are provided to help illustrate the numerical computations that are required to set up the prediction equations. Chapter 6 ends with a list of topics that were not covered here but which hopefully will be discussed in a future edition.

I am extremely grateful to many friends for their help in compiling the material for this work. In particular, I am indebted to Drs. John W. Gorman and R. Lymann Ott, Professors Irving John Good (Virginia Polytechnic Institute and State University) and Andre I. Khuri (University of Florida), with whom I have had the pleasure of working on research problems in mixtures, and to Dr. Hubert M. Hill (Tennessee Eastman Company), who introduced me to the subject of mixture experiments in the middle 1960s. I am very much indebted also to Professor J. Stuart Hunter (Princeton University), who reviewed the initial drafts of this book; his many thoughtful and detailed comments on the style and content were instrumental in its organization. I wish to thank the many authors and various publishers for permission to reproduce their papers and tables and to thank the staff at John Wiley and Sons, particularly Beatrice Shube, for her encouragement in completing this work. Finally, I would like to express my sincere appreciation to Donna Alexander for her excellent typing of the final manuscript.

JOHN A. CORNELL

University of Florida
Gainesville, Florida
August 1980

Contents

Experiments With Mixtures

CHAPTER 1

Introduction

Many products are formed by mixing together two or more ingredients. Some examples are as follows:

1. *Cake formulations* using baking powder, shortening, flour, sugar, and water.
2. *Building construction concrete* formed by mixing sand, water, and one or more types of cement.
3. *Railroad flares* which are the product of blending together proportions of magnesium, sodium nitrate, strontium nitrate, and binder.
4. *Fruit punch* consisting of juices from watermelon, pineapple, and orange.

In each of cases 1–4, one or more product properties is of interest to the manufacturer or experimenter who is responsible for mixing the ingredients. Properties such as (1) the fluffiness of the cake or the layer appearance of the cake where the fluffiness or layer appearance is related to the ingredient proportions; (2) the hardness or the strength (measured in psi's) of the concrete, where the hardness is a function of the percentages of cement and sand and water in the mix; (3) the illumination in foot-candles and the duration of the illumination of the flares; and (4) the fruitiness flavor of the punch which will depend on the percentages of watermelon, pineapple, and orange that are present in the punch. In each of examples 1–4, the measured property of the final product depended on each of the individual ingredients being present in the formulation.

Another reason for mixing together ingredients in blending experiments is to see if there exist blends of two or more ingredients that produce more desirable product properties than is obtainable with the single ingredients individually. For example, let us imagine we have three different gasoline stocks, labeled A, B, and C and that we are interested in comparing the antiknock quality of the three stocks, singly and in combination. In particular we would like to know if there are combinations of the stocks, such as a 50% : 50% blend of $A:B$, or a 33% : 33% : 33% blend of $A:B:C$, or a 25% : 75% blend of $B:C$, which yields a higher antiknock rating than is obtained from using A alone or from using B alone, or C alone. If so, we would probably select the particular blend of two or more gasoline stocks that produces the highest rating, assuming of course that all other factors such as the cost and availability of the blending ingredients remain fixed.

In each of cases 1–4 listed above, it is assumed that the properties of interest are functionally related to the product composition and that, by varying the composition through the changing of ingredient proportions, the properties of the product will vary or change also. From an experimental standpoint, often the reason for studying the functional relationship between the measured property or the measured response (such as the strength of the concrete) and the controllable variables (which in this case are the proportions of the ingredients of cement to sand to water) is to try to determine if some combination of the ingredients can be considered best in some sense. The best ingredient combination for the concrete would be the combination that produced the absolutely strongest concrete without increasing cost. In an attempt to determine the best combination of the ingredients (or combinations if more than one blend produces concrete samples having approximately equally high strengths), often one resorts to trial and error. Other attempts resemble "scattergun" procedures, where a large number of combinations of the ingredients are tried. The procedures can require large expenditures in terms of time and cost of experimentation and in most cases better methods can be employed. Procedures used in screening unimportant mixture ingredients are discussed in Section 5.7. Before we discuss some methods that have been developed for studying functional relations and are referred to as response surface methods, we introduce the general mixture problem.

1.1. THE GENERAL MIXTURE PROBLEM

To formulate our thinking about experiments involving mixtures, we simplify the gasoline-blending example mentioned earlier by considering only two gasoline stocks, which we label fuels A and B. Instead of discussing the antiknock rating, let us assume that the response of interest is the mileage obtained by driving a test car with the fuel where the mileage is recorded in units of the *average number of miles per gallon*. It is known ahead of time that fuel A normally yields 13 miles per gallon and fuel B normally yields only 7 miles per gallon. If the car is tested with

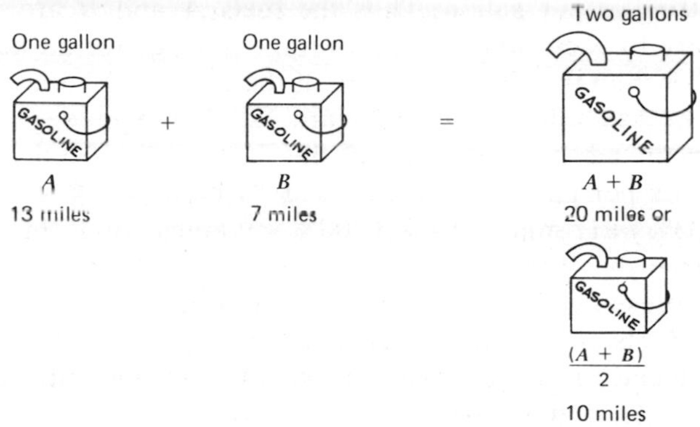

FIGURE 1.1. Summing the miles per gallon of fuels A and B.

each fuel separately by driving with 1 gallon of fuel A and then with 1 gallon of B, we would expect to drive $13+7=20$ miles on the 2 gallons or equivalently, we expect to average $20/2=10$ miles per gallon (Figure 1.1). The question we should like to answer therefore is, "If we combine or blend the two fuels and drive the same test car, is there a blend of A and B such as a 50% : 50% blend or a 33% : 67% blend of $A:B$ that yields a higher average number of miles per gallon than the 10 miles per gallon that was obtained by simply averaging A and B?"

To answer this question, an experiment is performed that consists of driving the test car containing a 50% : 50% blend of fuels A and B. A trial consists of driving the car with 2 gallons of fuel until the fuel is used up. Five trials were performed with the same car and the average mileage was calculated to be 12.0 miles per gallon. (See Table 1.1.)

The average number of miles per gallon for the blend is 12.0 miles per gallon and is higher than the simple average mileage of the two fuels

TABLE 1.1. The average mileage for each of five trials

Trial	Mileage from Two Gallons of 50% : 50% Blended Fuel	Average Mileage per Gallon
1	24.6	12.30
2	23.3	11.65
3	24.3	12.15
4	23.1	11.55
5	24.7	12.35
	Overall average	12.00

which was 10 miles per gallon. Thus the fuels A and B are said to be complementary to each other. If the average mileage for all blends of A and B is higher than the simple average of the two, this phenomena might be depicted by the solid curve in Figure 1.2. If the mileage figure per gallon is strictly additive, that is, if the 50% : 50% blend resulted in exactly 10 miles per gallon, or if a 33% : 67% blend of $A:B$ resulted in [(13 miles × 33%) + (7 miles × 67%)]/100% = 9 miles, and this additivity property is true of all possible $A:B$ blends, then this additive mileage property is represented by the straight line connecting the mileage values 13 and 7 for fuels A and B, respectively, in Figure 1.2. If the average mileage for blends is lower than the simple average mileage, this is represented by the dotted curve.

Definition In the general mixture problem, the response that is measured is a function only of the proportions of the ingredients present in the mixture and is not a function of the amount of the mixture.

In the previous fuel example, the measured response was the average number of miles per gallon and by putting the amount of fuel on a per gallon basis, we made the mileage dependent only on the proportions of the two fuels in the blends and not on the quantity of fuel used. On a slightly different note, if in the measurement of crop yields due to various mixtures of fertilizers the amount of fertilizer applied to the plots is allowed to vary, then the amount can greatly affect the yield. If we fix the amount of applied fertilizer to be constant on all plots, however, then the fertilizer trials can be considered a legitimate mixture problem because

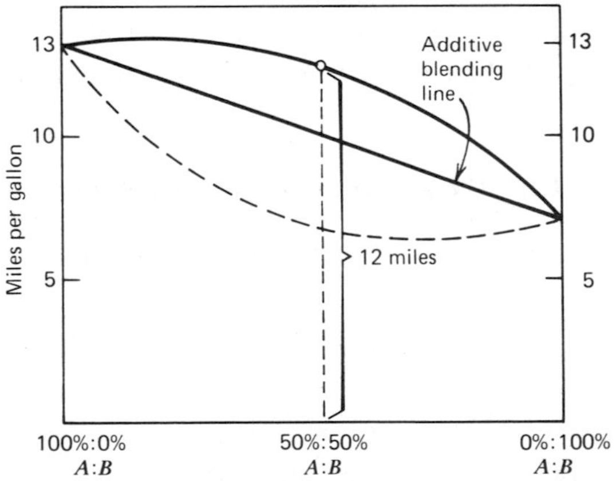

FIGURE 1.2. Plotting the mileage of the 50% : 50% blend of fuels A and B. The formula for the additive blending line is mileage = [(13 miles × A) + (7 miles × B)]/100%.

the crop yield per plot would then be a function only of the ingredient proportions.

The distinguishing feature of the mixture problem is that the independent or controllable factors (fuels A and B, or the fertilizer ingredients) will represent proportionate amounts of the mixture rather than unrestrained amounts where the proportions are by volume, by weight, or by mole fraction. These proportions must be nonnegative, and, if expressed as fractions of the mixture, they must sum to unity especially if they are the only ingredients to be studied comprising the mixtures. If the sum of the component proportions is less than unity, for example, if the sum is equal to 0.80 because 0.20 of the blend is held fixed, and we wish to work only with the variable proportions summing to 0.80, then the variable proportions are rewritten as scaled fractions so that the scaled fractions sum to unity. Clearly, if we let q represent the number of ingredients (or constituents) in the system under study and if we represent the proportion of the ith constituent in the mixture by x_i, then

$$x_i \geq 0, \qquad i = 1, 2, \ldots, q \tag{1.1}$$

and

$$\sum_{i=1}^{q} x_i = x_1 + x_2 + \cdots + x_q = 1.0 \tag{1.2}$$

According to Eq. (1.2), the sum of the nonnegative component proportions or fractions is unity. This latter condition (1.2) will be the fundamental restriction assigned to the proportions comprising the mixture experiment.

Satisfying the restrictions in Eqs. (1.1) and (1.2) means only that a mixture composition will be formed by adding together nonnegative quantities. Actually, since in Eq. (1.2) an individual proportion x_i could be unity, a mixture could be a single ingredient or constituent. Such a mixture is called a *pure* mixture. Pure mixtures are used mainly as a benchmark or as a standard against which multicomponent blends may be compared. We discuss the use of pure mixtures in the next chapter when introducing the simplex-lattice designs, but in Chapters 3–6, nearly all of the design points selected will require most of the constituents to be simultaneously present in the blends. Hereafter we shall call the x_i, $i = 1, 2, \ldots, q$, satisfying Eqs. (1.1) and (1.2) the *components* of the mixture.

By virtue of the constraints on the x_i shown in Eqs. (1.1) and (1.2) the geometric description of the factor space containing the q components consists of all points on or inside the boundaries (vertices, edges, faces, etc.) of a regular $(q-1)$-dimensional simplex. For $q = 2$ components, the simplex factor space is a straight line, represented by the horizontal axis in Figure 1.2. Each blend of the two fuels A and B is represented by a point on the line or axis.

With three components ($q = 3$), the simplex factor space is an equilateral triangle and for $q = 4$, the simplex is a tetrahedron. In Figure 1.3 is presented the factor space for the three components 1, 2, and 3, whose proportions are denoted by x_1, x_2, and x_3. Figure 1.4 is the tetrahedron for the four components whose proportions are x_1, x_2, x_3, and x_4. In Figure 1.3, we see that the vertices of the simplex which represent the pure mixtures are denoted by $x_i = 1$, $x_j = 0$ for $i, j = 1$, 2, and 3, $i \neq j$. The interior points of the triangle represent mixtures in which all of the component proportions are nonzero simultaneously, that is, $x_1 > 0$, $x_2 > 0$, and $x_3 > 0$.

Frequently situations exist where some of the proportions x_i are not allowed to range from 0 to 1.0. Instead, some, or possibly all, of the component proportions are restricted by either a lower bound and/or an upper bound. In the case of component i, these constraints might be written as

$$0 \leq a_i \leq x_i \leq b_i \leq 1.0, \qquad 1 \leq i \leq q$$

where a_i is the lower bound and b_i is the upper bound. As an example, in the production of a commercial laundry bleach which is to be used for removing ink dyes and which we describe as being comprised of the constituents bromine (x_1), dilute HCl (x_2), and hypochlorite powder (x_3),

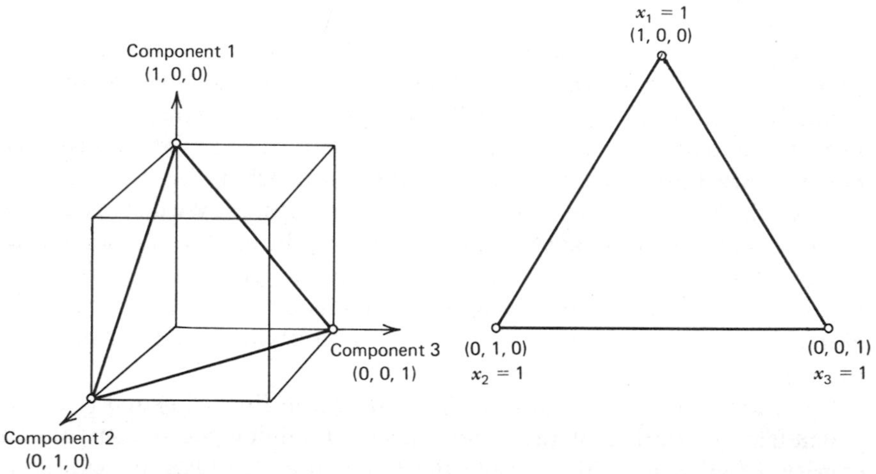

FIGURE 1.3. Three-component simplex region. All experimental points must lie on or inside the triangle whose equation is $x_1 + x_2 + x_3 = 1$.

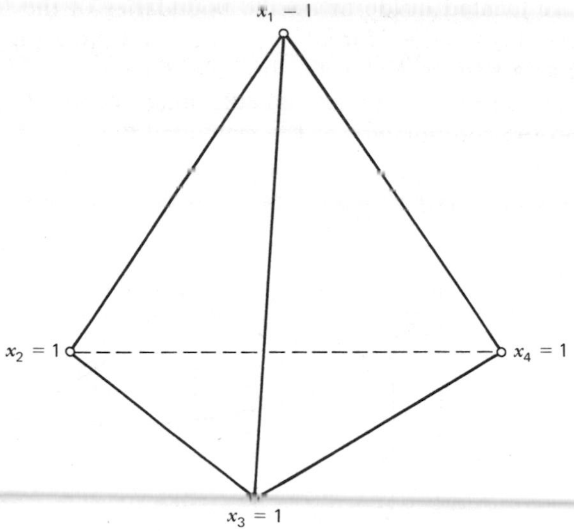

FIGURE 1.4. The four-component tetrahedron.

to be effective the bleach must contain in solution all three of the constituents. This means that each $a_i > 0$ and each $b_i < 1.0$. To be more exact, it might be necessary to require x_2 (dilute HCl) to take values in the interval $0.05 \leq x_2 \leq 0.09$, which therefore forces $a_2 = 0.05$ and $b_2 = 0.09$. Furthermore, the value of $a_2 = 0.05$ forces b_1 and b_3 to be at most equal to $1 - a_2 = 0.95$. Such mixtures where all of the components are simultaneously present are called *complete* mixtures.

1.2. GENERAL REMARKS ABOUT RESPONSE SURFACE METHODS

In much of the experimental work involving multicomponent mixtures, the emphasis is on studying the physical characteristics, such as the shape or the highest point, of the measured response surface. For example, let us assume we are making a fruit punch by blending proportions of orange juice (x_1), pineapple juice (x_2), and grapefruit juice (x_3). The response of interest is the fruitiness flavor of the punch quantified on a 1–10 scale as $1 =$ not fruity, $5 =$ average, $10 =$ extremely fruity. If the measured response or flavor rating in this case to any blend of the juices can be represented by the perpendicular height directly above the blend whose coordinates are located inside or on the boundaries of the triangle, then the locus of the flavor values for all one-, two-, and three-juice blends can be visualized as a surface above the triangle. One such surface, which is assumed to be continuous for all possible juice blends, is presented in Figure 1.5 and the contour plot of the estimated flavor surface is presented in Figure 1.6.

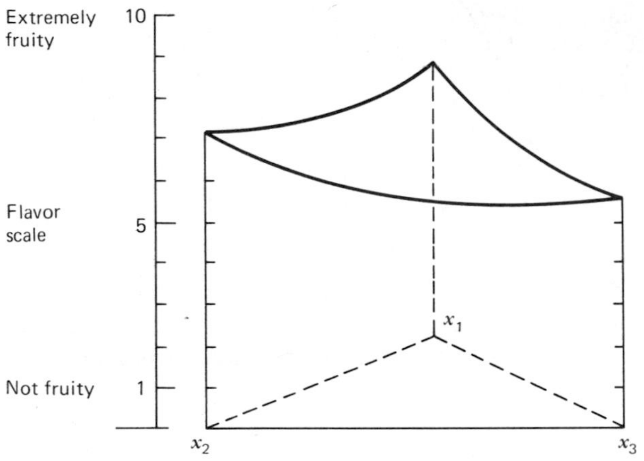

FIGURE 1.5. Fruitiness flavor surface of fruit punch.

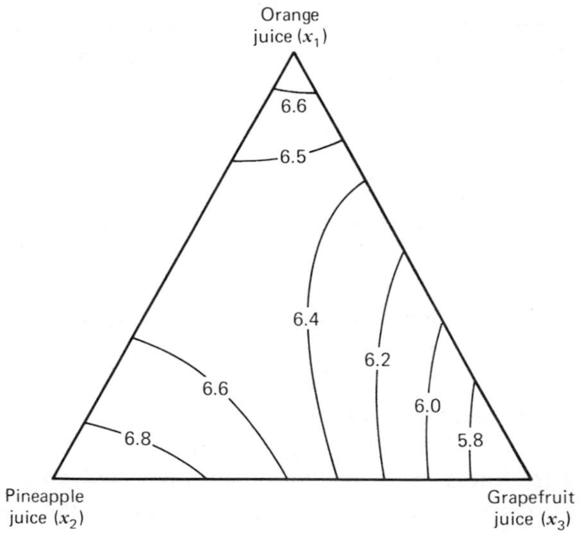

FIGURE 1.6. Contours of constant fruitiness flavor of the fruit punch surface.

The main considerations connected with the exploration of the response surface over the simplex region are (1) the choice of a proper model to approximate the surface over the region of interest, (2) the testing of the adequacy of the model in representing the response surface, and (3) a suitable design for collecting observations, fitting the model and for testing the adequacy of fit. To this end, we shall assume that there exists some functional relationship

$$\eta = \phi(x_1, x_2, \ldots, x_q) \tag{1.3}$$

which, in theory, exactly describes the surface. We shall write the quantity η to denote the response value which is dependent on the proportions $x_1, x_2, \ldots, x_q$ of the components. One very basic assumption that we are making here is that the response surface, represented by the function ϕ, is depicted to be a continuous function in the x_i, $i = 1, 2, \ldots, q$. This assumption might be questionable with some systems, for example, a gaseous system whose catalytic reactions breakdown with the addition or deletion of components. For these systems model forms other than the standard polynomial equations that we are working with initially will need to be considered. In Chapter 6 we present equations containing inverse terms for the purpose of modeling discontinuities of this type in the surface.

The problem of associating the measured properties (physical characteristics) of the response surface with the ingredient composition centers around the determining of the mathematical equation that adequately represents the function $\phi(\cdot)$ in Eq. (1.3). In general, polynomial functions are used to represent $\phi(x_1, \ldots, x_q)$, the justification being, mainly that one can expand $\phi(x_1, \ldots, x_q)$ using a Taylor series, and thus a polynomial can be used also as an approximation. Normally a low-degree polynomial such as the first-degree polynomial

$$\eta = \beta_0 + \sum_{i=1}^{q} \beta_i x_i \tag{1.4}$$

or the second-degree polynomial

$$\eta = \beta_0 + \sum_{i=1}^{q} \beta_i x_i + \sum_{i \leq j}^{q} \sum^{q} \beta_{ij} x_i x_j \tag{1.5}$$

are the kinds of models we believe to represent the surface. The low-degree polynomial equations are more conveniently handled than the higher-degree equations because the lower-degree polynomials contain fewer numbers of terms and therefore require fewer numbers of observed response values in order to estimate the parameters (the β's) in the equation. On those occasions when a very complicated system is being studied, we may feel the need to use a third-degree equation or some special form of a cubic or third-degree (Figure 1.7) equation (especially when even a transformation of the data values does not simplify the structure). Most of the time, however, we will try to be successful with at most the second-degree model.

Figure 1.7 shows contours of equal dielectric constant lines in the system $Pb(Co_{1/3}Nb_{2/3})O_3$–$PbTiO_3$–$PbZrO_3$ as estimated with a third-degree polynomial equation. As seen from the contours, the dielectric constants in the system increase with increasing proportion of $Pb(Co_{1/3}Nb_{2/3})O_3$ up to about 80% : 20% of $Pb(Co_{1/3}Nb_{2/3})O_3$: $PbTiO_3$ and then the values drop off as one approaches pure $Pb(Co_{1/3}Nb_{2/3})O_3$. Near the center of the system is a steep cliff which appears to drop off in the directions of pure $PbTiO_3$ and pure $PbZrO_3$. Contour plots as in Figures 1.6 and 1.7 are extremely helpful when studying a system.

While observing the response η during an experimental program consisting of N trials, it is natural to assume that the observed value that we denote by y_u for the uth trial ($u = 1, 2, \ldots, N$) varies about a mean of η_u with a common variance σ^2 for all $u = 1, 2, \ldots, N$. The observed value contains additive experimental error ϵ_u,

$$y_u = \eta_u + \epsilon_u, \qquad 1 \le u \le N \tag{1.6}$$

The experimental errors ϵ_u are assumed to be independently and identically distributed with zero mean and common variance σ^2. These properties

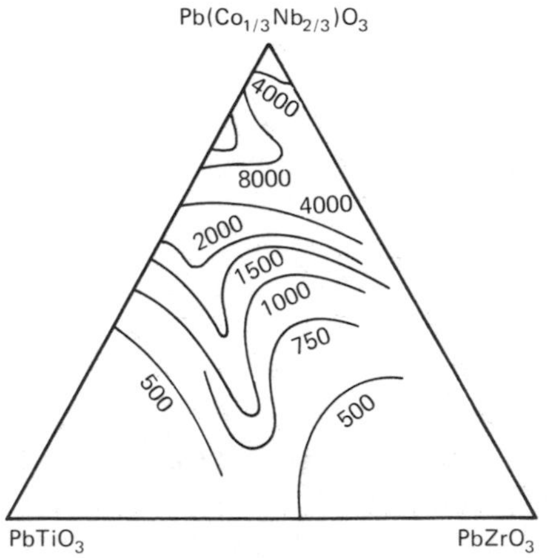

FIGURE 1.7. Equal dielectric constant lines in system $Pb(CO_{1/3}Nb_{2/3})$–$PbTiO_3$–$PbZrO_3$.

of the errors are defined, using an expectation operator $E(\cdot)$, as

$$E(\epsilon_u) = 0, \qquad E(\epsilon_u^2) = \sigma^2, \qquad E(\epsilon_u \epsilon_{u'}) = 0,$$

$$u \neq u', \qquad u, u' = 1, 2, \ldots, N$$

and therefore, the expected value for the observed value y_u is $E(y_u) = \eta_u$, for all $u = 1, 2, \ldots, N$.

In order to approximate the functional relationship $\eta = \phi(x_1, x_2, \ldots, x_q)$ with a polynomial or with any other form of model equation, some preselected number of experimental runs are performed at various predetermined combinations of the proportions of the q components. This *set* of combinations of the proportions (or blends of the ingredients) is referred to as the *experimental design*. Once the N observations are collected, the parameters in the model shown in Eqs. (1.4) or (1.5), are estimated by the *method of least squares*.

As an example, in Eq. (1.4) let us suppose we have $q = 2$ so that coupled with the structure of y_u in Eq. (1.6) we may write

$$y_u = (\beta_0 + \beta_1 x_1 + \beta_2 x_2)_u + \epsilon_u \tag{1.7}$$

With some number $N \geq 3$ of observations collected on y_u, we can obtain the estimates b_0, b_1, and b_2 of the parameters β_0, β_1, and β_2, respectively. If it is decided that the parameter estimates b_0, b_1, and b_2 are satisfactory in the sense that they are nonzero and therefore they relay information about the system we are modeling, then the unknown parameters in Eq. (1.7) are replaced by their respective estimates to give

$$\hat{y} = b_0 + b_1 x_1 + b_2 x_2 \tag{1.8}$$

where $\hat{y}$, read "y hat," denotes the predicted or estimated value of η for given values of x_1 and x_2. Of course, before any predictions are made with Eq. (1.8), we must determine that the prediction equation (1.8) does an adequate job of fitting the observed data. We discuss ways of testing the adequacy of empirical models fitted to data in Chapters 2 and 5.

The properties of the polynomials used to estimate the response function depend to a large extent on the specific program of experiments that we have called the experimental design. The experimental design also represents the range of interest of the experimenter. This is because the design may cover the entire simplex factor space if the experimenter's interest is with all the values of x_i ranging from 0 to 1.0 for all $i = 1, 2, \ldots, q$, or the design might cover only a subportion or smaller subspace within the simplex. This latter situation comes up in practice when additional constraints in the form of upper and/or lower bounds are placed on the component proportions, or, perhaps, when the experimenter is interested only in a group of mixtures which are located in some small region inside the simplex. Both of these cases are discussed in Chapters 4 and 3, respectively.

1.3. AN HISTORICAL PERSPECTIVE

Statistical research on mixture experiments, as represented by the number of papers which have appeared in the statistical literature, is a relatively new activity. Almost all of the theory and methodology that has emanated from the statistical community has surfaced during the last two decades. A few noticeable exceptions are the discussion on mixtures which appears in Quenouille's 1953 book, the designed experiment for administering joint dosages of hormones to mice by Claringbold in 1955, and the pioneering article in 1958 by H. Scheffé.

In the remainder of this book, we concentrate on experimental designs and techniques used in the analysis of mixture data that have evolved since the mid-1950s. (In fields such as in the cereal industry, the tire manufacturing industry, and the soap industry, for example, mixture experiments date back as far as the turn of the century.) In Table 1.2, a chronological sequence of papers that have appeared in the statistical literature is illustrated. The list includes authors of papers that have appeared during the period 1953–1979 in journals of statistical societies and associations and in textbooks on statistics, as well as in related periodicals, such as academic technical reports, industrial bulletins, and armed services reports. Within each year the authors are listed in alphabetical order. It was not possible to include the authors of all of the work that appeared in every journal during this time period for the obvious reason of lack of space. Several of the works that are omitted in Table 1.2, but which appeared in journals like *Cereal Chemistry* and *Food Technology*, are listed in the bibliography at the end of the book.

Several of the papers cited in Table 1.2 are particularly noteworthy for their content and also because of the time in which they appeared in the statistical literature. These papers are designated with an asterisk. Each of these papers is described briefly now.

TABLE 1.2. A chronological ordering of the statistical literature on mixtures.

1950	1953	1955	1958	1959	1961	1962	1963
	Quenouille	Claringbold*	Scheffé*	Quenouille	John and Gorman Scheffé	Gorman and Hinman* Wagner and Gorman	Kenworthy Scheffé* Wagner and Gorman

1964	1965	1966	1967	1968	1969
Myers Uranisi	Bounds, Kurotori and Cruise *Draper and Lawrence**	Box and Gardiner Cruise *Gorman* Kurotori* McLean and Anderson*	Diamond Drew	Becker *Lambrakis* Murty and Das Thompson and Myers*	Becker* Hewlett *Lambrakis* Watson

1970	1971	1973	1974	1975
Becker Cornell and Good Nigam	Cornell Cox Li Paku, Manson and Nelson Snee Van Schalkwyk	Cornell* Nigam Saxena and Nigam Snee*	Hare Marquardt and Snee Nigam Snee and Marquardt	Cornell Cornell and Ott Laake *Snee*

1976	1977	1978	1979
Snee and Marquardt	Cornell *Draper and St. John* *Galil and Kiefer* Hare and Brown Saxena and Nigam	Becker Cornell and Gorman Park	Cornell Cornell and Khuri Hare Snee

*Denotes paper that is referred to frequently.
Author names in italics denote that more than one paper appeared during the year.

Claringbold, P. J. (1955). *Use of the simplex design in the study of the joint action of related hormones.* The first paper to introduce a design on the three-component simplex and to present the corresponding fitted model and analysis of the data.

Scheffé, H. (1958). *Experiments with mixtures.* Introduced the simplex-lattice designs and the corresponding polynomial models. This paper is probably recognized as having done more than any other paper towards generating interest in the areas of design and analysis of mixture experiments.

Gorman, J. W. and J. E. Hinman (1962). *Simplex-lattice designs for multicomponent systems.* A lucid presentation on the use of simplex-lattice designs and Scheffé's polynomials.

Scheffé, H. (1963). *The simplex-centroid design for experiments with mixtures.* Introduced an alternative design to the $\{q, m\}$ simplex-lattice. The first paper to consider designs and models for experiments consisting of process variables and mixture components.

Draper, N. R. and W. E. Lawrence (1965). *Mixture designs for three factors. Mixture designs for four factors.* The first papers to suggest using designs which minimized the bias in the fitted model as well as the variance through minimizing the mean square error of the estimate of the response over the simplex region.

Kurotori, I. S. (1966). *Experiments with mixtures of components having lower bounds.* The first paper to use pseudocomponents to define a pseudocomponent simplex inside the original simplex space. This paper simplified the design problem when the components proportions are restricted by lower bounds.

McLean, R. A. and V. L. Anderson (1966). *Extreme vertices design of mixture experiments.* The first paper to recommend an algorithm to generate the coordinates of the vertices of a constrained factor space resulting from the placing of upper and lower bounds on some or all of the component proportions.

Thompson, W. O. and R. H. Myers (1968). *Response surface designs for experiments with mixtures.* The first paper to consider an ellipsoidal region of interest inside the simplex factor space and the conditions for using rotatable response surface designs for fitting polynomial models.

Becker, N. G. (1969). *Regression problems when the predictor variables are proportions.* The first paper to discuss adaptations to commonly used regression and response surface techniques, such as the method of steepest descent, reduction to canonical form, and so on, when working with proportions.

Cornell, J. A. (1973). *Experiments with mixtures: A review.* A complete review of nearly all of the published statistical papers on mixture designs and models.

Snee, R. D. (1973). *Techniques for the analysis of mixture data.* A lucid discussion of several fitted model forms as well as ways of analyzing mixture data.

1.4. REFERENCES AND RECOMMENDED READING

Cornell, J. A. (1973). Experiments with mixtures: A review, *Technometrics*, **15**, No. 3, 437–455.

Gorman, J. W. and J. E. Hinman (1962). Simplex-lattice designs for multicomponent systems. *Technometrics*, **4**, No. 4, 463–487.

Hare, L. B. (1974). Mixture designs applied to food formulation, *Food Technol.*, **28**, 50–62.

Scheffé, H. (1958). Experiments with mixtures. *J. R. Stat. Soc., B*, **20**, No. 2, 344–360.

Snee, R. D. (1973). Techniques for the analysis of mixture data. *Technometrics*, **15**, No. 3, 517–528.

QUESTIONS FOR CHAPTER 1

1.1. The functional relationship between the measured response and the ingredient proportions in a mixture problem is different in several ways from the standard regression functional relationship between a dependent variable and one or more independent variables. Explain.

1.2. List several experimental situations that fall under the heading of a mixture experiment.

1.3. Pure mixtures, binary mixtures, and complete mixtures are names given to blends of one or more ingredients or components. Distinguish between the various mixture types.

CHAPTER 2

The Original Mixture Problem

In this chapter we treat the most general description of the mixture problem, which is where the component proportions are to satisfy the constraints $x_i \geq 0, x_1 + x_2 + \cdots + x_q = 1.0$. Each component proportion x_i can take values from zero to unity and all blends among the ingredients are possible. We concentrate on the fitting of mathematical equations to model the response surface over the entire simplex factor space, so that the empirical prediction of the response to any mixture over the entire simplex is possible.

What is meant by modeling is that a model or an equation is postulated to represent the response surface. We then choose a design at whose points we may collect observations to which the equation can be fitted. Finally, we test the adequacy of the model. This final step is to ensure that our fitted equation is a prediction tool that we can feel comfortable with.

The modeling sequence just mentioned is altered slightly. First we discuss the simplex-lattice designs that were introduced by Scheffé in the early years (1958–1965) of the period in which research on mixture experiments was being developed. These designs are credited by many researchers to be the foundation upon which the theory of experimental designs for mixtures was built, and yet these designs are still very much in use today. We then present the associated polynomial models to be fitted to data recorded at the points of these designs.

2.1. THE SIMPLEX-LATTICE DESIGNS

To accommodate a polynomial equation to represent the response surface over the entire simplex region, a natural choice for a design would be one whose points are spread evenly over the whole simplex factor space. An ordered arrangement consisting of a uniformly spaced distribution of points on a simplex is known as a *lattice.* (We remark here that any similarity that exists between these lattices and the lattice squares and rectangular lattices which belong to the class of balanced incomplete block designs is purely coincidental. The name lattice is used here simply to make reference to an array of points.)

A lattice may have a special correspondence to a specific polynomial equation. For example, to support a polynomial model of degree m in q components over the simplex, the lattice, referred to as a $\{q, m\}$ simplex-lattice consists of points whose coordinates are defined by the following combinations of the component proportions; the proportions assumed by each component will take the $m+1$ equally spaced values from 0 to 1, that is,

$$x_i = 0, \frac{1}{m}, \frac{2}{m}, \ldots, 1 \qquad (2.1)$$

and the $\{q, m\}$ simplex-lattice consists of *all* possible combinations (mixtures) of the components where the proportions (2.1) for each component are used.

The listing of the specific component combinations comprising the $\{q, m\}$ simplex-lattice is illustrated as follows. Let us consider a $q = 3$ component system where the factor space for all blends is an equilateral triangle and for the mixture blends to be considered, let each component assume the proportions $x_i = 0, \frac{1}{2}$, and 1 for $i = 1, 2$, and 3. Setting $m = 2$ for the proportions, we are thinking of using a second-degree model to represent the response surface over the triangle. The $\{3, 2\}$ simplex-lattice will consist of the six points on the boundary of the triangle

$$(x_1, x_2, x_3) = (1, 0, 0), (0, 1, 0), (0, 0, 1), (\tfrac{1}{2}, \tfrac{1}{2}, 0), (\tfrac{1}{2}, 0, \tfrac{1}{2}), (0, \tfrac{1}{2}, \tfrac{1}{2})$$

The three points, which are defined as $(1, 0, 0)$ or $x_1 = 1$, $x_2 = x_3 = 0$; $(0, 1, 0)$ or $x_1 = x_3 = 0$, $x_2 = 1$; and $(0, 0, 1)$ or $x_1 = x_2 = 0$, $x_3 = 1$, each represent a pure mixture where an individual component is considered to be a mixture by itself, and these points are the three vertices of the triangle. The points $(\frac{1}{2}, \frac{1}{2}, 0)$, $(\frac{1}{2}, 0, \frac{1}{2})$, and $(0, \frac{1}{2}, \frac{1}{2})$ represent the binary blends or two-component mixtures $x_i = x_j = \frac{1}{2}$, $x_k = 0$, $k \neq i, j$, for which the nonzero component proportions are equal. The binary blends are located at the midpoints of the three edges of the triangle. The $\{3, 2\}$ simplex-lattice is shown in Figure 2.1.

Let us present another example. We set the number of equally spaced levels (or proportions) for each component to be four, that is, $x_i = 0, \frac{1}{3}, \frac{2}{3}, 1$. If we consider all possible blends of the three components with

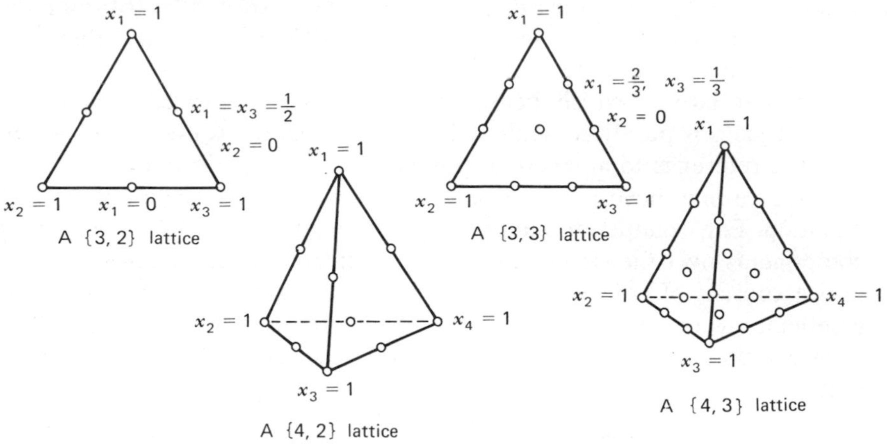

FIGURE 2.1. Some $\{3, m\}$ and $\{4, m\}$ simplex-lattice arrangements, $m = 2$ and $m = 3$.

these proportions, then the $\{3, m = 3\}$ simplex-lattice contains the following blending coordinates:

$$(x_1, x_2, x_3) = (1, 0, 0), (0, 1, 0), (0, 0, 1), (\tfrac{2}{3}, \tfrac{1}{3}, 0), (\tfrac{2}{3}, 0, \tfrac{1}{3}), (\tfrac{1}{3}, \tfrac{2}{3}, 0), (\tfrac{1}{3}, 0, \tfrac{2}{3}),$$
$$(\tfrac{1}{3}, \tfrac{1}{3}, \tfrac{1}{3}), (0, \tfrac{2}{3}, \tfrac{1}{3}), (0, \tfrac{1}{3}, \tfrac{2}{3})$$

We see that each of the proportions of the components in every blend or mixture is a fractional number and the sum of the fractions equals unity. When plotted as a lattice arrangement these points are an array that is symmetrical with respect to the orientation of the simplex (that is, symmetrical with respect to the vertices and the sides of the simplex). The arrangement of the 10 points of a $\{3, 3\}$ simplex-lattice is presented in Figure 2.1.

Before proceeding further, we remark that throughout this book the coordinate system that we are using with the mixture components is called a simplex coordinate system. With three components, for example, the triangular coordinate system is represented by the fractional values in parentheses (x_1, x_2, x_3) where each $0 \leq x_i \leq 1$, $i = 1, 2$ and 3 and $x_1 + x_2 + x_3 = 1$. Several points of composition are chosen in the triangular system presented in Figure 2.2. When there appears to be no chance for confusion, the composition $(x_1 = a_1, x_2 = a_2, x_3 = a_3)$ is denoted by (a_1, a_2, a_3).

The number of design points in the $\{q, m\}$ simplex-lattice is $\binom{q+m-1}{m} = (q + m - 1)!/m!(q - 1)!$ where $m!$ is "m factorial" and $m! = m(m-1)(m-2) \cdots (2)(1)$. [The symbol $\binom{a}{b}$ is the combinational symbol

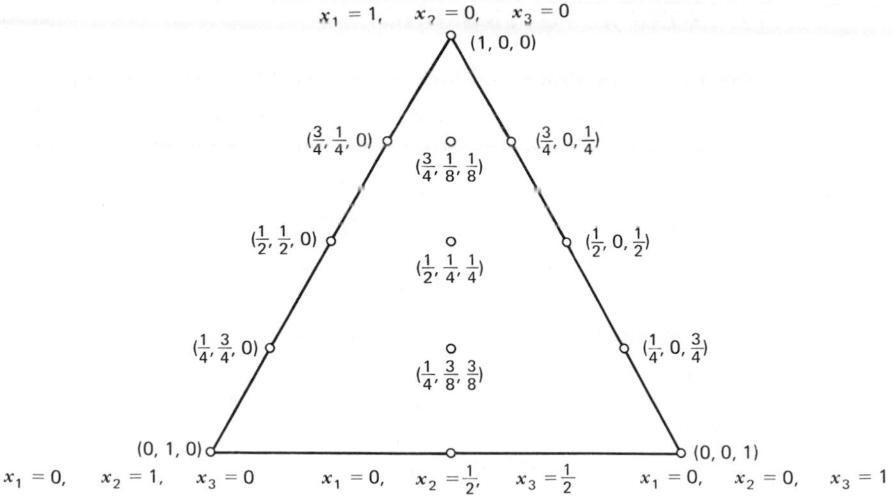

FIGURE 2.2. Triangular coordinates (x_1, x_2, x_3).

for the number of ways a things can be taken b at a time and $\binom{a}{b} = a!/b!(a-b)!$] In the $\{3, 2\}$ simplex-lattice, for example, the number of points is $\binom{3+2-1}{2} = 4!/2!2! = 6$, while the $\{3, 3\}$ simplex-lattice consists of $\binom{5}{2} = 10$ points. In the $\{q, m\}$ simplex-lattice then, the points correspond to pure or single-component mixtures, to binary or two-component mixtures, to ternary or three-component blends and so on, up to mixtures consisting of at most m components. In Figure 2.1 the $\{4, 2\}$ and $\{4, 3\}$ simplex-lattices are shown. In Table 2.1 is listed the number of points in a $\{q, m\}$ simplex-lattice for values of q and m from $3 \le q \le 10$, $1 \le m \le 4$.

TABLE 2.1. Number of points in the $\{q, m\}$ simplex-lattice for $3 \le q \le 10$, $1 \le m \le 4$ where the number of levels for each component is $m + 1$

Degree of Model		Number of Components							
m	$q =$	3	4	5	6	7	8	9	10
1		3	4	5	6	7	8	9	10
2		6	10	15	21	28	36	45	55
3		10	20	35	56	84	120	165	220
4		15	35	70	126	210	330	495	715

2.2. THE CANONICAL POLYNOMIALS

A general form of regression function that can be fitted to data collected at the points of a $\{q, m\}$ simplex-lattice is derived using the following procedure. First, we recall that the equation of an mth-degree polynomial is written as

$$\eta = \beta_0 + \sum_{i=1}^{q} \beta_i x_i + \sum_{i \le j}^{q} \sum^{q} \beta_{ij} x_i x_j + \sum_{i \le j \le k}^{q} \sum^{q} \sum^{q} \beta_{ijk} x_i x_j x_k + \cdots \tag{2.2}$$

where the terms up to the mth degree are included. The number of terms in Eq. (2.2) is $\binom{q+m}{m}$, but because the terms in Eq. (2.2) have meaning for us only subject to the restriction $x_1 + x_2 + \cdots + x_q = 1$, we know that the parameters $\beta_i, \beta_{ij}, \beta_{ijk}, \ldots$ associated with the terms are not unique. However, we may make the substitution

$$x_q = 1 - \sum_{i=1}^{q-1} x_i \tag{2.3}$$

in Eq. (2.2), thereby removing the dependency among the x_i terms, and this will not affect the degree of the polynomial. The effect of substituting Eq. (2.3) into Eq. (2.2) is that η becomes a polynomial of degree m in $q-1$ components $x_1, x_2, \ldots, x_{q-1}$ with $\binom{q+m-1}{m}$ terms. And although the resulting formula after the substitution is simpler in form because it contains fewer components and fewer terms, the effect of component q is obscured by this substitution because the component is not included in the equation. Since we do not wish to sacrifice information on component q, we do not use Eq. (2.3). Instead we use another approach to derive an equation in place of Eq. (2.2) to represent the surface.

An alternative equation to Eq. (2.2) for a polynomial of degree m in q components, subject to the restriction on the x_i's in Eq. (1.2), is derived by multiplying some of the terms in Eq. (2.2) by the identity $(x_1 + x_2 + \cdots + x_q) = 1$ and simplifying. The resulting equation has been called the "canonical" polynomial or the "canonical form of the polynomial" or simply the $\{q, m\}$ polynomial. (The name $\{q, m\}$ polynomial is given to these equations by some authors because this polynomial form is used often in conjunction with the $\{q, m\}$ simplex-lattice.) The $\{q, m\}$ polynomial contains $\binom{q+m-1}{m}$ terms which is the same number of points that make up the associated $\{q, m\}$ simplex-lattice design. For example, for $m = 1$ and from Eq. (1.4),

$$\eta = \beta_0 + \sum_{i=1}^{q} \beta_i x_i$$

and upon multiplying the β_0 term by $(x_1 + x_2 + \cdots + x_q) = 1$, the resulting equation is

$$\eta = \beta_0\left(\sum_{i=1}^{q} x_i\right) + \sum_{i=1}^{q} \beta_i x_i = \sum_{i=1}^{q} \beta_i^* x_i \tag{2.4}$$

where $\beta_i^* = \beta_0 + \beta_i$ for all $i = 1, 2, \ldots, q$. The number of terms in Eq. (2.4) is q, which is the number of points in the $\{q, 1\}$ lattice. The parameters $\beta_i^*, i = 1, 2, \ldots, q$, have simple and clear meanings in terms of describing the shape of the response surface over the simplex region.

The general second-degree polynomial in q variables is

$$\eta = \beta_0 + \sum_{i=1}^{q} \beta_i x_i + \sum_{i=1}^{q} \beta_{ii} x_i^2 + \sum_{i<j}\sum^{q} \beta_{ij} x_i x_j \tag{2.5}$$

and if we apply the identities $x_1 + x_2 + \cdots + x_q = 1$ and

$$x_i^2 = x_i\left(1 - \sum_{\substack{j=1\\ j\neq i}}^{q} x_j\right) \tag{2.6}$$

then for $m = 2$

$$\begin{aligned}\eta &= \beta_0\left(\sum_{i=1}^{q} x_i\right) + \sum_{i=1}^{q} \beta_i x_i + \sum_{i=1}^{q} \beta_{ii} x_i\left(1 - \sum_{j\neq i}^{q} x_j\right) + \sum_{i<j}\sum^{q} \beta_{ij} x_i x_j \\ &= \sum_{i=1}^{q} (\beta_0 + \beta_i + \beta_{ii}) x_i - \sum_{i=1}^{q} \beta_{ii} x_i \sum_{j\neq i}^{q} x_j + \sum_{i<j}\sum^{q} \beta_{ij} x_i x_j \\ &= \sum_{i=1}^{q} \beta_i^* x_i + \sum_{i<j}\sum^{q} \beta_{ij}^* x_i x_j\end{aligned} \tag{2.7}$$

The number of terms in Eq. (2.7) is $q + q(q-1)/2 = q(q+1)/2$.

A comparison made between Eqs. (2.5) and (2.7) reveals that the parameters in Eq. (2.7) are simple functions of the parameters in Eq. (2.5), that is, $\beta_i^* = \beta_0 + \beta_i + \beta_{ii}$ and $\beta_{ij}^* = \beta_{ij} - \beta_{ii} - \beta_{jj}$, $i, j = 1, 2, \ldots, q, i < j$. Furthermore, Eq. (2.7) can be written in the homogeneous form

$$\eta = \sum_{i=1}^{q} \delta_{ii} x_i^2 + \sum_{i<j}\sum^{q} \delta_{ij} x_i x_j = \sum_{i\leq j}\sum^{q} \delta_{ij} x_i x_j \tag{2.8}$$

which resulted from multiplying $\Sigma_{i=1}^{q} \beta_i^* x_i$ in Eq. (2.7) by the identity $(x_1 + x_2 + \cdots + x_q) = 1$ and then simplifying the terms.

The two models in Eqs. (2.7) and (2.8) are equivalent in the sense that one was derived from the other, Eq. (2.8) from Eq. (2.7), without changing the degree of the polynomial nor reducing the number of components. Owing to the restriction $x_1 + x_2 + \cdots + x_q = 1$ on the component proportions, an infinite number of regression functions can be derived from Eq. (2.5), and these equations are equivalent to Eqs. (2.7) and (2.8) when all of the component proportions are included. This is seen by realizing that for all functions ϕ, the linear equations $\beta_0 - \phi + \Sigma_{i=1}^{q}(\beta_i + \phi)x_i$ are equivalent when $\Sigma_{i=1}^{q} x_i = 1$.

The formula for the third-degree polynomial can be derived by multiplying the identity $(x_1 + x_2 + \cdots + x_q) = 1$ as well as the restriction in Eq. (2.6) to the terms of the general third-degree polynomial in Eq. (2.2) and then simplifying the terms as was done for the $\{q, 2\}$ polynomial. The full cubic or $\{q, 3\}$ polynomial is

$$\eta = \sum_{i=1}^{q} \beta_i^* x_i + \sum_{i<j}^{q}\sum \beta_{ij}^* x_i x_j + \sum_{i<j}^{q}\sum \delta_{ij} x_i x_j (x_i - x_j) + \sum_{i<j<k}^{q}\sum\sum \beta_{ijk}^* x_i x_j x_k \tag{2.9}$$

A simpler formula for a special case of the cubic polynomial where the terms $\delta_{ij}x_i x_j(x_i - x_j)$ are not considered is the special cubic polynomial

$$\eta = \sum_{i=1}^{q} \beta_i^* x_i + \sum_{i<j}^{q}\sum \beta_{ij}^* x_i x_j + \sum_{i<j<k}^{q}\sum\sum \beta_{ijk}^* x_i x_j x_k \tag{2.10}$$

Hereafter we shall remove the asterisks from β_i^*, β_{ij}^*, and β_{ijk}^* and use β_i, β_{ij} and β_{ijk} in all of the $\{q, m\}$ polynomials. The asterisks were assigned to the parameters only to keep the parameters in the general polynomial Eq. (2.2) separate from the parameters in the derived $\{q, m\}$ polynomials. In three components, the models of Eqs. (2.4), (2.7), (2.9), and (2.10), respectively, are to appear hereafter as

$$\eta = \beta_1 x_1 + \beta_2 x_2 + \beta_3 x_3$$

$$\eta = \beta_1 x_1 + \beta_2 x_2 + \beta_3 x_3 + \beta_{12} x_1 x_2 + \beta_{13} x_1 x_3 + \beta_{23} x_2 x_3$$

$$\begin{aligned}\eta = \beta_1 x_1 + \beta_2 x_2 + \beta_3 x_3 + \beta_{12} x_1 x_2 + \beta_{13} x_1 x_3 + \beta_{23} x_2 x_3 + \delta_{12} x_1 x_2 (x_1 - x_2) \\ + \delta_{13} x_1 x_3 (x_1 - x_3) + \delta_{23} x_2 x_3 (x_2 - x_3) + \beta_{123} x_1 x_2 x_3\end{aligned}$$

$$\eta = \beta_1 x_1 + \beta_2 x_2 + \beta_3 x_3 + \beta_{12} x_1 x_2 + \beta_{13} x_1 x_3 + \beta_{23} x_2 x_3 + \beta_{123} x_1 x_2 x_3$$

The number of terms in the $\{q, m\}$ polynomials is a function of m, the degree of the equation, as well as the number of components q. The numbers of terms for several values of q are listed in Table 2.2.

The terms $\beta_i x_i$ and $\beta_{ij} x_i x_j$ in the $\{q, 1\}$ and $\{q, 2\}$ polynomial equations have simple interpretations. At the vertex corresponding to pure com-

TABLE 2.2. The number of terms in the canonical polynomials

Number of Components q	Linear	Quadratic	Special Cubic	Full Cubic
2	2	3	—	—
3	3	6	7	10
4	4	10	14	20
5	5	15	25	35
6	6	21	41	56
7	7	28	63	84
8	8	36	92	120
⋮	⋮	⋮	⋮	⋮
q	q	$q(q+1)/2$	$q(q^2+5)/6$	$q(q+1)(q+2)/6$

ponent i, for example, if with either of the two models Eq. (2.4) or Eq. (2.7), we set $x_i = 1$ which thus forces $x_j = 0$ for all $j \neq i$, then $\eta = \beta_i$. The parameter β_i therefore represents the expected response to pure component i and pictorially, β_i is the height of the surface above the simplex at the vertex, where $x_i = 1$ for $i = 1, 2, \ldots, q$. (As heights, the β_i are usually nonnegative quantities unless side conditions are imposed on the values of the β_i. A set of conditions will be imposed on the β_i in Sections 6.7 and 6.8 when discussing Cox's polynomial model.) When Eq. (2.4) defines the response surface exactly, which is the case when the blending or mixing among the components is strictly linear or additive, then the surface is depicted by a plane over the simplex. An example for $q = 3$ is presented in Figure 2.3.

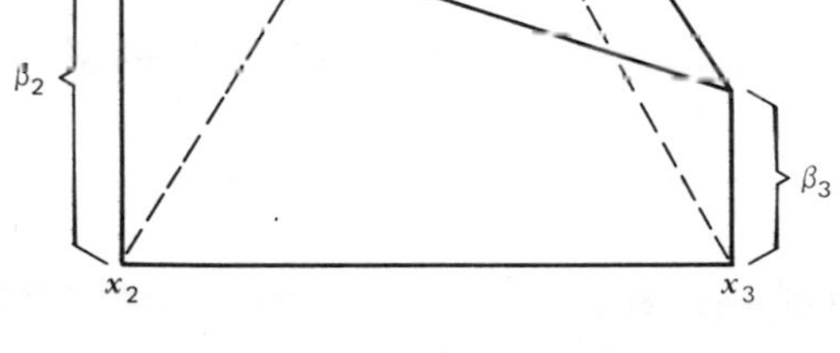

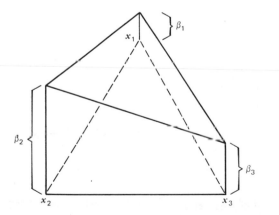

FIGURE 2.3. A planar surface above the three-component triangle.

When the blending among the components is assumed to be linear (Figure 2.3), the response to the binary mixture of components i and j in the proportions x_i and x_j is given by Eq. (2.4) to be $\eta = \beta_i x_i + \beta_j x_j$, since all of the other x_k values are set to zero. If on the other hand the true response to the binary mixture of components i and j is more correctly represented by Eq. (2.7), which is $\eta = \beta_i x_i + \beta_j x_j + \beta_{ij} x_i x_j$, then an excess exists. The excess, which is represented by the term $\beta_{ij} x_i x_j$, is found by taking the difference between the models in Eqs. (2.7) and (2.4). If with a positive response the quantity β_{ij} is positive, the excess is called the *synergism* of the binary mixture, and β_{ij} is the quadratic or second-order coefficient of the binary synergism. (See Figure 2.4.) The opposite of synergism (that is, when β_{ij} is negative) is called *antagonism* of the binary mixture.

If the cubic formula (2.9) is the truest representation of the surface, the excess or synergism of the binary mixture would include the additional term $\delta_{ij} x_i x_j (x_i - x_j)$ where δ_{ij} is the cubic coefficient of the binary synergism. The term $\beta_{ijk} x_i x_j x_k$ in Eq. (2.9), represents the ternary synergism of the components i, j, and k.

Another plausible way of understanding the separate terms in the first- and second-degree models in Eqs. (2.4) and (2.7) is to consider how the terms individually contribute towards the description of the shape of the mixture surface. The term $\beta_i x_i$ contributes to the model only when the value of $x_i > 0$ and since β_i represents the height of the surface above the simplex at the vertex $x_i = 1$, the term $\beta_i x_i$ contributes most (i.e., the value of $\beta_i x_i = \beta_i$ is greatest when $\beta_i > 0$) at $x_i = 1$. A term $\beta_{ij} x_i x_j$ in Eq. (2.7) contributes to the model everywhere in the simplex where both $x_i > 0$ and $x_j > 0$. On the edge joining the vertices corresponding to components i and j, in fact, the value of the term $\beta_{ij} x_i x_j$ is maximum when $\beta_{ij} > 0$ at $x_i = x_j = \frac{1}{2}$ where it is equal to $\beta_{ij} x_i x_j = \beta_{ij}/4$.

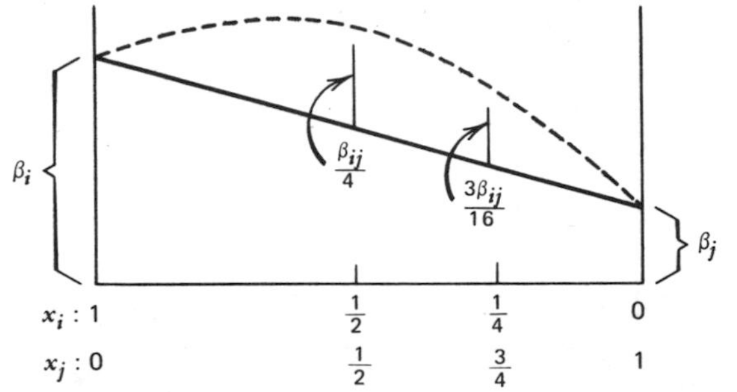

FIGURE 2.4. Synergisms of binary mixtures.

Returning to the example at the beginning of Chapter 1 in which fuels A and B were blended together to see if the automobile mileage could be increased, we recall that with fuel $x_1 = A$, the mileage figure was 13 miles per gallon but with fuel $x_2 = B$, the mileage was only 7 miles per gallon. The 50% : 50% blend of the two fuels resulted in an average mileage figure of 12 miles per gallon. Now, the simple average of the two fuels used individually is $(13+7)/2 = 10$ miles per gallon and since the 50% : 50% blend average of 12 miles per gallon exceeds the simple average of the fuels, the fuels are said to be *synergistic*. Hence, when blending two components which are synergistic to one another, we anticipate a more desirable yield than would be expected by taking the average of the yields of the two pure blends. A mileage figure less than 10 miles per gallon would imply that the fuels are *antagonistic* to one another.

2.3. THE POLYNOMIAL COEFFICIENTS AS FUNCTIONS OF THE RESPONSES AT THE POINTS OF THE LATTICES

We mentioned previously that the $\{q, m\}$ simplex-lattice has a special correspondence to the $\{q, m\}$ polynomial equation. This correspondence is that the parameters in the polynomial can be expressed as simple functions of the expected responses which are measured at the points of the $\{q, m\}$ simplex-lattice. In order to show this, we shall reintroduce the response nomenclature which was first proposed by Scheffé in his 1958 paper on mixtures.

Let the response to pure component i be denoted by η_i; the response to the binary mixture with equal proportions (50% : 50%) of components i and j be denoted by η_{ij} and the response to the ternary mixture with equal proportions of components i, j, and k by η_{ijk}. In Figure 2.5 the response nomenclature is illustrated at the points of the $\{3, 2\}$ and $\{3, 3\}$ simplex-lattices, respectively. The subscripts on a response designate three characteristics:

1. The number of subscripts equals the denominator in the fractions used in the mixture (the two subscripts in η_{ij} implies the response to a mixture where the component proportions are $\frac{1}{2}$, that is, where $x_i = \frac{1}{2}$ and $x_j = \frac{1}{2}$).
2. The number of distinct numbers or letters indicates how many components are present in nonzero proportions in the mixture.
3. The number of times a number or letter appears indicates the relative proportion assumed by the corresponding component in the mixture.

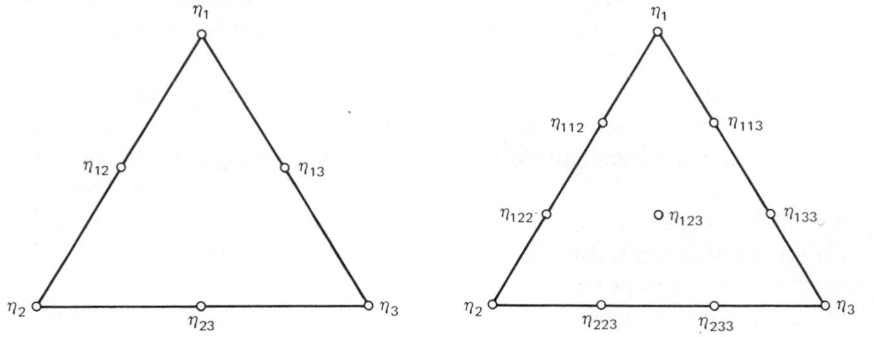

FIGURE 2.5. Response nomenclature at the points of the {3, 2} and {3, 3} simplex-lattices.

As an example, the response η_{112} has three subscripts, implying the composition at which η_{112} is measured consists of component fractions of size $\frac{1}{3}$. The two distinct numbers, 1 and 2, indicate two separate components are present in nonzero proportions. Since the subscript consists of two 1's and one 2 the proportions are $\frac{2}{3}$ for component 1 and $\frac{1}{3}$ for component 2. Expected responses at other points of composition are listed in Table 2.3.

TABLE 2.3. Scheffé's response nomenclature for some three- and four-component mixtures

	Mixture Composition						
Response	x_1	x_2	x_3	x_1	x_2	x_3	x_4
η_1	1	0	0	1	0	0	0
η_2	0	1	0	0	1	0	0
η_{12}	$\frac{1}{2}$	$\frac{1}{2}$	0	$\frac{1}{2}$	$\frac{1}{2}$	0	0
η_{23}	0	$\frac{1}{2}$	$\frac{1}{2}$	0	$\frac{1}{2}$	$\frac{1}{2}$	0
η_{123}	$\frac{1}{3}$	$\frac{1}{3}$	$\frac{1}{3}$	$\frac{1}{3}$	$\frac{1}{3}$	$\frac{1}{3}$	0
η_{112}	$\frac{2}{3}$	$\frac{1}{3}$	0	$\frac{2}{3}$	$\frac{1}{3}$	0	0
η_{233}	0	$\frac{1}{3}$	$\frac{2}{3}$	0	$\frac{1}{3}$	$\frac{2}{3}$	0
η_{1112}	$\frac{3}{4}$	$\frac{1}{4}$	0	$\frac{3}{4}$	$\frac{1}{4}$	0	0
η_{1223}	$\frac{1}{4}$	$\frac{1}{2}$	$\frac{1}{4}$	$\frac{1}{4}$	$\frac{1}{2}$	$\frac{1}{4}$	0
η_{1144}				$\frac{1}{2}$	0	0	$\frac{1}{2}$
η_{2344}				0	$\frac{1}{4}$	$\frac{1}{4}$	$\frac{1}{2}$
η_{1234}				$\frac{1}{4}$	$\frac{1}{4}$	$\frac{1}{4}$	$\frac{1}{4}$

The equations for expressing the parameters in the polynomial models in terms of η_i, η_{ij}, and η_{ijk}, are given by solving $\binom{q+m-1}{m}$ equations simultaneously. This number corresponds not only to the number of parameters in the $\{q, m\}$ polynomial equation but also to the number of lattice points and therefore to the number of expected responses η_i, η_{ij}, measured at the points in the $\{q, m\}$ simplex lattice as well. For example, if the second-degree model in Eq. (2.7) is to be used for a three-component system and we have the measured responses at the points of the $\{3, 2\}$ simplex-lattice design, then for the equation

$$\eta = \beta_1 x_1 + \beta_2 x_2 + \beta_3 x_3 + \beta_{12} x_1 x_2 + \beta_{13} x_1 x_3 + \beta_{23} x_2 x_3 \tag{2.11}$$

if we substitute

$$\eta_i \quad \text{at } x_i = 1, \quad x_j = 0, \quad i, j = 1, 2, 3, \quad j \neq i$$

$$\eta_{ij} \quad \text{at } x_i = \tfrac{1}{2}, \quad x_j = \tfrac{1}{2}, \quad x_k = 0 \quad i < j, \quad k \neq i, j$$

into Eq. (2.11), the following $\binom{3+2-1}{2} = 6$ equations result

$$\eta_1 = \beta_1, \qquad \eta_2 = \beta_2, \qquad \eta_3 = \beta_3$$

$$\eta_{12} = \beta_1(\tfrac{1}{2}) + \beta_2(\tfrac{1}{2}) + \beta_{12}(\tfrac{1}{4})$$

$$\eta_{13} = \beta_1(\tfrac{1}{2}) + \beta_3(\tfrac{1}{2}) + \beta_{13}(\tfrac{1}{4})$$

$$\eta_{23} = \beta_2(\tfrac{1}{2}) + \beta_3(\tfrac{1}{2}) + \beta_{23}(\tfrac{1}{4})$$

Solving the six equations simultaneously, and this is possible since the number of equations is equal to the number of unknown parameters, we find that the formulas for the parameters β_i and β_{ij}, $i, j = 1, 2$, and 3, $i < j$, are

$$\begin{aligned} \beta_1 &= \eta_1, & \beta_{12} &= 4\eta_{12} - 2\eta_1 - 2\eta_2 \\ \beta_2 &= \eta_2, & \beta_{13} &= 4\eta_{13} - 2\eta_1 - 2\eta_3 \\ \beta_3 &= \eta_3, & \beta_{23} &= 4\eta_{23} - 2\eta_2 - 2\eta_3 \end{aligned} \tag{2.12}$$

The parameter β_i represents the response to pure component i, and β_{ij} is a contrast that compares the response at the midpoint of the edge connecting the vertices of components i and j with the responses at the vertices of components i and j. Thus in the six-term polynomial Eq. (2.11), the sum $\beta_1 x_1 + \beta_2 x_2 + \beta_3 x_3$ represents linear or additive blending of the three components while the extra terms $\beta_{ij} x_i x_j$, $i < j$, are said to represent measures of the deviations from the plane of the second-degree surface resulting from the nonadditive blending of the components.

Equations (2.12) are derived using only three components for reasons of convenience. It is easy to display the expected responses at the points of the $\{3, 2\}$ simplex-lattice and only six equations are necessary for setting up the formulas for the β_i and β_{ij}, $i, j = 1, 2$, and 3, $i < j$. For the general case of q components where the second-degree model of Eq. (2.7) contains $q(q+1)/2$ terms and the expected responses are measured at the points of a $\{q, 2\}$ simplex-lattice design, the formulas for expressing the parameters β_i and β_{ij} in terms of the η_i and η_{ij}, are identical to Eqs. (2.12). In other words, for general q, where $i, j = 1, 2, \ldots, q$, $i < j$,

$$\beta_i = \eta_i, \qquad \beta_{ij} = 4\eta_{ij} - 2(\eta_i + \eta_j) \tag{2.13}$$

For higher degree cases $m > 2$, the formulas can be derived in a manner similar to that for the second-degree model. Gorman and Hinman (1962) present the formulas for the parameters in the cubic and quartic polynomial equations. Also, see Appendix 2B.

2.4. ESTIMATING THE PARAMETERS IN THE $\{q, m\}$ POLYNOMIALS

The parameters in the $\{q, m\}$ polynomials are expressible as simple functions of the expected responses at the points of the $\{q, m\}$ simplex-lattice designs. Thus one might conjecture that to estimate the parameters in the models using observed values of the response at the lattice points, the computing formulas for b_i and b_{ij}, the estimates of β_i and β_{ij}, respectively, will be identical to Eqs. (2.13) with the observed values substituted in Eq. (2.13) in place of η_i and η_{ij}. To show that this is the case, we shall consider the fitting of the three-component second-degree model in Eq. (2.11) to data values collected at the points of a $\{3, 2\}$ simplex-lattice design.

We recall from Section 1.2 that the observed value of the response in the uth trial, $1 \le u \le N$, denote by y_u, is expressible in terms of the expected or true response η_u in the form $y_u = \eta_u + \epsilon_u$ where the ϵ_u for all $u \le 1 \le N$, are independent and identically distributed random errors assumed to have a zero mean and a variance σ^2. Now let us alter this notation temporarily by writing the observed response with the same nomenclature that was used for the expected response, that is, we denote the observed value of the response to the pure component i (i.e., at $x_i = 1$, $x_j = 0$, $j \ne i$) by y_i and the observed value of the response to the 50% : 50% binary mixture ($x_i = \frac{1}{2}$, $x_j = \frac{1}{2}$, $x_k = 0$ for all $i < j \ne k$) of components i and j by y_{ij}. Replacing the η_i and η_{ij} with y_i and y_{ij}, respectively, in Eq. (2.13) and letting b_i and b_{ij} denote the estimates of β_i and β_{ij}, respectively, we find that

$$b_i = y_i, \qquad i = 1, 2, \ldots, q$$
$$b_{ij} = 4y_{ij} - 2(y_i + y_j), \qquad i, j = 1, 2, \ldots, q, \qquad i < j \tag{2.14}$$

or

$$\frac{b_{ij}}{4} = y_{ij} - \frac{y_i + y_j}{2}$$

This last contrast says that the quantity $b_{ij}/4$ represents the amount the surface deviates, at the blend $x_i = x_j = \frac{1}{2}$, from the plane that connects the vertices corresponding to components i and j. Furthermore, if r_i and r_{ij} replicate observations are collected at $x_i = 1$, $x_j = 0$, $j \neq i$, and at $x_i = x_j = \frac{1}{2}$, $x_k = 0$, $i < j$, $k \neq i, j$, respectively, then the averages $\bar{y}_i$ and $\bar{y}_{ij}$ are substituted into Eq. (2.14) and the calculating formulas for the parameter estimates become

$$b_i = \bar{y}_i, \qquad i = 1, 2, \ldots, q$$
$$b_{ij} = 4\bar{y}_{ij} - 2(\bar{y}_i + \bar{y}_j), \qquad i, j = 1, 2, \ldots, q, \qquad i < j \tag{2.15}$$

The formulas for calculating the estimates of the parameters in the cubic model and the quartic or fourth-degree polynomial equation are presented in Appendix 2B at the end of this chapter.

Equations (2.15) for the estimates b_i and b_{ij} are the solutions to the normal equations that are shown as Eqs. (7.6) in Chapter 7. The matrix formula for the least squares estimates is Eq. (7.7). The properties of the estimates depend on the distributional properties of the random errors ϵ_u. We have assumed that the errors ϵ_u, $1 \leq u \leq N$, are independent and identically distributed with mean zero and a variance of σ^2. Thus the means and the variances of the distributions of the estimates b_i and b_{ij}, given that the observations were collected at the points of the lattice only, are

$$E(b_i) = E(\bar{y}_i) = \beta_i$$

$$\text{var}(b_i) = \text{var}(\bar{y}_i) = \frac{\sigma^2}{r_i}$$

$$E(b_{ij}) = E[4\bar{y}_{ij} - 2(\bar{y}_i + \bar{y}_j)] = \beta_{ij} \tag{2.16}$$

$$\text{var}(b_{ij}) = \text{var}[4\bar{y}_{ij} - 2(\bar{y}_i + \bar{y}_j)] = \frac{16\sigma^2}{r_{ij}} + \frac{4\sigma^2}{r_i} + \frac{4\sigma^2}{r_j}$$

and

$$\text{cov}(b_i, b_j) = E[\bar{y}_i(\bar{y}_j)] - E(\bar{y}_i)E(\bar{y}_j) = 0, \qquad i \neq j$$

$$\begin{aligned}\text{cov}(b_i, b_{ij}) &= E[\bar{y}_i(4\bar{y}_{ij} - 2\bar{y}_i - 2\bar{y}_j)] - E(\bar{y}_i)E[4\bar{y}_{ij} - 2\bar{y}_i - 2\bar{y}_j] = \\ &\quad - 2E(\bar{y}_i^2) + 2(E\bar{y}_i)^2 = \frac{-2\sigma^2}{r_i}\end{aligned} \tag{2.17}$$

$$\text{cov}(b_{ij}, b_{ik}) = \frac{4\sigma^2}{r_i}, \qquad j \neq k$$

Furthermore, if the errors are assumed to be normally distributed, that is, if the $\epsilon_u \sim \text{normal}(0, \sigma^2)$, and if an equal number of replicates $r_i = r_{ij} = r$ is collected at each of the design points, then $b_i \sim \text{normal}(\beta_i, \sigma^2/r)$ and $b_{ij} \sim \text{normal}(\beta_{ij}, 24\sigma^2/r)$, where $\sim$ denotes "is distributed as."

Once the parameters in the second-degree model in Eq. (2.7) have been estimated, the estimates b_i and b_{ij} are substituted into Eq. (2.7) for the β_i and β_{ij} respectively, and the fitted equation is

$$\hat{y} = \sum_{i=1}^{q} b_i x_i + \sum_{i<j}^{q}\sum b_{ij} x_i x_j \tag{2.18}$$

An estimate of the value of the response at the point $\mathbf{x} = (x_1, x_2, \ldots, x_q)'$ in the simplex region is found by substituting the values of the x_i into Eq. (2.18). We denote the estimate of η at $\mathbf{x}$ by $\hat{y}(\mathbf{x})$.

2.5. PROPERTIES OF THE ESTIMATE OF THE RESPONSE $\hat{y}(\mathbf{x})$

Since the estimates b_i and b_{ij} are linear functions of random variables (the y_i and y_{ij}) and are therefore themselves random variables, the estimate $\hat{y}(\mathbf{x})$ of the response at $\mathbf{x}$ is a random variable. When the estimates b_i and b_{ij} are unbiased, which is the case when the fitted model is of the same degree in the x_i's as the true surface, then the expectation of $\hat{y}(\mathbf{x})$ is $E[\hat{y}(\mathbf{x})] = \eta$.

A formula for the variance of the estimate $\hat{y}(\mathbf{x})$ can be written in terms of the variances and covariances of the b_i and b_{ij}, which are given in Eqs. (2.16) and (2.17). An easier method for obtaining the variance of $\hat{y}(\mathbf{x})$, on the other hand, is to replace the parameter estimates b_i and b_{ij} by their respective linear combinations of the averages $\bar{y}_i$ and $\bar{y}_{ij}$, which are defined in Eq. (2.15). The variance of $\hat{y}(\mathbf{x})$ can then be written as a function of the variances of the $\bar{y}_i$ and $\bar{y}_{ij}$.

Written as a function of the averages $\bar{y}_i$ and $\bar{y}_{ij}$ at the lattice points, the estimate of the response is

$$\begin{aligned}\hat{y}(\mathbf{x}) &= \sum_{i=1}^{q} b_i x_i + \sum_{i<j}^{q}\sum b_{ij}x_i x_j \\ &= \sum_{i=1}^{q} \bar{y}_i x_i + \sum_{i<j}^{q}\sum (4\bar{y}_{ij} - 2\bar{y}_i - 2\bar{y}_j)x_i x_j \\ &= \sum_{i=1}^{q} \bar{y}_i \left[x_i - 2x_i\left(\sum_{j\neq i}^{q} x_j\right)\right] + \sum_{i<j}^{q}\sum 4\bar{y}_{ij}x_i x_j \\ &= \sum_{i=1}^{q} a_i \bar{y}_i + \sum_{i<j}^{q}\sum a_{ij}\bar{y}_{ij} \end{aligned} \tag{2.19}$$

where $a_i = x_i(2x_i - 1)$ and $a_{ij} = 4x_i x_j$, $i, j = 1, 2, \ldots, q, i < j$. In the coefficients a_i and a_{ij}, the values of the x_i are specified by the values in $\mathbf{x} = (x_1, \ldots, x_q)'$ and thus are fixed without error. Since the $\bar{y}_i$ and $\bar{y}_{ij}$ are averages of r_i and r_{ij} observations, respectively, then the variance of $\hat{y}(\mathbf{x})$ in Eq. (2.19) can be written as

$$\text{var}[\hat{y}(\mathbf{x})] = \sigma^2\left\{\sum_{i=1}^{q} \frac{a_i^2}{r_i} + \sum_{i<j}^{q}\sum \frac{a_{ij}^2}{r_{ij}}\right\} \tag{2.20}$$

Of course, when there is an equal number of observations, r, at each lattice point, the formula for the variance of the estimate of the response at the point $\mathbf{x}$, is simplified to

$$\text{var}[\hat{y}(\mathbf{x})] = \frac{\sigma^2}{r}\left\{\sum_{i=1}^{q} a_i^2 + \sum_{i<j}^{q}\sum a_{ij}^2\right\} \tag{2.21}$$

In Eq. (2.21) the quantity σ^2/r is dependent on the precision (through σ^2) of the experimental observations, while $\{\Sigma_{i=1}^{q} a_i^2 + \Sigma\Sigma_{i<j}^{q} a_{ij}^2\}$ is dependent only on the composition (through the a_i's and thus the x_i's) of the mixture at which the estimate $\hat{y}(\mathbf{x})$ is being considered. When σ^2 is unknown, then an estimate, s^2, is calculated from the r_i and r_{ij} replicate observations. An estimate of var$[\hat{y}(\mathbf{x})]$ is obtained by substituting s^2 for σ^2 in Eq. (2.20) and is written as $\widehat{\text{var}}[\hat{y}(\mathbf{x})]$. Finally, if, at $\mathbf{x}$, a $(1-\alpha)\times 100$ percent confidence interval for η is desired, then the interval is

$$\Pr[\hat{y}(\mathbf{x}) - \Delta < \eta < \hat{y}(\mathbf{x}) + \Delta] = 1 - \alpha \tag{2.22}$$

where $\Delta = [t_{f,\alpha/2}]\sqrt{\widehat{\text{var}}[\hat{y}(\mathbf{x})]}$, $p =$ number of terms in the model, $f =$ number of degrees of freedom for estimating σ^2 in Eq. (2.21), and $t_{f,\alpha/2}$ is the tabled t-value with f degrees of freedom at the $\alpha/2$ level of significance. The general formula for the variance of $\hat{y}(\mathbf{x})$, using matrix notation, is presented in Appendix 2A at the end of this chapter.

We now illustrate the fitting of the second-degree polynomial using data collected from a three-component yarn-manufacturing experiment. We shall use the example data to introduce methods for determining how well the fitted model represents the response surface.

2.6. A THREE-COMPONENT YARN EXAMPLE USING A {3, 2} SIMPLEX-LATTICE DESIGN

Three constituents—polyethylene (x_1), polystyrene (x_2), and polypropylene (x_3)—were blended together and the resulting fiber material was spun to form yarn for draperies. Only pure blends and binary blends are studied in this example where the response of interest is the thread elongation of the yarn measured in kilograms of force applied. Averages of the thread elongation values are presented in Figure 2.6 at the points of the {3, 2} simplex-lattice design where the averages were calculated from two replicate samples collected from each of the pure blends and three replicate samples collected on the binary or two-component blends. The mixture settings, the observed thread elongation values and the averages of the elongation values are presented in Table 2.4.

The fitted model in the three components is of the form

$$\hat{y}(\mathbf{x}) = b_1x_1 + b_2x_2 + b_3x_3 + b_{12}x_1x_2 + b_{13}x_1x_3 + b_{23}x_2x_3$$

where from Eq. (2.15), the estimates are

$$\begin{aligned} b_1 &= \bar{y}_1 = (11.0 + 12.4)/2 = 11.7, \qquad b_2 = 9.4, \qquad b_3 = 16.4 \\ b_{12} &= 4\bar{y}_{12} - 2(\bar{y}_1 + \bar{y}_2) = 4(15.3) - 2(11.7 + 9.4) = 61.2 - 2(21.1) = 19.0 \\ b_{13} &= 4\bar{y}_{13} - 2(\bar{y}_1 + \bar{y}_3) = 4(16.9) - 2(11.7 + 16.4) = 11.4, \qquad b_{23} = -9.6 \end{aligned} \tag{2.23}$$

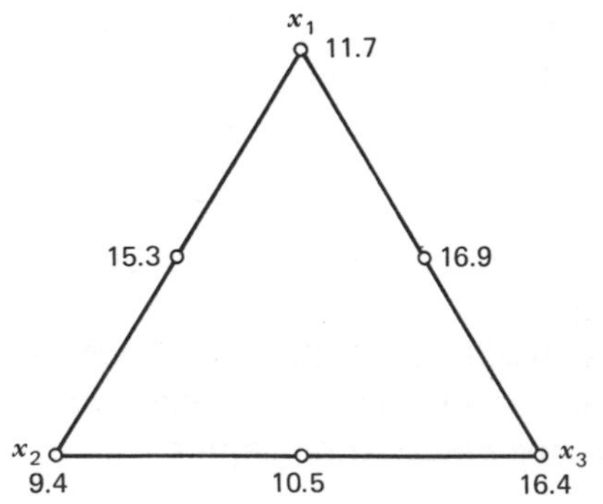

FIGURE 2.6. Average thread elongation values at the design points.

TABLE 2.4. Observed yarn elongation values

Design Point	Component Proportions x_1	x_2	x_3	Observed Elongation Values (y_u)	Average Elongation Value ($\bar{y}$)
1	1	0	0	11.0, 12.4	11.7
2	$\frac{1}{2}$	$\frac{1}{2}$	0	15.0, 14.8, 16.1	15.3
3	0	1	0	8.8, 10.0	9.4
4	0	$\frac{1}{2}$	$\frac{1}{2}$	10.0, 9.7, 11.8	10.5
5	0	0	1	16.8, 16.0	16.4
6	$\frac{1}{2}$	0	$\frac{1}{2}$	17.7, 16.4, 16.6	16.9

An estimate of the error variance σ^2 is obtained from the replicate observations at the lattice points. The estimate is

$$s^2 = \sum_{l=1}^{6} \sum_{u=1}^{2 \text{ or } 3} \frac{(y_{lu} - \bar{y}_l)^2}{\sum_{l=1}^{6} (r_l - 1)}$$

$$= \frac{(11.0-11.7)^2 + (12.4-11.7)^2 + (15.0-15.3)^2 + \cdots + (16.6-16.9)^2}{1+2+1+2+1+2}$$

$$s^2 = \frac{6.56}{9} = 0.73$$

where in the formula for s^2, $\bar{y}_l$ is the average of the r_l observations at the lth design point. The number of degrees of freedom for the estimate s^2 is 9 and this number appears in the denominator of the quotient for s^2. Estimates of the variances of the parameter estimates in (2.23) are obtained from Eqs. (2.16) along with $s^2 = 0.73$, to give

$$\widehat{\text{var}}(b_i) = \frac{s^2}{r_i} = \frac{0.73}{2} = 0.37, \qquad i = 1, 2 \text{ and } 3$$

$$\widehat{\text{var}}(b_{ij}) = s^2\left\{\frac{16}{r_{ij}} + \frac{4}{r_i} + \frac{4}{r_j}\right\} = 0.73\left\{\frac{16}{3} + 4\right\} = 6.81$$

The estimated standard error of each parameter estimate is the positive square root of the estimated variance of the estimate. The estimated standard errors are est. s.e.$(b_i) = \sqrt{\widehat{\text{var}}(b_i)}$ and est. s.e.$(b_{ij}) = \sqrt{\widehat{\text{var}}(b_{ij})}$, respectively. The values of the estimates of the standard errors are est. s.e.$(b_i) = 0.61$ and est. s.e.$(b_{ij}) = 2.61$. We shall adopt the practice of placing the estimates of the standard errors in parentheses directly below the corresponding parameter estimate in the fitted equation.

The fitted second-degree polynomial equation is

$$\hat{y}(\mathbf{x}) = \underset{(0.61)}{11.7}x_1 + \underset{(0.61)}{9.4}x_2 + \underset{(0.61)}{16.4}x_3 + \underset{(2.61)}{19.0}x_1x_2 + \underset{(2.61)}{11.4}x_1x_3 - \underset{(2.61)}{9.6}x_2x_3 \quad (2.24)$$

If we can assume that the fitted model in Eq. (2.24) is an adequate representation of the yarn elongation surface, we can draw the following conclusions from the magnitudes of the parameter estimates:

$$b_3 > b_1 > b_2$$

Of the three pure ingredients, component 3 (polypropylene) produced yarn with the highest elongation, followed by component 1 and then component 2.

$$b_{12} > 0, \qquad b_{13} > 0$$

Components 1 and 2, and components 1 and 3, have binary synergistic effects, that is, the binary blends with component 1 produced higher elongation values than would be expected by simply averaging the elongation values of the pure blends. Components 2 and 3 when blended produced yarn whose average elongation was lower than the average of the two pure blends.

Our conclusions then are, if yarn with high elongation is desirable and a single component blend is wanted, use component 3. If a binary blend is desired because either component 3 costs more than the others or because of the lack of availability of component 3, use component 1 with either of the other two components. A plot of the elongation surface is presented in Figure 2.7.

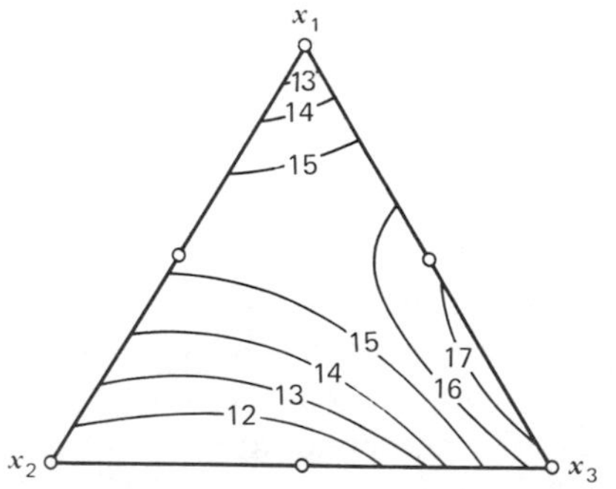

FIGURE 2.7. Contours of constant estimated thread elongation values obtained with the second-degree model.

The formula for the estimated variance of $\hat{y}(\mathbf{x})$ at the point $\mathbf{x}$ in the three-component triangle is, from Eq. (2.20),

$$\widehat{\text{var}}[\hat{y}(\mathbf{x})] = s^2\left\{\sum_{i=1}^{3}\frac{a_i^2}{r_i} + \sum_{i<j}^{3}\sum\frac{a_{ij}^2}{r_{ij}}\right\} = 0.73\left\{\sum_{i=1}^{3}\frac{a_i^2}{2} + \sum_{i<j}^{3}\sum\frac{a_{ij}^2}{3}\right\} \quad (2.25)$$

where $a_i = x_i(2x_i - 1)$ and $a_{ij} = 4x_ix_j$. At the point $(x_1 = \frac{2}{3}, x_2 = \frac{1}{3}, x_3 = 0)$, for example, the coefficients are $a_1 = (\frac{2}{3})(\frac{1}{3}) = \frac{2}{9}$, $a_2 = (\frac{1}{3})(-\frac{1}{3}) = -\frac{1}{9}$, $a_3 = (0)$, $a_{12} = 4(\frac{2}{3})(\frac{1}{3}) = \frac{8}{9}$, $a_{13} = 0$, $a_{23} = 0$, and the estimated variance of $\hat{y}(\frac{2}{3}, \frac{1}{3}, 0)$ is

$$\widehat{\text{var}}[\hat{y}(\mathbf{x})] = 0.73\left\{\frac{(\frac{2}{9})^2 + (\frac{1}{9})^2 + (0)^2}{2} + \frac{(\frac{8}{9})^2 + (0)^2 + (0)^2}{3}\right\} = 0.73\{\tfrac{143}{486}\} = 0.21$$

The estimate of the response at the point $\mathbf{x} = (\frac{2}{3}, \frac{1}{3}, 0)'$ is

$$\begin{aligned}\hat{y}(\mathbf{x}) &= 11.7(\tfrac{2}{3}) + 9.4(\tfrac{1}{3}) + 16.4(0) + 19.0(\tfrac{2}{3})(\tfrac{1}{3}) + 11.4(\tfrac{2}{3})(0) - 9.6(\tfrac{1}{3})(0) \\ &= 15.2\end{aligned}$$

and a 95% confidence interval for η using Eq. (2.22) and letting $\alpha = 0.05$ so that $\alpha/2 = 0.05/2 = 0.025$, is

$$\Pr[15.2 - \Delta < \eta < 15.2 + \Delta] = 0.95 = \Pr[14.2 < \eta < 16.2]$$

where $\Delta = [t_{9,(0.025)} = 2.262]\sqrt{0.21} = 1.0$.

2.7. THE ANALYSIS OF VARIANCE TABLE

In the last section a second-degree polynomial equation was fitted to observed yarn elongation values and the objective was to try to understand how the variation or differences in the elongation values could be explained (or accounted for) in terms of the effects of the pure components and in terms of the binary effects of pairs of components. We shall now show how to partition or separate the overall variation in the elongation values into two assigned sources; the first being the variation among the averages of the elongation values attributed to the different blends and the second being the measure of variation among the replicate samples within each blend.

The variation among the six blends in the yarn elongation example is measured by computing the differences between each average blend and the overall average, squaring the differences, weighting the squared quantities by the number of replicates in each average blend and summing the weighted quantities. Computationally, we would use the equation

$$\text{Sum of squares among blends} = \sum_{l=1}^{6} r_l(\bar{y}_l - \bar{y})^2 \tag{2.26}$$

where r_l is the number of replicate observations of the lth blend and $\bar{y}_l$ is the average of the r_l replicate observations of the lth blend. The sum of squares in Eq. (2.26) has $6-1=5$ degrees of freedom associated with it. If there had been p different blends selected over the simplex-lattice, then $l=1,2,\ldots,p$ and the sum of squares among blends would have $p-1$ degrees of freedom.

Since the $\{q, m\}$ polynomial was fitted to the data collected at the points of the $\{q, m\}$ simplex-lattice design, the number of terms in the model must equal the number of different blends defined by the design. This number is $\binom{q+m-1}{m}$ which in our example is $\binom{3+2-1}{2}=6$. Thus the variation in the observations explained by the fitted model and which is called the "sum of squares due to regression" or "sum of squares due to the fitted model" is the same as the sum of squares among the blends in Eq. (2.26). An alternative formula to Eq. (2.26) for calculating the sum of squares due to regression, designated as SSR, is

$$\text{SSR} = \sum_{u=1}^{N} (\hat{y}_u - \bar{y})^2 \tag{2.27}$$

where $\hat{y}_u$ is the predicted value of y_u using the fitted model (actually $\hat{y}_u$ is the estimate of η at the uth setting of the mixture components obtained with the fitted model) and $\bar{y}$ is the overall average of the observations.

The variation among the replicate observations within the blends is not accounted for (or explained) by the differences among the blends (nor by the fitted model) and is referred to as the residual variation. The formula for the residual sum of squares, denoted by SSE, is

$$\text{SSE} = \sum_{u=1}^{N} (y_u - \hat{y}_u)^2 \tag{2.28}$$

and SSE has $N-p$ degrees of freedom where p is the number of different blends or, as in this case, the number of terms in the model.

Using Eqs. (2.27) and (2.28), the *analysis of variance* table for the fitted model, containing p terms, is shown as Table 2.5.

TABLE 2.5. The analysis of variance table

Source of Variation	Degrees of Freedom	Sum of Squares	Mean Square
Regression (Fitted Model)	$p-1$	$SSR = \sum_{u=1}^{N} (\hat{y}_u - \bar{y})^2$	$SSR/(p-1)$
Residual	$N-p$	$SSE = \sum_{u=1}^{N} (y_u - \hat{y}_u)^2$	$SSE/(N-p)$
Total	$N-1$	$SST = \sum_{u=1}^{N} (y_u - \bar{y})^2$	

2.8. ANALYSIS OF VARIANCE CALCULATIONS OF THE YARN ELONGATION DATA

The analysis of variance table for the yarn elongation example of the previous section is Table 2.7 where the deviations used in calculating the sums of squares quantities are listed in Table 2.6.

Fitted model:

$$\hat{y}(\mathbf{x}) = 11.7x_1 + 9.4x_2 + 16.4x_3 + 19.0x_1x_2 + 11.4x_1x_3 - 9.6x_2x_3$$

Overall average:

$$\bar{y} = \sum_{u=1}^{15} y_u/15 = 203.1/15 = 13.5$$

The value of the F-ratio in Table 2.7 is found by comparing the mean squares for regression and residual.

The mean squares for regression and for residual are functions of the squares of the observed y_u random variables and as such are themselves random variables. It can be shown that if the assumed model is correct, the expected values of the mean squares for regression and for residual are

$$E(\text{mean square residual}) = \sigma^2$$
$$E(\text{mean square regression}) = \sigma^2 + f(\beta_1, \beta_2, \ldots, \beta_{23})$$

where the quantity $f(\beta_1, \beta_2, \ldots, \beta_{23})$ equals zero if the surface $\eta = \beta_1x_1 + \beta_2x_2 + \cdots + \beta_{23}x_2x_3 = \beta$ is a level plane above the simplex (i.e.,

TABLE 2.6. Sum of squared deviations for yarn elongation data

Observed Values (y_u)	Predicted Value ($\hat{y}_u$)	Differences or Residuals ($y_u - \hat{y}_u$)	Deviations ($y_u - \bar{y}$)	Regression Deviations ($\hat{y}_u - \bar{y}$)
11.0	11.7	−0.7	−2.5	−1.8
12.4	11.7	0.7	−1.1	−1.8
15.0	15.3	−0.3	1.5	1.8
14.8	15.3	−0.5	1.3	1.8
16.1	15.3	0.8	2.6	1.8
10.0	9.4	0.6	−3.5	−4.1
8.8	9.4	−0.6	−4.7	−4.1
10.0	10.5	−0.5	−3.5	−3.0
9.7	10.5	−0.8	−3.8	−3.0
11.8	10.5	1.3	−1.7	−3.0
16.8	16.4	0.4	3.3	2.9
16.0	16.4	−0.4	2.5	2.9
17.7	16.9	0.8	4.2	3.4
16.4	16.9	−0.5	2.9	3.4
16.6	16.9	−0.3	3.1	3.4
		$\sum_{u=1}^{15} (y_u - \hat{y}_u)^2 = 6.56$	$\sum_{u=1}^{15} (y_u - \bar{y})^2 = 134.88$†	$\sum_{u=1}^{15} (\hat{y}_u - \bar{y})^2 = 128.32$

†When deviations are not rounded to tenths, SST = 134.86.

TABLE 2.7. The analysis of variance table for the yarn elongation example

Source of Variation	Degrees of Freedom	Sum of Squares	Mean Square	F-ratio
Regression	5	128.32	25.66	25.66/0.73 = 35.1
Residual	9	6.56	0.73	
Total	14	134.88		

$\beta_1 = \beta_2 = \beta_3 = \beta$, $\beta_{12} = \beta_{13} = \beta_{23} = 0$). For the F-ratio test in Table 2.7, we assume that the errors ϵ_u are independent, normal$(0, \sigma^2)$ variables. Then if the null hypothesis $H_0: \beta_1 = \beta_2 = \beta_3 = \beta$, $\beta_{12} = \beta_{13} = \beta_{23} = 0$ is true, implying the surface above the simplex or triangle is a level plane whose height is the same at all points, the ratio

$$\frac{(p-1) \times (\text{mean square regression})}{\{\sigma^2 + f(\beta_1, \beta_2, \ldots, \beta_{23})\} = \sigma^2} \sim \chi^2_{(p-1)} \tag{2.29}$$

is distributed as a chi square random variable with $p-1$ degrees of freedom, and further

$$\frac{(N-p)\times(\text{mean square residual})}{\sigma^2} \sim \chi^2_{(N-p)} \tag{2.30}$$

and the two chi square random variables in Eqs. (2.29) and (2.30) are independent. Since the ratio of two independent χ^2's, over their respective degrees of freedom, is an F random variable, we have

$$F = \frac{(p-1)\times(\text{mean square regression})/(p-1)\sigma^2}{(N-p)\times(\text{mean square residual})/(N-p)\sigma^2} \tag{2.31}$$

and the F-distribution has $(p-1)=5$ and $(N-p)=9$ degrees of freedom in the numerator and denominator, respectively. The value of the F-ratio in Eq. (2.31) is compared with the $100(1-\alpha)\%$ point of the tabulated $F_{(p-1,N-p,\alpha)}$ distribution at the end of the book to see if the null hypothesis should be rejected or not. Since the F-value in Table 2.7 is $F=35.1$ and this value exceeds $F_{(5,9,\alpha=0.01)}=6.06$, we reject $H_0:\beta_1=\beta_2=\beta_3=\beta$, $\beta_{12}=\beta_{13}=\beta_{23}=0$ at the $\alpha=0.01$ level in favor of at least one equality being false. We shall present some additional discussion on testing hypotheses in Chapter 5.

The F-test performed on the mean squares in Table 2.7 prompted us to reject the hypothesis that the elongation surface was a planar surface of constant height above the triangle. It is not clear by the rejection of the level plane whether or not the surface is a plane. It might be a plane that is tilted such as shown in Figure 2.3, or the surface may possess curvature and not be a plane at all. If the surface is a tilted plane whose heights above the three vertices are unequal, then $H_0:\beta_1=\beta_2=\beta_3=\beta$, $\beta_{12}=\beta_{13}=\beta_{23}=0$ is rejected in favor of $H_A:\beta_{12}=\beta_{13}=\beta_{23}=0$ and $\beta_1\neq\beta_2=\beta_3$ or $\beta_1=\beta_2\neq\beta_3$ or $\beta_1\neq\beta_2\neq\beta_3$. If the surface possess curvature, then $H_0:\beta_1=\beta_2=\beta_3=\beta$, $\beta_{12}=\beta_{13}=\beta_{23}=0$ is rejected in favor of $H_A:\beta_{12}=\beta_{13}=\beta_{23}\neq 0$ and $\beta_1=\beta_2=\beta_3$ or $\beta_1\neq\beta_2=\beta_3$ or $\beta_1=\beta_2\neq\beta_3$ or $\beta_1=\beta_2\neq\beta_3$. Also, another statistic which we shall find useful quite often when checking the closeness of the fitted model to the observed data values is the adjusted multiple correlation coefficient R_A where $R_A^2=1-[(N-1)\text{SSE}/(N-p)\text{SST}]$, see Section 7.7. With the yarn elongation data, the value of R_A^2 with the fitted second-degree model is $R_A^2=1-[(14)6.56/(9)134.86]=0.924$.

2.9. THE PLOTTING OF INDIVIDUAL RESIDUALS

A measure of the closeness of the predicted surface by the proposed model to observed values of the response at the design points can be obtained by computing the differences $y_u - \hat{y}_u$, $u = 1, 2, \ldots, N$, where y_u is the observed value of the response and $\hat{y}_u$ is the predicted value of the response for the uth trial. These differences are called *residuals*. An approximate average of the squares of the residuals was used to obtain an estimate of the observation or error variance when SSE in Eq. (2.28) was divided by $N - p$ but often it is equally important to examine the residuals individually as well as jointly for detecting inadequacies in the model. This is because the sizes of the residuals can be made large by factors which were not accounted for in the proposed model as well as by not utilizing the strengths of the important components to the correct degree with the fitted model.

In order that the individual residuals $y_u - \hat{y}_u$, $u = 1, 2, \ldots, N$, be meaningful, the number of distinct design points, say t, must exceed the number of terms in the model. This is because when the number of different parameters to be estimated is equal to the number of distinct design points, the model is fitted to the average response at each design point forcing the predicted value $\hat{y}_u$ at design point l to be equal to $\bar{y}_l$, the average response at design point l, $l = 1, 2, \ldots, t$. When it happens that $\hat{y}_u = \bar{y}_l$, then the residuals $y_u - \hat{y}_u = y_u - \bar{y}_l$ at design point l are correlated with one another. The presence of correlation between residuals is seen by looking at the residuals in Table 2.6, where at design point 1, for example, $r_1 = -0.7$, and $r_2 = -r_1 = 0.7$.

Disturbances by outside factors are more likely to be detected when they affect the behavior of uncorrelated residuals than when they affect the behavior of correlated residuals. Thus we shall strive to work with uncorrelated residuals as often as possible when studying the sizes of the residuals to judge the adequacy of the fitted model.

Residuals may be plotted in various ways, using different types of plotting paper. First a general inspection of the signs and of the sizes can be carried out with just a dot diagram on standard grid paper. (See Box, Hunter, and Hunter, 1978, Chapter 6.) If the assumption is true that the residuals are independently and identically distributed normal$(0, \sigma^2)$, the plot will have roughly the appearance of a sample from a normal distribution centered at zero. Of course, unless the number of residuals is sufficiently large (say $N \geq 30$ for a rough rule of thumb), it is difficult to envision any shape tendencies in the appearance of the distribution. Daniel and Wood (1971) suggest several techniques for studying residual patterns, but state that none of the techniques can be expected to work well with fewer than 20 observations.

We shall not present a plot of the sizes of the residuals from Table 2.6 at this time but a dot diagram is presented in the next section. As we shall see, the use of the dot diagram can be an effective way of comparing the fits of competing models. In fact we compare the fit of the second-degree model of Eq. (2.24) to the elongation data with the fit of a first-degree model to the same data by noting the reduction in the sizes or spread of the residuals with the two models. A diagram of residuals is presented in Figure 2.9.

Plots of residuals $y_u - \hat{y}_u$ versus the individual values of y_u, or of $\hat{y}_u$, or values of $y_u - \hat{y}_u$ versus the sizes of the individual values of the x_i or linear combinations of the x_i are other techniques that are often useful. Still another statistic, the C_p statistic, can be used to compare the fits of competing models. This statistic measures the sum of the squared biases, plus the squared random errors or residuals at all data points. The form of C_p is given in Question 2.12. (See Daniel and Wood, 1971, Chapter 4.)

2.10. TESTING THE DEGREE OF THE FITTED MODEL: A QUADRATIC MODEL OR PLANAR MODEL?

The fitting of the second-degree yarn elongation model of Eq. (2.24) provided information on each of the components individually as well as on pairs of components. At the conclusion of the example, component 1 was said to blend in a synergistic manner with each of components 2 and 3, while components 2 and 3 were said to blend in an antagonistic way. These claims were made because of the sizes and signs of the estimates b_{12}, b_{13}, and b_{23}.

The question we ask ourselves now is, "If we refit the data using only a first-degree model in x_1, x_2, and x_3, would the first-degree model fit the 15 elongation values as well as the second-degree equation (2.24)?" That is to say, is there enough evidence to reject $H_0: \beta_{12} = \beta_{13} = \beta_{23} = 0$ because of the falsehood of one or more of the equality signs?

The least squares estimates of the parameters in the first-degree polynomial $\eta = \beta_1 x_1 + \beta_2 x_2 + \beta_3 x_3$ are easily calculated using matrix algebra, even when the data is collected at the points of the {3, 2} simplex-lattice. (See Section 2.13 on the use of axial designs, or Appendix 2A at the end of this chapter, or Section 7.3 for a discussion of the matrix formulas.) The first-degree model fitted to the 15 yarn elongation values is

$$\hat{y}(\mathbf{x}) = \underset{(1.41)}{15.0x_1} + \underset{(1.41)}{9.8x_2} + \underset{(1.41)}{15.8x_3} \qquad (2.32)$$

where the numbers in parentheses below the parameter estimates are the estimated standard errors of the estimates. The surface contours estimated with the fitted model (2.32) are shown in Figure 2.8.

The sums of squares quantities associated with the fitted first-degree model of Eq. (2.32) are

$$SSR = \sum_{u=1}^{N=15} (\hat{y}_u - \bar{y})^2 = 57.63$$

$$SSE = \sum_{u=1}^{N=15} (y_u - \hat{y}_u)^2 = 77.23$$

$$SST = \sum_{u=1}^{N=15} (y_u - \bar{y})^2 = 134.86$$

A test for comparing the first- and second-degree models in terms of how well they account for or explain the variation in the response values uses the residual sum of squares associated with each of the models. The test statistic is an F-ratio of the form (see Section 7.7).

$$F = \frac{(SSE_{reduced} - SSE_{complete})/r}{SSE_{complete}/(N-p)} \tag{2.33}$$

where $SSE_{reduced}$ = the residual sum of squares associated with the reduced model (the first-degree equation of Eq. (2.32))

$SSE_{complete}$ = the residual sum of squares associated with the complete model (the second-degree equation of Eq. (2.24))

r = the difference in the number of parameters in the complete and reduced models ($r = 6 - 3 = 3$), and p is the number of parameters in the complete model ($p = 6$)

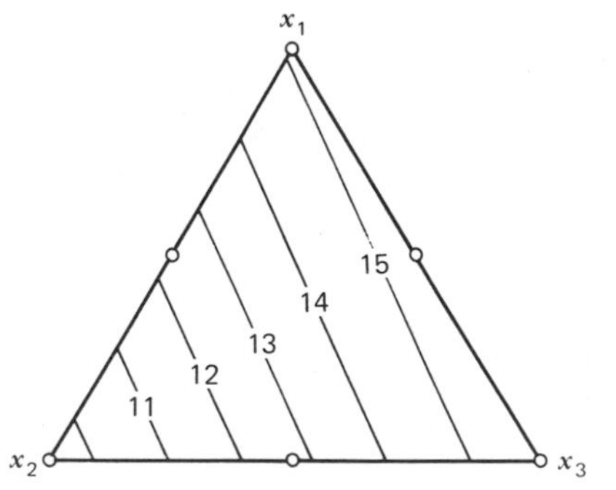

FIGURE 2.8. Surface contours of the estimated planar surface of Eq. (2.32).

In the F-test of Eq. (2.33), the reduction in the sum of squared residuals (or unexplained variation) that is achieved when going from the first-degree to the second-degree model, divided by the number of additional terms in the second-degree model, is compared to the estimate of the observation or error variance found with the complete second-degree model. With the yarn elongation data, the value of the F-ratio in Eq. (2.33) is

$$F = \frac{(77.23 - 6.56)/3}{6.56/9} = 32.2 \tag{2.34}$$

and since $F = 32.2$ exceeds $F_{(3,9,0.01)} = 6.99$, we conclude that the drop in the residual variation is large enough, relative to $s^2 = 0.73$, to justify using the second-degree model. The decrease in the sizes of the individual residuals from fitting the second-degree model is illustrated in Figure 2.9 where the dot diagram is a plot of the residuals from both models.

Having decided from the F-test in Eq. (2.34) or from the plot of the residuals in Figure 2.9 that the second-degree model does a better job than the first-degree model in terms of reducing the amount of unexplained variation, tests of hypotheses on the individual parameters β_{12}, β_{13}, and β_{23} might be performed to determine if each parameter is important. Tests of the individual parameters are discussed in Section 5.1.

We have seen that the $\{q, m\}$ simplex-lattice design, for $m = 2$ and $m = 3$, gives an equally spaced distribution of points over the simplex and that it provides just enough points to enable the associated $\{q, m\}$ polynomial in the x_i to be fitted. A possible objection to the simplex-lattice design is that, while it is generally intended for the prediction of the response to mixtures of q components, of $q - 1$ components, and of $q - 2$ components at least, the model is fitted to observations collected on mixtures of at most m components. This means that the parameters in the

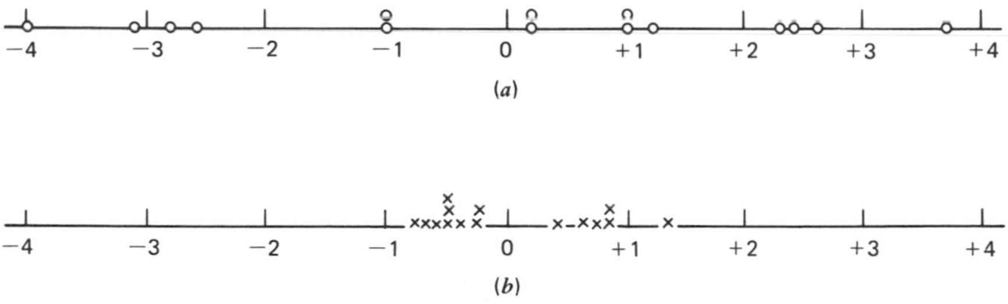

FIGURE 2.9. Plots of the sizes of the residuals with the yarn elongation data arising from the fit of the first-degree model (○) and the fit of the second-degree model (×). Sizes of the residuals = $y_u - \hat{y}_u$.

second-degree model ($m = 2$) are estimated from data collected on mixtures consisting of at most pairs of components. A partial solution to the use of too few component mixtures is provided by the simplex-centroid design.

2.11. THE SIMPLEX-CENTROID DESIGN AND THE ASSOCIATED POLYNOMIAL MODEL

In a q-component simplex-centroid design, the number of distinct points is $2^q - 1$. These points correspond to q permutations of $(1, 0, 0, \ldots, 0)$ or q pure blends, the $\binom{q}{2}$ permutations of $(\frac{1}{2}, \frac{1}{2}, 0, \ldots, 0)$ or all binary mixtures, the $\binom{q}{3}$ permutations of $(\frac{1}{3}, \frac{1}{3}, \frac{1}{3}, 0, \ldots, 0), \ldots$, and so on, with finally the overall centroid point $(1/q, 1/q, \ldots, 1/q)$ or q-nary mixture. In other words, the design consists of every (nonempty) subset of the q components, but only on mixtures in which the components present appear in equal proportions. Such mixtures are located at the centroid of the $(q-1)$-dimensional simplex and at the centroids of all the lower-dimensional simplices contained within the $(q-1)$-dimensional simplex. Presented in Figure 2.10 are the three-component and four-component simplex-centroid designs.

At the points of the simplex-centroid design, data on the response are collected and a polynomial is fitted that has the same number of terms (or parameters to be estimated) as there are points in the associated design. The polynomial equation is

$$\eta = \sum_{i=1}^{q} \beta_i x_i + \sum_{i<j}^{q} \sum \beta_{ij} x_i x_j + \sum_{i<j<k}^{q} \sum \sum \beta_{ijk} x_i x_j x_k + \cdots + \beta_{12\ldots q} x_1 x_2 \cdots x_q \tag{2.35}$$

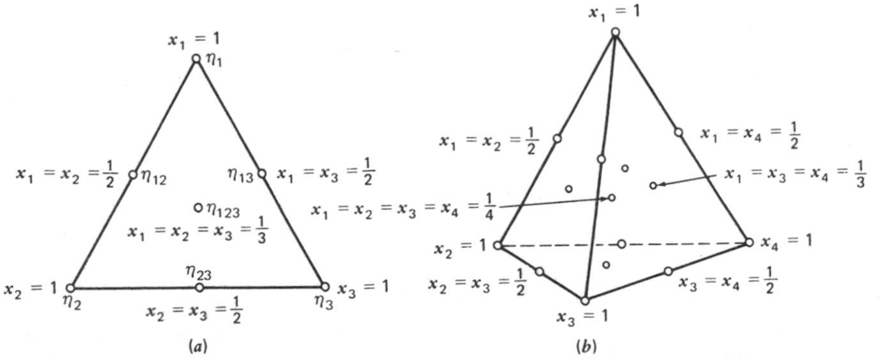

FIGURE 2.10. Simplex-centroid designs for (a) three components, and (b) four components.

As in the case of the previous models which were expressed in the canonical forms in Eqs. (2.4) and (2.7), the parameter β_i in Eq. (2.35) represents the expected response to the pure component i and is called the linear blending value of component i, and β_{ij} is the coefficient of the nonadditive blending of components i and j. The other β_{ijk}'s are defined similarly.

The $2^q - 1$ parameters in the polynomial Eq. (2.35) are expressible as linear functions of the expected responses at the points of the simplex-centroid design. (See Section 2.2 with the simplex-lattices.) For example, if we substitute η_i, η_{ij}, and η_{ijk} into Eq. (2.35) for the responses to $x_i = 1$, $x_j = 0$, $j \neq i$, to $x_i = x_j = \frac{1}{2}$, and to $x_i = x_j = x_k = \frac{1}{3}$, respectively, for all i, j, and k, then the parameters are

$$\beta_i = \eta_i, \qquad \beta_{ij} = 2\{2^1\eta_{ij} - 1^1(\eta_i + \eta_j)\}$$
$$\beta_{ijk} = 3\{3^2\eta_{ijk} - 2^2(\eta_{ij} + \eta_{ik} + \eta_{jk}) + 1^2(\eta_i + \eta_j + \eta_k)\} \tag{2.36}$$

More generally, if we write Sr to denote any subset $\{i_1, i_2, \ldots, i_r\}$ of r elements of $\{1, 2, \ldots, q\}$, then the general formula for the model parameters is

$$\beta_{Sr} = r\{r^{r-1}L_r(Sr) - (r-1)^{r-1}L_{r-1}(Sr) + \cdots + (-1)^{r-1}t^{r-1}L_1(Sr)\}$$
$$= r\left\{\sum_{t=1}^{r} (-1)^{r-t}t^{r-1}L_t(Sr)\right\} \tag{2.37}$$

where $L_t(Sr)$ is the sum of the responses of *all* $\binom{r}{t}$ of the t-nary mixtures with equal proportions formed from the r components in Sr.

The formulas for the estimates of the parameters in Eq. (2.35) are the formulas in Eq. (2.37) with the averages $\bar{y}_i, \bar{y}_{ij}, \bar{y}_{ijk}, \ldots$ substituted for the respective $\eta_i, \eta_{ij}, \eta_{ijk}$, in Eq. (2.37). To show this, let us refer to Figure 2.10(a) where the expected responses are located at the points of the three-component simplex-centroid design. If we replace η_i with $\bar{y}_i$, η_{ij} with $\bar{y}_{ij}$, and η_{123} with $\bar{y}_{123}$, then b_{12} which is the estimate of β_{12} is written by setting $Sr = 12$ and $r = 2$, so that from Eq. (2.37)

$$b_{12} = 2[(-1)^{2-1}1^{2-1}L_1(12) + (-1)^{2-2}2^{2-1}L_2(12)]$$
$$= 2[-(\bar{y}_1 + \bar{y}_2) + 2\bar{y}_{12}]$$

or

$$b_{12} = 4\bar{y}_{12} - 2(\bar{y}_1 + \bar{y}_2) \tag{2.38}$$

since $L_1(12) = \bar{y}_1 + \bar{y}_2$ and $L_2(12) = \bar{y}_{12}$. The formula of Eq. (2.38) is identical to Eq. (2.14) or Eq. (2.15), which was used previously to calculate b_{12} with the simplex-lattice design. The equality of the formulas in Eqs. (2.38) and (2.11) means that with the more elaborate simplex-centroid design, an observation that is collected at the centroid of the face of the simplex-centroid design is not used to estimate the binary coefficient β_{ij}, nor is it used to estimate β_i or β_j.

The magnitudes of the variances and covariances of the parameter estimates using Eq. (2.37), with the y_i and y_{ij} replacing η_i and η_{ij}, respectively, are obtained as follows. Let us consider any two nonempty subsets Sr and Sr' of $\{1, 2, \ldots, q\}$, of r and r' elements, respectively, and let h be the number of elements that Sr and Sr' have in common. For example, if $Sr' = 123$ and $Sr = 12$, then $h = 2$. Now, if $h = 0$, that is, when estimating β_{Sr} and $\beta_{Sr'}$ in Eq. (2.37), if none of the observed responses are used for both b_{Sr} and $b_{Sr'}$, then the coefficient estimates are independent.

If $h > 0$, and we set $r \leq r'$, then $Sr = \{1, \ldots, r\}$ and $Sr' = \{1, \ldots, h, r+1, \ldots, r+r'-h\}$. Let us denote the observed responses which appear in both b_{Sr} and $b_{Sr'}$ by y_i, y_{ij}, and so on. Common to both estimates b_{Sr} and $b_{Sr'}$ there are h y_i for which $1 \leq i \leq h$; there are $\binom{h}{2}$ y_{ij} for which $1 \leq i < j \leq h; \ldots$; there are $\binom{h}{t}$ responses $y_{j_1 j_2 \cdots j_t}$ for which $1 \leq j_1 < \cdots < j_t \leq h$, and, the coefficient of these $y_{j_1 j_2 \cdots j_t}$ is, from Eq. (2.37), $r(-1)^{r-t}t^{r-1}$ in b_{Sr} and $r'(-1)^{r'-t}t^{r'-1}$ in $b_{Sr'}$. Hence

$$\operatorname{cov}(b_{Sr}, b_{Sr'}) = \sum_{t=1}^{h} \binom{h}{t} r(-1)^{r-t}t^{r-1} r'(-1)^{r'-t}t^{r'-1}\sigma^2$$

$$= rr'\{f(r+r', h)\}\sigma^2$$

where

$$f(s, h) = (-1)^s \sum_{t=1}^{h} \binom{h}{t} t^{s-2}$$

If $h = 0$, we define $f(s, 0) = 0$ and if $r = r' = h$, then b_{Sr} and $b_{Sr'}$ are the

TABLE 2.8. Standard error $[\tilde{g}(r)]^{1/2}$ of the regression coefficients estimated from the simplex-centroid design, $\sigma^2 = 1$.

r	$\tilde{g}(r)$	$[\tilde{g}(r)]^{1/2}$	$r^{-r}[\tilde{g}(r)]^{1/2}$
1	1	1.00	1.00
2	24	4.90	1.23
3	1188	34.47	1.28
4	1.184×10^5	344.09	1.34
5	1.966×10^7	4434.19	1.42
6	4.895×10^9	69967.54	1.50
7	1.706×10^{12}	1306062.79	1.59

same, in which case,

$$\text{var}(b_{Sr}) = \tilde{g}(r)\sigma^2 = r^2 f(2r, r)\sigma^2 = r^2 \sum_{t=1}^{r} \binom{r}{t} t^{2r-2}\sigma^2 \tag{2.39}$$

Some examples of the sizes of the variances of the parameter estimates and covariances between pairs of estimates in the $q = 3$ component case are

$$\begin{array}{llllll}
b_i: & i = 1, 2, 3: \text{cov}(b_i, b_j) = \sigma^2 & i = j, & h = 1, & r = r' = h & \\
 & \qquad = 0 & i \neq j, & h = 0, & r = 1, & r' = 1 \\
b_{ij}: & i < j: \text{cov}(b_i, b_{i'j}) = -2\sigma^2 & i = i', & h = 1, & r = 1, & r' = 2 \\
 & \qquad = 0 & i \neq i', & h = 0, & r = 1, & r' = 2 \\
 & \text{cov}(b_{ij}, b_{i'k}) = 4\sigma^2 & i = i', & j < k, & h = 1, & r = 2, \; r' = 2 \\
 & \qquad = 24\sigma^2 & i = i', & j = k, & h = 2, & r = r' = h \\
 & \qquad = 0 & i \neq i', & j \neq k, & h = 0, & r = 2, \; r' = 2 \\
b_{123}: & \text{cov}(b_i, b_{123}) = 3\sigma^2 & h = 1, & r = 1, & r' = 3 & \\
 & \text{cov}(b_{ij}, b_{123}) = -60\sigma^2 & h = 2, & r = 2, & r' = 3 & \\
 & \text{cov}(b_{123}, b_{123}) = 1188\sigma^2 & h = 3, & r = r' = h & &
\end{array} \tag{2.40}$$

In Table 2.8 are tabulated values of $\tilde{g}(r)$ in Eq. (2.39) up to $r = 7$. One notices immediately how quickly the variance of b_{Sr} increases with the number r of subscripts in Sr. However, it must be remembered that the term $x_{i_1}x_{i_2} \cdots x_{i_r}$ in the model that multiplies b_{Sr} decreases rapidly with increasing r since each $x_i \leq 1, i = 1, 2, \ldots, q$. In fact, the maximum value of the term $x_{i_1}x_{i_2} \cdots x_{i_r}$ over the simplex is r^{-r}. Hence the maximum value of the standard deviation of the term $b_{Sr}x_{i_1}x_{i_2} \cdots x_{i_r}$ increases as $r^{-r}[\tilde{g}(r)]^{1/2}$, which is seen from Table 2.8 to be rather slowly.

2.12. AN APPLICATION OF A FOUR-COMPONENT SIMPLEX-CENTROID DESIGN. BLENDING CHEMICAL PESTICIDES FOR CONTROL OF MITES

Four chemical pesticides Vendex (x_1), Omite (x_2), Kelthane (x_3), and Dibrom (x_4) were sprayed on strawberry plants in an attempt to control the mite population. Each chemical was applied individually and in combination with each of the others to comprise the four pure blends, six binary blends, four ternary blends, and the four chemicals together. Each of the 15 chemical treatments was sprayed on three plants in each of four blocks of 45 plants. The response of interest was the average percentage of the mites per treatment over the four replications 7 days after spraying relative to the mite population counted on the plants just prior to spraying. This response approximates percentage survival.

The average relative percentages and the component proportions are presented in Table 2.9. The fitted model is

$$\begin{aligned}\hat{y}(\mathbf{x}) = {} & \underset{(7.4)}{1.8x_1} + 25.4x_2 + 28.6x_3 + 38.5x_4 - \underset{(36.4)}{34.8x_1x_2} - 48.4x_1x_3 + 34.2x_1x_4 \\ & - 94.4x_2x_3 + 21.8x_2x_4 - 91.4x_3x_4 + \underset{(255.8)}{624.6x_1x_2x_3} - 584.7x_1x_2x_4 \\ & - 238.5x_1x_3x_4 - 40.8x_2x_3x_4 - \underset{(2554.2)}{1439.2x_1x_2x_3x_4} \end{aligned} \tag{2.41}$$

and the estimate of the error variance is $s^2 = 220.4$. The quantities in parentheses below the parameter estimates in Eq. (2.41) are the estimated standard errors of the parameter estimates. The estimated standard errors

TABLE 2.9. Average percentage of mites on plants relative to initial numbers 7 days after spraying chemical treatments

Chemical Blend				
x_1	x_2	x_3	x_4	Average Percentage ($\bar{y}$)
1	0	0	0	1.8
0	1	0	0	25.4
0	0	1	0	28.6
0	0	0	1	38.5
0.5	0.5	0	0	4.9
0.5	0	0.5	0	3.1
0.5	0	0	0.5	28.7
0	0.5	0.5	0	3.4
0	0.5	0	0.5	37.4
0	0	0.5	0.5	10.7
0.33	0.33	0.33	0	22.0
0.33	0.33	0	0.33	2.4
0.33	0	0.33	0.33	2.6
0	0.33	0.33	0.33	11.1
0.25	0.25	0.25	0.25	0.8

are the positive square roots of the variance estimates where the latter are

$$\widehat{\text{var}}(b_i) = \frac{s^2}{4} = 55.1$$

$$\widehat{\text{var}}(b_{ij}) = \frac{24s^2}{4} = 1322.4$$

$$\widehat{\text{var}}(b_{ijk}) = \frac{1188s^2}{4} = 65.5 \times 10^3$$

$$\widehat{\text{var}}(b_{1234}) = \frac{118400s^2}{4} = 65.2 \times 10^5$$

In this experiment where the response is the average percentage of mites on the plants seven days after treatment relative to the number counted on the plants just prior to spraying, treatments with low ($\bar{y}_u \leq 20$) average response values are preferred to (or are considered more effective than) treatments with high ($\bar{y}_u \geq 50$) average response values. According to the parameter estimates in the fitted model in Eq. (2.41), the following inferences might be made about the chemical blends used in the experiment. With the seven-days posttreatment data, there is some evidence that the most effective blends are as follows:

Particular Blend	Reason for Inference	
V	The estimate b_1 appears to be significantly lower than the other pure blend estimates	
VO is better than O	$b_{12} < 0$ (?) where $<$ means "is	
VK is better than K	$b_{13} < 0$ (?) significantly less than"	
OK is better than either O or K	$b_{23} < 0$	
KD is better than K or D	$b_{34} < 0$	
VOD is the best three-component *blend*	$b_{124} < 0$	(2.42)

In order to validate the inferences made in (2.42), it would be necessary to perform tests of hypotheses on the specific model parameters that are listed in the reasons for making the inferences. Hypotheses tests are discussed in Section 5.1.

Plotting estimated response contours for systems with four or more components is not an easy exercise. This is because to graphically represent the surface described by the fitted model in Eq. (2.41) in two dimensions, one of the component proportions, say x_i, must be fixed so that the values of the other x_j can be varied over the range $0 \leq x_j \leq 1 - x_i$. For example, let us assume that surface contours across the values of the component proportions x_1, x_2, and x_3 are desired at three levels 0, 0.2, and 0.6 of x_4. The estimated response equation (2.41) is, with each of these cases,

$$x_4 = 0 \quad \hat{y}(\mathbf{x}) = 1.8x_1 + 25.4x_2 + 28.6x_3 - 34.8x_1x_2 - 48.4x_1x_3 - 94.4x_2x_3 + 624.6x_1x_2x_3 \tag{2.43}$$

$$\begin{aligned} x_4 = 0.2 \quad \hat{y}(\mathbf{x}) &= 1.8x_1 + 25.4x_2 + 28.6x_3 + 38.5(0.2) - 34.8x_1x_2 - 48.4x_1x_3 \\ (0 \leq x_i \leq 0.8) \quad &+ 34.2x_1(0.2) - 94.4x_2x_3 + 21.8x_2(0.2) - 91.4x_3(0.2) \\ i = 1, 2, 3 \quad &+ 624.6x_1x_2x_3 - 584.7x_1x_2(0.2) - 238.5x_1x_3(0.2) \\ &- 40.8x_2x_3(0.2) - 1439.2x_1x_2x_3(0.2) \\ &= 7.7 + (1.8 + 6.8)x_1 + (25.4 + 4.4)x_2 + (28.6 - 18.3)x_3 \\ &- (34.8 + 116.9)x_1x_2 - (48.4 + 47.7)x_1x_3 - (94.4 + 8.2)x_2x_3 \\ &+ (624.6 - 287.8)x_1x_2x_3 \\ &= 7.7 + 8.6x_1 + 29.8x_2 + 10.3x_3 - 151.7x_1x_2 - 96.1x_1x_3 \\ &- 102.6x_2x_3 + 336.8x_1x_2x_3 \end{aligned} \tag{2.44}$$

$$\begin{aligned} x_4 = 0.6 \quad \hat{y}(\mathbf{x}) &= 23.1 + 22.3x_1 + 38.5x_2 - 26.2x_3 - 385.6x_1x_2 - 191.5x_1x_3 \\ (0 \leq x_i \leq 0.4) \quad &- 118.9x_2x_3 - 238.9x_1x_2x_3 \\ i = 1, 2, 3 \quad & \end{aligned} \tag{2.45}$$

The plots of the surface contours for Eqs. (2.43)–(2.45) are presented in Figure 2.11.

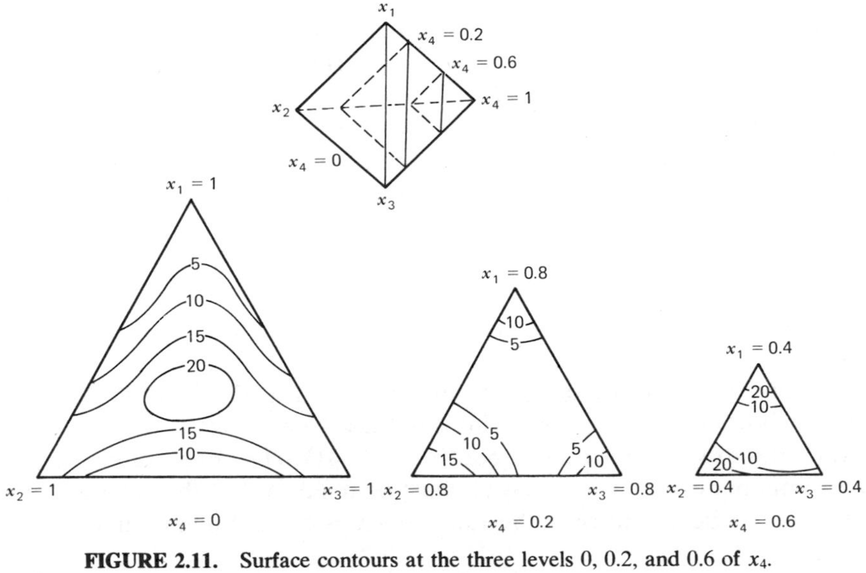

FIGURE 2.11. Surface contours at the three levels 0, 0.2, and 0.6 of x_4.

2.13. AXIAL DESIGNS

The $\{q, m\}$ simplex-lattice and q-component simplex-centroid designs are boundary designs in that, with the exception of the overall centroid, the points of these designs are positioned on the boundaries (vertices, edges, faces, etc.) of the simplex factor space. Axial designs on the other hand, are designs consisting mainly of complete mixtures or q-component blends where most of the points are positioned inside the simplex. Axial designs have been recommended for use when component effects are to be measured and in screening experiments (Section 5.7), particularly when first-degree models are to be fitted. To define an axial design, we quote the following.

Definition: The axis of component i is the imaginary line extending from the base point $x_i = 0$, $x_j = 1/(q-1)$ for all $j \neq i$, to the vertex where $x_i = 1$, $x_j = 0$ all $j \neq i$.

The base point is the centroid of the $(q-2)$-dimensional boundary (sometimes called a $(q-2)$-flat) which is opposite the vertex $x_i = 1$, $x_j = 0$, all $j \neq i$. The length of the axis is the shortest distance from the opposite $(q-2)$-dimensional boundary to the vertex. This distance is defined as one unit in the values of the x_i's. Figure 2.12 presents the axes for components 1, 2, and 3 in the three-component triangle.

An *axial* design's points are positioned only on the component axes. (With three components, both the $\{3, 2\}$ simplex-lattice and simplex-centroid design are outer extreme point axial designs.) The simplest form of axial design is one whose points are positioned equidistant from the centroid $(1/q, 1/q, \ldots, 1/q)$ toward each of the vertices. The distance from the centroid, as measured in the units of x_i, is denoted by Δ and the maximum value for Δ is $(q-1)/q$. Such a design has been suggested in Cornell (1975). A three-component axial design is shown in Figure 2.13.

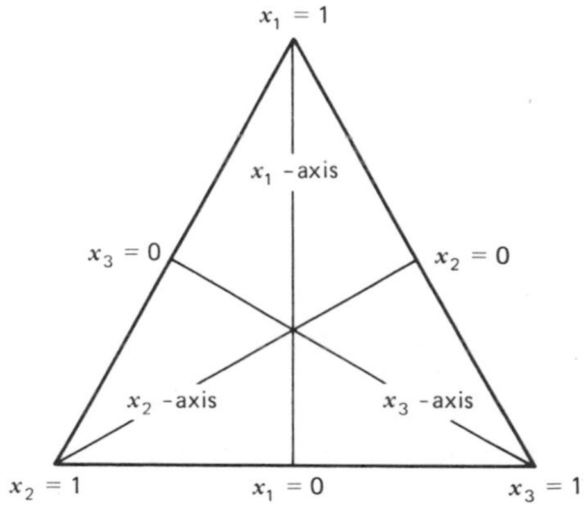

FIGURE 2.12. The x_i-axes, $i = 1, 2,$ and 3.

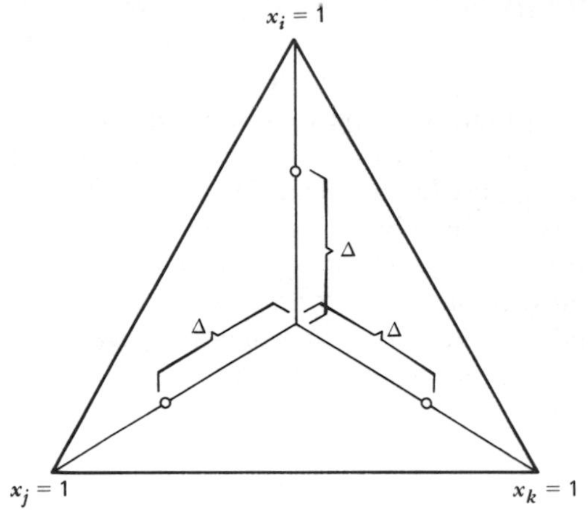

FIGURE 2.13. A three-component axial design where the distance from the center of the simplex to the points is Δ.

Let us write the matrix form of the first-degree model $\eta = \Sigma_{i=1}^{q} \beta_i x_i$ as $\eta = \mathbf{x}'\boldsymbol{\beta}$ where $\mathbf{x} = (x_1, x_2, \ldots, x_q)'$. With an axial design of the form shown in Figure 2.13 if r observations are collected at each of the points, the form of the variance-covariance matrix of the vector $\mathbf{b} = (b_1, b_2, \ldots, b_q)'$ of estimates of the parameters in the model is

$$\text{var}(\mathbf{b}) = \{d\mathbf{I} + e\mathbf{J}\}\sigma^2 \tag{2.46}$$

where

$$d = \frac{(q-1)^2}{r\Delta^2 q^2}, \qquad e = \frac{\Delta^2 q^2 - (q-1)^2}{r\Delta^2 q^3} \tag{2.47}$$

$\mathbf{I}$ is an identity matrix of order q and $\mathbf{J}$ is a $q \times q$ matrix of ones. Given the form of the matrix var(**b**) in Eq. (2.46), the variance of each parameter estimate b_i is

$$\text{var}(b_i) = (d+e)\sigma^2 = \left\{\frac{(q-1)^3 + \Delta^2 q^2}{r\Delta^2 q^3}\right\}\sigma^2 \tag{2.48}$$

and the covariance between pairs of estimates b_i and b_j is $\text{cov}(b_i, b_j) = e\sigma^2$. Now, the larger the value of Δ, the smaller the values of $\text{var}(b_i)$ and $\text{cov}(b_i, b_j)$. In fact, if the points are positioned at the vertices, then $\Delta = (q-1)/q$, $\text{cov}(b_i, b_j) = 0$ and

$$\text{var}(b_i) = \left\{\frac{(q-1)^3 + (q-1)^2}{r(q-1)^2 q}\right\}\sigma^2 = \frac{\sigma^2}{r}$$

If the points are positioned midway between the centroid of the simplex

TABLE 2.10. Variances and covariances of parameter estimates associated with the simple axial design, $\Delta = a/q$

Number of Components $q =$	3	4	4	5	5	5	6	6	6	6	7	7	7	7	7
Size of $\Delta = a/q$ where $a =$	1	1	2	1	2	3	1	2	3	4	1	2	3	4	5
$\mathrm{var}(b_i)r/\sigma^2$	3	7	$\frac{31}{16}$	13	$\frac{17}{5}$	$\frac{73}{45}$	21	$\frac{43}{8}$	$\frac{67}{27}$	$\frac{47}{32}$	31	$\frac{55}{7}$	$\frac{25}{7}$	$\frac{58}{28}$	$\frac{241}{175}$
$\mathrm{cov}(b_i, b_j)r/\sigma^2$	-1	-2	$-\frac{5}{16}$	-3	$-\frac{3}{5}$	$-\frac{7}{45}$	-4	$-\frac{7}{8}$	$-\frac{8}{27}$	$-\frac{3}{32}$	-5	$-\frac{8}{7}$	$-\frac{3}{7}$	$-\frac{5}{28}$	$-\frac{11}{175}$

and each of the vertices so that $\Delta = (q-1)/2q$, then $\text{cov}(b_i, b_j) = -3\sigma^2/rq$, and

$$\text{var}(b_i) = \frac{(4q-3)}{rq}\sigma^2$$

In Table 2.10, values of $\text{var}(b_i)r/\sigma^2$ and $\text{cov}(b_i, b_j)r/\sigma^2$ are listed for increasing values of Δ for $q = 3$, 4, 5, 6, and 7. Incremental values of $\Delta = a/q$ range from $a = 1$ to $a = q - 2$. With the larger values of q or larger numbers of components, the faster the values of $\text{var}(b_i)$ and $\text{cov}(b_i, b_j)$ decrease for increasing values of Δ. This means that when fitting the first-degree model to an axial design of the type considered, the greater the number of components, the more spread the design should be in order to increase the precision of the parameter estimates and reduce the correlations between the estimates. Here precision refers to the reciprocal of the variance and correlation between pairs of estimates is directly related to the covariance between the pairs. Additional discussion on the use of axial designs is presented in Chapters 5 and 6 when topics such as the screening of components, the measuring of component effects using Cox's mixture model, and the measuring of the slope of the response surface along the component axes are introduced.

2.14. SUMMARY OF CHAPTER 2

In this chapter the $\{q, m\}$ simplex-lattice designs and the q-component simplex-centroid designs were introduced. Accompanying both classes of designs are their associated polynomial models. These designs and models are appropriate when the entire simplex region is to be explored.

Formulas for estimating the parameters in the models using data collected at the design points were presented. Data from a three-component yarn experiment, where a $\{3, 2\}$ simplex-lattice design was set up, were used to illustrate the calculations performed in obtaining the second-degree fitted model coefficients as well as to show how to calculate the sums of squares quantities for the analysis of variance table. Some discussion on choosing between a fitted first-degree model and a fitted second-degree model was presented and a plot of the residuals from both models was constructed to aid in the decision making.

A four-chemical pesticide experiment where mite survival percentages were recorded after spraying was the application of the simplex-centroid design. Axial designs, consisting of points positioned on the component axes, were introduced at the end of the chapter. These latter designs have been recommended for situations where the relative effects of the components are to be studied and for screening experiments where the most important components out of a large number of components are sought. We refer to axial designs later, in Chapters 5 and 6.

2.15. REFERENCES AND RECOMMENDED READING

Becker, N. G. (1968). Models for the response of a mixture. *J. R. Stat. Soc., B*, **30**, 349–358.

Box, G. E. P., W. G. Hunter, and J. S. Hunter (1978). *Statistics For Experimenters. An Intro to Design, Data Analysis, and Model Building.* Wiley, New York.

Cornell, J. A. (1975). Some comments on designs for Cox's mixture polynomial. *Technometrics*, **17**, 25–35.

Cornell, J. A. and J. W. Gorman (1978). On the detection of an additive blending component in multicomponent mixtures. *Biometrics*, **34**, No. 2, 251–263.

Daniel, C. and F. S. Wood (1971). *Fitting Equations to Data.* Wiley, New York.

Gorman, J. W. and J. E. Hinman (1962). Simplex-lattice designs for multicomponent systems. *Technometrics*, **4**, 463–487.

Marquardt, D. W. and R. D. Snee (1974). Test statistics for mixture models. *Technometrics*, **16**, 533–537.

Scheffé, H. (1958). Experiments with mixtures. *J. R. Stat. Soc., B*, **20**, No. 2, 344–360.

Scheffé, H. (1963). Simplex-centroid design for experiments with mixtures. *J. R. Stat. Soc., B*, **25**, No. 2, 235–263.

Snee, R. D. (1971). Design and analysis of mixture experiments. *J. Qual. Technol.*, **3**, 159–169.

Snee, R. D. (1973). Techniques for the analysis of mixture data. *Technometrics*, **15**, 517–528.

QUESTIONS FOR CHAPTER 2

2.1. The simplex is known in geometrical terms as a regular figure. What is a regular figure?

2.2. List the component proportions comprising the blends in a two component {2, 4} simplex-lattice. List the 20 blends that comprise a {4, 3} simplex-lattice?

2.3. Mileage figures for each of the individual fuels A and B as well as the 50% : 50% blend of the two fuels $A:B$ in five separate cases are as follows:

Case	A	B	$A:B$
1	17	10	15
2	12	18	15
3	6	6	4
4	10	20	12
5	9	12	12

In which of the cases are the fuels synergistic? In which of the cases are the fuels antagonistic and when are the fuels neither synergistic nor antagonistic. In this last situation, the fuels are said to be ________.

2.4. Show for $q = 3$ that when the special cubic model in Eq. (2.10) is fitted to the expected responses at the points of a design which is the $\{3, 2\}$ simplex-lattice augmented with a point at the centroid $x_1 = x_2 = x_3 = \frac{1}{3}$ of the triangle, then $\beta_{123} = 27\eta_{123} - 12(\eta_{12} + \eta_{13} + \eta_{23}) + 3(\eta_1 + \eta_2 + \eta_3)$ where η_{123} is the response at the centroid. How do the formulas for the parameters β_i and β_{ij} differ from Eq. (2.13) if at all, by the addition of the centroid response η_{123}?

2.5. Refer to the three-component yarn elongation example in Section 2.6. Predict the elongation value of yarn produced from the blend whose proportions are $\mathbf{x} = (0.40, 0.30, 0.30)'$ and set up a 95% confidence interval for the true elongation value η at $\mathbf{x}$.

2.6. With the three-component yarn elongation example of Section 2.6, suppose an extra blend whose component proportions are $\mathbf{x} = (0.40, 0.50, 0.10)'$ was formulated and the elongation values from two samples of the spun yarn were observed to be 18.4 and 17.6. Set up a test statistic and test the hypothesis that says this blend conforms to the second-degree model presented as Eq. (2.24). State any assumptions you make prior to performing the test. (Hint: see Scheffé, 1958, Section 6.)

2.7. The following six fitted models resulted from using data collected at the points of a simplex-centroid design in three components 1, 2, and 3. The models produced the following six contour plots. Match each model with the associated contour plot and define the surface as planar, second-degree, or special cubic.

Model	**Plot**
1. $\hat{y}(\mathbf{x}) = 5.4x_1 + 5.9x_2 + 7.2x_3 + 2.7x_1x_2 + 0.1x_1x_3 - 1.9x_2x_3$	____
2. $\hat{y}(\mathbf{x}) = 5.6x_1 + 6.2x_2 + 7.5x_3$	____
3. $\hat{y}(\mathbf{x}) = 1.84x_1 + 0.67x_2 + 1.51x_3 + 0.14x_1x_2 - 1.01x_1x_3 + 0.27x_2x_3 + 8.68x_1x_2x_3$	____
4. $\hat{y}(\mathbf{x}) = 4.85x_1 + 1.28x_2 + 2.74x_3 - 3.80x_1x_2 - 3.66x_1x_3 + 0.58x_2x_3$	____
5. $\hat{y}(\mathbf{x}) = 1.69x_1 + 0.68x_2 + 1.29x_3$	____
6. $\hat{y}(\mathbf{x}) = 4.29x_1 + 4.57x_2 + 4.43x_3 + 2.84x_1x_2 + 0.84x_1x_3 - 2.56x_2x_3 - 11.19x_1x_2x_3$	____

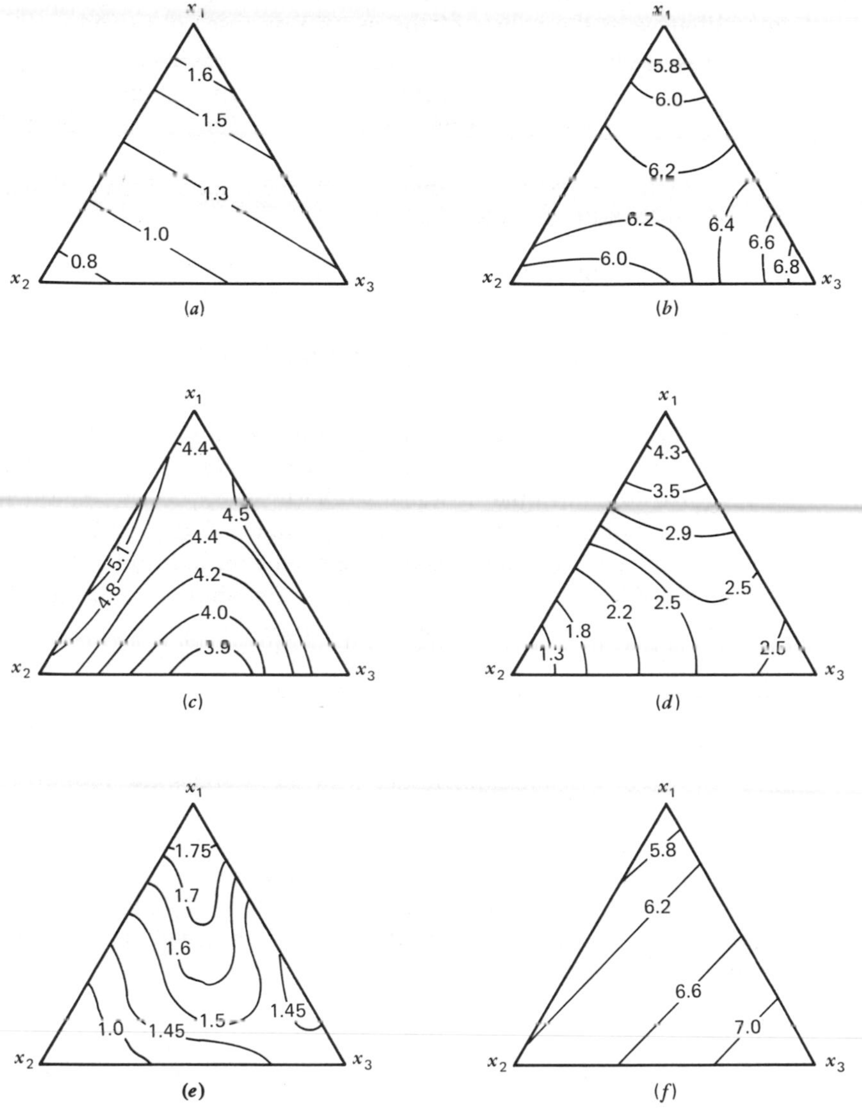

2.8. The following parameter estimates were calculated from multiple observations collected at the points of a $\{3, 2\}$ simplex-lattice.

$$b_1 = 5.5, \qquad b_2 = 7.0, \qquad b_3 = 8.0$$

$$\tfrac{1}{4}b_{12} = 8.75, \qquad \tfrac{1}{4}b_{13} = 2.25, \qquad \tfrac{1}{4}b_{23} = 4.5$$

a. What were the values of the mean responses observed at the lattice points?

b. Suppose three observations were collected at each binary blend ($x_i = x_j = 0.50$, $x_k = 0$) and four observations were collected at each vertex. Which of the quadratic coefficients are significantly ($\alpha = 0.05$) greater than zero if an estimate of the variance of each observation is $s^2 = 3.0$?

2.9. In the special cubic model

$$\hat{y}(\mathbf{x}) = \sum_{i=1}^{3} b_i x_i + \sum_{i<j}^{3}\sum b_{ij} x_i x_j + b_{123} x_1 x_2 x_3$$

the size of the quadratic coefficient b_{ij} must be approximately how many times the size of the planar coefficient b_i in order for the term $b_{ij}x_i x_j$ to contribute as much as the term $b_i x_i$ to the model at the lattice points $x_i = x_j = \frac{1}{2}$. The magnitude of the ratio of the cubic coefficient to the planar coefficient b_{123}/b_i must be approximately how large for the terms to contribute equally to the model at the centroid point. How large must the ratio b_{123}/b_{ij} be for the special cubic term $b_{123}x_1x_2x_3$ to contribute about the same to the model as the quadratic term $b_{ij}x_i x_j$ at the centroid?

2.10. Designs A and B are to be compared based on the simplicity of the coefficient estimation formulas with the designs, the properties of the coefficient estimates and of $\hat{y}(\mathbf{x})$, and any other design properties that are of interest to an experimenter. Design A is the $\{3, 2\}$ simplex-lattice, while design B is a binary blend design derived by clockwise rotating the points of a $\{3, 2\}$ lattice ever so slightly. The coordinates of the designs and observed values of the response at the design points are as follows:

Design A:	(1, 0, 0)	(0.5, 0, 0.5)	(0, 0, 1)	(0, 0.5, 0.5)	(0, 1, 0)
$y_u =$	10.4	15.0	7.0	11.0	8.5
Design A:	(0.5, 0.5, 0)				
$y_u =$	5.0				
Design B:	(0.9, 0, 0.1)	(0.4, 0, 0.6)	(0, 0.1, 0.9)	(0, 0.6, 0.4)	
$y_u =$	10.2	14.6	6.8	10.0	
Design B:	(0.1, 0.9, 0)	(0.6, 0.4, 0)			
$y_u =$	8.0	4.6			

With each design fit a second-degree model to the data observed at the respective points and comment on which of the two designs you prefer and why? If an estimate of the observation variance is $s^2 = 0.40$, are the fitted surfaces with the two designs approximately the same?

2.11. The first six sets of numbers in the following data set were collected at the points (points 1–6) of a $\{3, 2\}$ simplex-lattice. The points 7–10 were chosen after the fitted second-degree model was obtained and observations were collected at these points to check the fit of the second-degree model.

Design Point	Components Proportions (x_1, x_2, x_3)	Observed Response (y_u)	Check Point	(x_1, x_2, x_3)	Observed Response (y_u)
1	$(1, 0, 0)$	4.80, 4.90	7	$(\frac{1}{3}, \frac{1}{3}, \frac{1}{3})$	2.75
2	$(0, 1, 0)$	1.38, 1.18	8	$(\frac{2}{3}, \frac{1}{6}, \frac{1}{6})$	2.80
3	$(0, 0, 1)$	2.58, 2.90	9	$(\frac{1}{6}, \frac{2}{3}, \frac{1}{6})$	1.73
4	$(\frac{1}{2}, \frac{1}{2}, 0)$	1.92, 1.78, 1.76	10	$(\frac{1}{6}, \frac{1}{6}, \frac{2}{3})$	2.20
5	$(0, \frac{1}{2}, \frac{1}{2})$	2.25, 2.43, 2.46			
6	$(\frac{1}{2}, 0, \frac{1}{2})$	2.86, 3.16, 3.25			

a. Fit a second-degree model to the data observed at points 1–6.

b. Predict the response value at the centroid (point 7) using the fitted model of **a** and compare the observed response value at point 7 with the estimate. Are you satisfied with the form of the fitted model? What next course of action do you recommend?

c. With the fitted model in **a**, predict the values of the response at points 8, 9, and 10 and compare each predicted value with the corresponding observed value. Suggest a procedure for testing the adequacy of the fitted model in **a** which uses all four check point observations simultaneously.

d. Use all 19 observations to obtain a prediction equation for this response system.

2.12. With a fitted model containing p terms, the C_p statistic is defined as

$$C_p = \frac{\text{RSS}_p}{s^2} - (N - 2p)$$

where RSS_p is the residual sum of squares, and s^2 is an estimate of the error variance. [This RSS_p has been denoted by SSE in Eq.

(2.28).] If the fitted model has negligible bias, then the expectation of C_p is $E[C_p \text{ given zero bias}] = p$. Using as an estimate of the error variance $s^2 = 0.05$, refer to the data at the design points 1–6 in Question 2.11 and fit a first-degree model and compute C_3, and then compute C_6 with the model 2.11**a**. What is the reduction in the value of C_6 compared to C_3 the result of?

APPENDIX 2A. LEAST SQUARES ESTIMATION FORMULAS FOR THE POLYNOMIAL COEFFICIENTS AND THEIR VARIANCES: MATRIX NOTATION

In Section 7.3, a general review of least squares is presented. In this appendix, the variance properties of the parameter estimates and of $\hat{y}(\mathbf{x})$ are reviewed for the case where the second-degree model is fitted to data at the points of the $\{q, 2\}$ simplex-lattice design.

The general form of the mixture model is $\mathbf{y} = \mathbf{X}\boldsymbol{\beta} + \boldsymbol{\epsilon}$, where $\mathbf{y}$ is an $N \times 1$ vector of observations, $\mathbf{X}$ is an $N \times p$ matrix whose elements are the mixture component proportions and functions (such as pairwise products) of the component proportions, $\boldsymbol{\beta}$ is a $p \times 1$ vector of parameters and $\boldsymbol{\epsilon}$ is an $N \times 1$ vector of random errors. When the model $\mathbf{y} = \mathbf{X}\boldsymbol{\beta} + \boldsymbol{\epsilon}$ is of the first degree, then $p = q$ and

$$\mathbf{y} = \begin{bmatrix} y_1 \\ y_2 \\ \vdots \\ y_N \end{bmatrix}, \quad \mathbf{X} = \begin{bmatrix} x_{11} & x_{12} & \cdots & x_{1q} \\ x_{21} & x_{22} & \cdots & x_{2q} \\ \vdots & \vdots & & \vdots \\ x_{N1} & x_{N2} & \cdots & x_{Nq} \end{bmatrix}, \quad \boldsymbol{\beta} = \begin{bmatrix} \beta_1 \\ \beta_2 \\ \vdots \\ \beta_q \end{bmatrix}, \quad \boldsymbol{\epsilon} = \begin{bmatrix} \epsilon_1 \\ \epsilon_2 \\ \vdots \\ \epsilon_N \end{bmatrix}$$

In fitting the model $\mathbf{y} = \mathbf{X}\boldsymbol{\beta} + \boldsymbol{\epsilon}$ over the N observations, suppose $r_i \geq 1$ observations are collected at the vertex $x_i = 1$, $x_j = 0$, $j \neq i$, $i = 1, 2, \ldots, q$, and $\sum_{i=1}^{q} r_i = N$. The normal equations (7.6) used for estimating the elements of the parameter vector $\boldsymbol{\beta}$ are

$$\mathbf{X'Xb} = \mathbf{X'y}$$

$$\begin{bmatrix} r_1 & & & \\ & r_2 & & \mathbf{0} \\ & & \ddots & \\ \mathbf{0} & & & r_q \end{bmatrix} \begin{bmatrix} b_1 \\ b_2 \\ \vdots \\ b_q \end{bmatrix} = \begin{bmatrix} \sum x_{u1}y_u \\ \sum x_{u2}y_u \\ \vdots \\ \sum x_{uq}y_u \end{bmatrix} \tag{2A.1}$$

where all of the summations $\sum_{u=1}^{N} x_{ui}y_u$ are over $u = 1, 2, \ldots, N$. The solutions to the normal equations provide the estimates

$$b_i = \sum_{u=1}^{N} x_{ui}y_u/r_i = \bar{y}_i, \qquad i = 1, 2, \ldots, q \tag{2A.2}$$

where $\bar{y}_i$ is the average of the r_i observations collected at the vertex $x_i = 1$, $x_l = 0$, $j \neq i$.

Let us consider the case where a second-degree polynomial is fitted to the observations on a $\{q, 2\}$ simplex-lattice where r observations are collected at each vertex $x_i = 1$, and r observations are collected at the midpoint $x_i = x_j = \frac{1}{2}$, $x_l = 0$, $l \neq i, j$ of the edge connecting the vertices corresponding to components i and j. The second-degree polynomial is

$$y_u = \beta_1 x_{u1} + \beta_2 x_{u2} + \cdots + \beta_q x_{uq} + \beta_{12} x_{u1} x_{u2} + \cdots + \beta_{q-1,q} x_{uq-1} x_{uq} + \epsilon_u \qquad u = 1, 2, \ldots, N. \tag{2A.3}$$

If we set $q = 3$ and rearrange the $x_i x_j$ terms in the model, the normal equations are

$$\underset{\mathbf{X'X}}{\begin{bmatrix} A & B & B & C & D & D \\ B & A & B & D & C & D \\ B & B & A & D & D & C \\ C & D & D & E & F & F \\ D & C & D & F & E & F \\ D & D & C & F & F & E \end{bmatrix}} \underset{\mathbf{b}}{\begin{bmatrix} b_1 \\ b_2 \\ b_3 \\ b_{23} \\ b_{13} \\ b_{12} \end{bmatrix}} = \underset{\mathbf{X'y}}{\begin{bmatrix} \Sigma x_{u1} y_u \\ \Sigma x_{u2} y_u \\ \Sigma x_{u3} y_u \\ \Sigma x_{u2} x_{u3} y_u \\ \Sigma x_{u1} x_{u3} y_u \\ \Sigma x_{u1} x_{u2} y_u \end{bmatrix}} \tag{2A.4}$$

where

$$A = \sum x_{ui}^2 = \frac{3r}{2}, \qquad B = \sum x_{ui} x_{uj} = \frac{r}{4}, \qquad C = \sum x_{u1} x_{u2} x_{u3} = 0,$$

$$D = \sum x_{ui}^2 x_{uj} = \frac{r}{8}, \qquad E = \sum x_{ui}^2 x_{uj}^2 = \frac{r}{16}, \qquad F = \sum x_{ui}^2 x_{uj} x_{uk} = 0,$$

$$i, j, k = 1, 2, \text{ and } 3, \qquad i \neq j \neq k \tag{2A.5}$$

and all summations are over $u = 1, 2, \ldots, N$. Furthermore, if a $\{3, m\}$ lattice is considered where $m > 2$, then both C and F will be nonzero.

The matrix $\mathbf{X'X}$ in Eq. (2A.4) is of the composite form $\left[\begin{array}{c|c} \mathbf{U} & \mathbf{V} \\ \hline \mathbf{V} & \mathbf{W} \end{array}\right]$ where each partition matrix is of the form $a\mathbf{I} + b\mathbf{J}$. Hence the inverse matrix $(\mathbf{X'X})^{-1}$ will also be of the composite form $(\mathbf{X'X})^{-1} = \begin{bmatrix} \mathbf{M} & \mathbf{O} \\ \mathbf{O} & \mathbf{P} \end{bmatrix}$ where

$$\mathbf{M} = \mathbf{WQ}, \qquad \mathbf{O} = -\mathbf{VQ}, \qquad \mathbf{P} = \mathbf{UQ}, \qquad \mathbf{Q} = (\mathbf{UW} - \mathbf{V}^2)^{-1} \tag{2A.6}$$

In particular, from Eq. (2A.5),

$$\mathbf{U} = \begin{bmatrix} \frac{6r}{4} & \frac{r}{4} & \frac{r}{4} \\ \frac{r}{4} & \frac{6r}{4} & \frac{r}{4} \\ \frac{r}{4} & \frac{r}{4} & \frac{6r}{4} \end{bmatrix} = \frac{5r}{4}\mathbf{I} + \frac{r}{4}\mathbf{J}, \qquad \mathbf{V} = -\frac{r}{8}\mathbf{I} + \frac{r}{8}\mathbf{J}, \qquad \mathbf{W} = \frac{r}{16}\mathbf{I}^*$$

*Note that if a matrix $\mathbf{T}$ can be expressed as a sum of the matrices $\mathbf{I}$ and $\mathbf{J}$ each of order q in the form $\mathbf{T} = a\mathbf{I} + b\mathbf{J}$, then $\mathbf{T}^{-1} = c\mathbf{I} + d\mathbf{J}$ where $c = 1/a$ and $d = -b/a(a + bq)$.

and therefore,

$$\mathbf{Q} = \frac{16}{r^2}\mathbf{I}, \qquad \mathbf{M} = \frac{1}{r}\mathbf{I}, \qquad \mathbf{O} = \frac{2}{r}\mathbf{I} - \frac{2}{r}\mathbf{J}, \qquad \mathbf{P} = \frac{20}{r}\mathbf{I} + \frac{4}{r}\mathbf{J} \tag{2A.7}$$

On the {3, 2} simplex-lattice, the right-hand side $\mathbf{X}'\mathbf{y}$ of the normal equations in (2A.4) take on the following values,

$$\begin{aligned}
&\sum x_{u1}y_u = r[\bar{y}_1 + \tfrac{1}{2}(\bar{y}_{12} + \bar{y}_{13})], && \sum x_{u2}x_{u3}y_u = r(\bar{y}_{23})/4 \\
&\sum x_{u2}y_u = r[\bar{y}_2 + \tfrac{1}{2}(\bar{y}_{12} + \bar{y}_{23})], && \sum x_{u1}x_{u2}y_u = r(\bar{y}_{12})/4 \\
&\sum x_{u3}y_u = r[\bar{y}_3 + \tfrac{1}{2}(\bar{y}_{13} + \bar{y}_{23})], && \sum x_{u1}x_{u3}y_u = r(\bar{y}_{13})/4
\end{aligned}$$

and upon substituting the partitions **M**, **O**, and **P** in $(\mathbf{X}'\mathbf{X})^{-1}$, the estimates of the coefficients, or, the solutions to the normal equations (2A.4) become

$$\mathbf{b} = \begin{bmatrix} b_1 \\ b_2 \\ b_3 \\ b_{23} \\ b_{12} \\ b_{13} \end{bmatrix} = (\mathbf{X}'\mathbf{X})^{-1}\mathbf{X}'\mathbf{y} = \begin{bmatrix} \bar{y}_1 + \tfrac{1}{2}(\bar{y}_{12} + \bar{y}_{13}) - \tfrac{1}{2}(\bar{y}_{12} + \bar{y}_{13}) \\ \bar{y}_2 \\ \bar{y}_3 \\ 4\bar{y}_{23} - 2(\bar{y}_2 + \bar{y}_3) \\ 4\bar{y}_{12} - 2(\bar{y}_1 + \bar{y}_2) \\ 4\bar{y}_{13} - 2(\bar{y}_1 + \bar{y}_3) \end{bmatrix} \tag{2A.8}$$

Furthermore, if a measure of the error variance σ^2 is available, then the variance–covariance matrix of the coefficient estimates

$$\begin{aligned}
\text{var–cov}(\mathbf{b}) &= (\mathbf{X}'\mathbf{X})^{-1}\sigma^2 \\
&= \left[\begin{array}{c:c} \frac{1}{r}\mathbf{I} & \frac{2}{r}\mathbf{I} - \frac{2}{r}\mathbf{J} \\ \hdashline \frac{2}{r}\mathbf{I} - \frac{2}{r}\mathbf{J} & \frac{20}{r}\mathbf{I} + \frac{4}{r}\mathbf{J} \end{array}\right]\sigma^2
\end{aligned} \tag{2A.9}$$

The predicted value of the response at a point $\mathbf{x} = (x_1, x_2, \ldots, x_q)'$ in the experimental region is expressed in matrix notation as

$$\hat{y}(\mathbf{x}) = \mathbf{x}_p'\mathbf{b}$$

where $\mathbf{x}_p'$ is a $1 \times p$ vector whose elements correspond to the elements in a row of the matrix **X**. A measure of the precision of the estimate $\hat{y}(\mathbf{x})$, at the point **x**, is defined as the variance of $\hat{y}(\mathbf{x})$ and is expressed in matrix notation as

$$\begin{aligned}
\text{var}[\hat{y}(\mathbf{x})] &= \text{var}[\mathbf{x}_p'\mathbf{b}] \\
&= \mathbf{x}_p' \,\text{var}[\mathbf{b}]\mathbf{x}_p \\
&= \mathbf{x}_p'(\mathbf{X}'\mathbf{X})^{-1}\mathbf{x}_p\sigma^2
\end{aligned}$$

Since the estimates of the parameters in $\boldsymbol{\beta}$ are expressible as linear functions of the average responses at the lattice points, as shown in (2A.8), $\hat{y}(\mathbf{x})$ can be written in terms of the averages. Thus the variance of $\hat{y}(\mathbf{x})$ can be written in terms of the variances of the averages as follows:

Quadratic model: $$\hat{y}(\mathbf{x}) = \sum_{i=1}^{q} a_i \bar{y}_i + \sum_{i<j}^{q}\sum a_{ij}\bar{y}_{ij}$$

where

$$a_i = x_i(2x_i - 1), \qquad a_{ij} = 4x_i x_j$$

and

$$\operatorname{var}[\hat{y}(\mathbf{x})] = \sigma^2\left[\sum_{i=1}^{q} a_i^2/r_i + \sum_{i<j}^{q}\sum a_{ij}^2/r_{ij}\right]$$

Special cubic model: $$\hat{y}(\mathbf{x}) = \sum_{i=1}^{q} b_i \bar{y}_i + \sum_{i<j}^{q}\sum b_{ij}\bar{y}_{ij} + \sum_{i<j<k}^{q}\sum\sum b_{ijk}\bar{y}_{ijk}$$

where

$$b_i = \frac{x_i}{2}(6x_i^2 - 2x_i + 1) - 3\sum_{j\neq i}^{q} x_j^2$$

$$b_{ij} = 4x_i x_j(3x_i + 3x_j - 2), \qquad b_{ijk} = 27x_i x_j x_k$$

and

$$\operatorname{var}[\hat{y}(\mathbf{x})] = \sigma^2\left[\sum_{i=1}^{q} b_i^2/r_i + \sum_{i<j}^{q}\sum b_{ij}^2/r_{ij} + \sum_{i<j<k}^{q}\sum\sum b_{ijk}^2/r_{ijk}\right]$$

Similar expressions corresponding to the full cubic model and quartic model can be found in Gorman and Hinman (1962).

APPENDIX 2B. CUBIC AND QUARTIC POLYNOMIALS AND FORMULAS FOR THE ESTIMATES OF THE COEFFICIENTS

In q components, the *cubic* model would be fitted to the points of a $\{q, 3\}$ simplex-lattice. The model and the expressions for the coefficient estimates in terms of the observed responses at the lattice points are,

$$\eta = \sum_{i=1}^{q} \beta_i x_i + \sum_{i<j}^{q}\sum \beta_{ij}x_i x_j + \sum_{i<j}^{q}\sum \gamma_{ij}x_i x_j(x_i - x_j) + \sum_{i<j<k}^{q}\sum\sum \beta_{ijk}x_i x_j x_k \tag{2B.1}$$

Denoting the observed responses at the vertices by y_i, at the point $x_i = \frac{2}{3}$, $x_j = \frac{1}{3}$, $x_k = 0$, $k \neq i, j$ by y_{iij} and at the centroids $x_i = x_j = x_k = \frac{1}{3}$, $x_l = 0$ of the faces by y_{ijk}, respectively, we have for the coefficient estimates

$$
\begin{aligned}
b_i &= y_i \qquad i = 1, 2, \ldots, q \\
b_{ij} &= \tfrac{9}{4}(y_{iij} + y_{ijj} - y_i - y_j) \qquad i<j \\
g_{ij} &= \tfrac{9}{4}(3y_{iij} - 3y_{ijj} - y_i + y_j) \qquad i<j \\
b_{ijk} &= 27y_{ijk} - \tfrac{27}{4}(y_{iij} + y_{ijj} + y_{iik} + y_{ikk} + y_{jjk} + y_{jkk}) + \tfrac{9}{2}(y_i + y_j + y_k)
\end{aligned} \tag{2B.2}
$$

If replicate observations are collected at some or all of the lattice points, we replace y_i, y_{iij}, and so on in Eq. (2B.2) with the averages $\bar{y}_i$, $\bar{y}_{iij}$, etc. of the replicates.

To the points of a $\{q, 4\}$ simplex-lattice, the *quartic* model to be fitted is

$$
\begin{aligned}
\eta = {} & \sum_{i=1}^{q} \beta_i x_i + \sum_{i<j}\sum^{q} \beta_{ij} x_i x_j + \sum_{i<j}\sum^{q} \gamma_{ij} x_i x_j (x_i - x_j) + \sum_{i<j}\sum^{q} \delta_{ij} x_i x_j (x_i - x_j)^2 \\
& + \sum_{i<j<k}\sum\sum^{q} \beta_{iijk} x_i^2 x_j x_k + \sum_{i<j<k}\sum\sum^{q} \beta_{ijjk} x_i x_j^2 x_k + \sum_{i<j<k}\sum\sum^{q} \beta_{ijkk} x_i x_j x_k^2 \\
& + \sum_{i<j<k<l}\sum\sum\sum^{q} \beta_{ijkl} x_i x_j x_k x_l
\end{aligned} \tag{2B.3}
$$

The formulas for the estimates of the coefficients, in terms of y_i, y_{ij}, and so on, are

$$
\begin{aligned}
b_i &= y_i \\
b_{ij} &= 4y_{ij} - 2(y_i + y_j) \\
g_{ij} &= \tfrac{8}{3}(2y_{iiij} - 2y_{ijjj} - y_i + y_j) \\
d_{ij} &= \tfrac{8}{3}(4y_{iiij} - 6y_{ij} + 4y_{ijjj} - y_i = y_j) \\
b_{iijk} &= 32(3y_{iijk} - y_{ijjk} - y_{ijkk}) + \tfrac{8}{3}(6y_i - y_j - y_k) - 16(y_{ij} + y_{ik}) \\
&\quad - \tfrac{16}{3}(5y_{iiij} + 5y_{iiik} - 3y_{ijjj} - 3y_{ikkk} - y_{jjjk} - y_{jkkk}) \\
b_{ijjk} &= 32(3y_{ijjk} - y_{iijk} - y_{ijkk}) + \tfrac{8}{3}(6y_j - y_i - y_k) - 16(y_{ij} + y_{jk}) \\
&\quad - \tfrac{16}{3}(5y_{ijjj} + 5y_{jjjk} - 3y_{iiij} - 3y_{jkkk} - y_{iiik} - y_{ikkk}) \\
b_{ijkk} &= 32(3y_{ijkk} - y_{iijk} - y_{ijjk}) + \tfrac{8}{3}(6y_k - y_i - y_j) - 16(y_{ik} + y_{jk}) \\
&\quad - \tfrac{16}{3}(5y_{ikkk} + 5y_{jkkk} - 3y_{iiik} - 3y_{jjjk} - y_{iiij} - y_{ijjj}) \\
b_{ijkl} &= 256y_{ijkl} - 32(y_{iijk} + y_{iijl} + y_{iikl} + y_{ijjk} + y_{ijjl} + y_{jjkl} + y_{ijkk} + y_{ikkl} + y_{jkkl} + y_{ijll} \\
&\quad + y_{jkll} + y_{ikll}) + \tfrac{32}{3}(y_{iiij} + y_{iiik} + y_{iiil} + y_{ijjj} + y_{jjjk} + y_{jjjl} + y_{ikkk} + y_{jkkk} + y_{kkkl} \\
&\quad + y_{illl} + y_{jlll} + y_{klll})
\end{aligned} \tag{2B.4}
$$

It is interesting to notice that the formula (2B.3) for the quartic model does not contain any of the third-degree terms $\beta_{ijk} x_i x_j x_k$ that appear in the cubic model (2B.1). Furthermore, none of the face centroids $x_i = x_j = x_k = \frac{1}{3}$, $x_l = 0$, $l \neq i, j, k$ are included as points of the $\{q, 4\}$ simplex-lattice and therefore the parameters β_{ijk} are not estimable if included in the model (2B.3) unless additional points are added to the design.

CHAPTER 3

The Use of Independent Variables

One of the very basic assumptions made when employing response surface methods is that the response surface can be approximated reasonably well within the range of the data by a low-order polynomial (a first-degree or a second-degree polynomial). In most of the experimental situations that we have considered thus far, the range of the data covered all of the points in the $(q-1)$-dimensional simplex and the low-order polynomial approximation was satisfactory. When it is discovered that the second-degree model does not fit the original data very well and it is suspected that a third-degree model will be needed, often a transformation of the observed response values will improve the fitted performance of the second-degree model to the point that we are satisfied with it.

While searching for a model to depict the shape of the surface over the simplex factor space, some researchers have chosen to work not with the q mixture components (whose values or settings are dependent on the values of the others) but instead with an alternative system consisting of $q-1$ mathematically independent variables. Two reasons cited by those who choose to work with $q-1$ independent variables is the mathematical convenience in knowing how to set up model forms and interpret parameter estimates in the independent variables, and the familiarity with standard designs for fitting models with independent variables as well as familiarity with design optimality criteria. Almost all of the transformations (or mappings) that have been suggested in the literature for going from the mixture component space to the independent variable space are equivalent in theory; the main difference in using the various transformations or mappings is only in the final form of the relationship between the mixture components and the transformed variables. Hereafter when we mention the use of transformations in this chapter we are referring to transformations performed on the mixture component proportions (x_i) and not to transformations made on the data values. We discuss a mixture component transformation now.

3.1. TRANSFORMING FROM THE q MIXTURE COMPONENTS TO $q-1$ MATHEMATICALLY INDEPENDENT VARIABLES

The factor space of the q mixture component proportions is represented by a $(q-1)$-dimensional regular simplex defined according to the restriction $x_1+x_2+\cdots+x_q=1$ (see Figure 3.1 for $q=3$). Instead of working directly with the q linearly dependent mixture components $x_1, x_2, \ldots, x_q$, let us redefine the system in terms of $q-1$ mathematically independent variables $w_1, w_2, \ldots, w_{q-1}$. To do so we suggest the following two-step procedure.

***Step* 1:** First, define the location of the origin of the new system to be at the centroid of the simplex by introducing the intermediate variables $\tilde{x}_i$, where

$$\tilde{x}_i = q\left(x_i - \frac{1}{q}\right) = qx_i - 1 \tag{3.1}$$

The centroid of the original simplex $x_i = 1/q$, $i = 1, 2, \ldots, q$, is expressed now as $\tilde{x}_i = 0$ in the intermediate variables and, the vertex where $x_i = 1$,

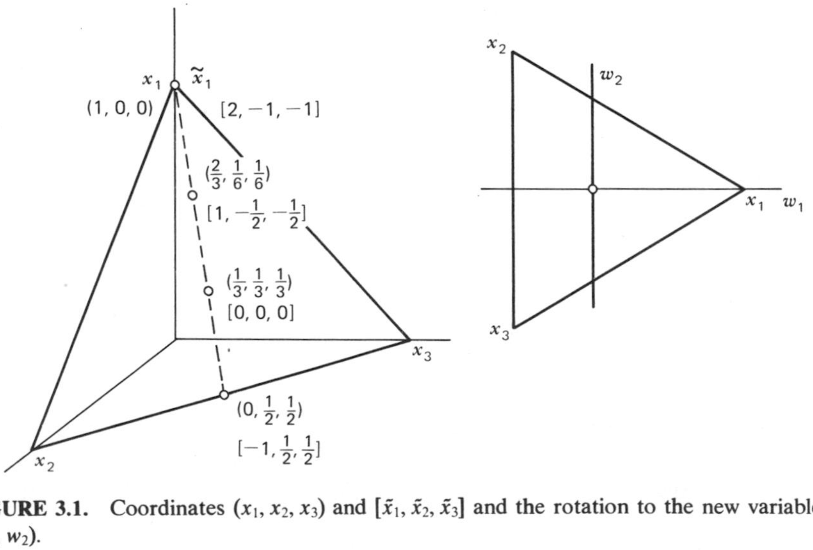

FIGURE 3.1. Coordinates (x_1, x_2, x_3) and $[\tilde{x}_1, \tilde{x}_2, \tilde{x}_3]$ and the rotation to the new variables (w_1, w_2).

$i = 1, 2, \ldots, q$ in the original components is defined now as $\tilde{x}_i = q - 1$. The values for the $\tilde{x}_i$ range from -1 to $q-1$, for example, in Figure 3.1, the values for $\tilde{x}_1$ are shown as -1, 0, 1, and 2 for $x_1 = 0, \frac{1}{3}, \frac{2}{3}$, and 1, respectively.

Step 2: Second, the axes of the original components are rotated to make the axis of component q orthogonal to the simplex. (Component q then is removed from further consideration.) This is accomplished by multiplying the vector of intermediate variates with an orthogonal matrix $\mathbf{\Theta}$ and defining new variates, in vector form, as

$$\mathbf{w}' = \frac{q(q-1)}{q}\tilde{\mathbf{x}}'\mathbf{\Theta} \tag{3.2}$$

where $\mathbf{w}'$ and $\tilde{\mathbf{x}}'$ are $1 \times q$ vectors and the $q \times q$ matrix $\mathbf{\Theta}$ is expressed as

$$\mathbf{\Theta} = \frac{1}{\sqrt{q(q-1)}} \begin{bmatrix} q-1 & 0 & 0 & \cdots & 0 & s \\ -1 & (q-2)l & 0 & \cdots & 0 & s \\ -1 & -l & (q-3)m & & 0 & s \\ -1 & -l & -m & & 0 & s \\ \vdots & \vdots & \vdots & \ddots & \vdots & \vdots \\ -1 & -l & -m & & t & s \\ -1 & -l & -m & & -t & s \end{bmatrix} \tag{3.3}$$

The elements $l, m, \ldots, t$ and s in $\mathbf{\Theta}$ are defined in a manner to force the sum of squares of the elements in each column to be $q(q-1)$. This will produce the orthogonality property of the matrix $\mathbf{\Theta}$, that is, $\mathbf{\Theta}'\mathbf{\Theta} = \mathbf{I}$. The $N \times q$ matrix $\mathbf{W}$ where $\mathbf{w}'$ in Eq. (3.2) is a row, is expressible in terms of the $N \times q$ matrix $\mathbf{X}$ and $\mathbf{\Theta}$ as $\mathbf{W} = (q-1)\{q\mathbf{X} - \mathbf{J}\}\mathbf{\Theta}$.

To illustrate the transformation just described let $q = 3$ and let us assume for our design that we have the seven points on the simplex consisting of the vertices, the midpoints of the three edges, and the overall centroid. In the original components, this design is the simplex-centroid configuration. The values of the elements l and s in the 3×3 matrix $\mathbf{\Theta}$ in Eq. (3.3) are $l = \sqrt{3}$, $s = \sqrt{2}$ and the resulting transformations in Eqs. (3.1) and (3.2) are

$$\tilde{x}_i = 3x_i - 1$$

$$(w_1, w_2, w_3)' = 2(\tilde{x}_1, \tilde{x}_2, \tilde{x}_3)\mathbf{\Theta} = 2(\tilde{x}_1, \tilde{x}_2, \tilde{x}_3)\frac{1}{\sqrt{6}}\begin{bmatrix} 2 & 0 & \sqrt{2} \\ -1 & \sqrt{3} & \sqrt{2} \\ -1 & -\sqrt{3} & \sqrt{2} \end{bmatrix}$$

and $w_3 = 0$ because $\tilde{x}_1 + \tilde{x}_2 + \tilde{x}_3 = 0$. The 7×3 design matrix $\mathbf{W}$ in the transformed variables, corresponding to the 7×3 design matrix $\mathbf{X}$ in the original components, is

$$
\mathbf{X} =
\begin{array}{ccc}
x_1 & x_2 & x_3 \\
\end{array}
\begin{bmatrix}
1 & 0 & 0 \\
\frac{1}{2} & \frac{1}{2} & 0 \\
0 & 1 & 0 \\
0 & \frac{1}{2} & \frac{1}{2} \\
0 & 0 & 1 \\
\frac{1}{2} & 0 & \frac{1}{2} \\
\frac{1}{3} & \frac{1}{3} & \frac{1}{3}
\end{bmatrix},
\quad
\tilde{\mathbf{X}} =
\begin{array}{ccc}
\tilde{x}_1 & \tilde{x}_2 & \tilde{x}_3 \\
\end{array}
\begin{bmatrix}
2 & -1 & -1 \\
\frac{1}{2} & \frac{1}{2} & -1 \\
-1 & 2 & -1 \\
-1 & \frac{1}{2} & \frac{1}{2} \\
-1 & -1 & 2 \\
\frac{1}{2} & -1 & \frac{1}{2} \\
0 & 0 & 0
\end{bmatrix},
\quad
\mathbf{W} = \sqrt{6}
\begin{array}{ccc}
w_1 & w_2 & w_3 \\
\end{array}
\begin{bmatrix}
2 & 0 & 0 \\
\frac{1}{2} & \sqrt{3}/2 & 0 \\
-1 & \sqrt{3} & 0 \\
-1 & 0 & 0 \\
-1 & -\sqrt{3} & 0 \\
\frac{1}{2} & -\sqrt{3}/2 & 0 \\
0 & 0 & 0
\end{bmatrix}
$$

$$(3.4)$$

where the w_3-axis is perpendicular to the plane of the simplex (and as such the w_3 values are all zero in the plane of w_1 and w_2). The locations of the design points, expressible in the new system (w_1, w_2), are pictured in Figure 3.2a.

At the points of the design in the $w_1, w_2, \ldots, w_{q-1}$ system, observed values of the response are collected and are used for the estimation of the parameters in the vector $\boldsymbol{\gamma}$ in the model

$$y = \mathbf{W}_A \boldsymbol{\gamma} + \boldsymbol{\epsilon} \tag{3.5}$$

The matrix $\mathbf{W}_A$ is the matrix $\mathbf{W}$ in Eq. (3.4) with the qth column removed and augmented with an $N \times 1$ column of ones and with columns whose elements represent terms in the model equation of degree higher than one. The vector

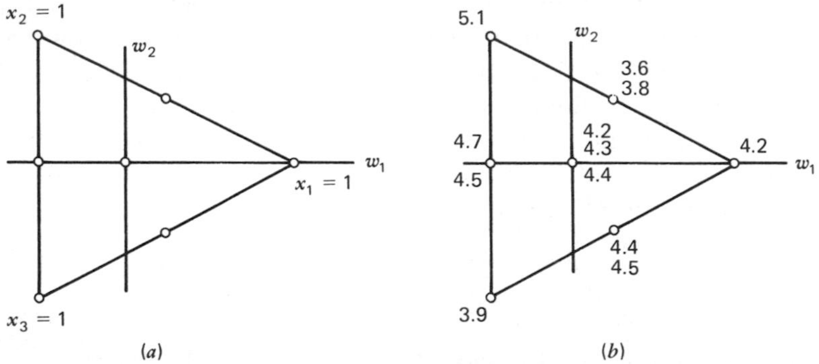

FIGURE 3.2. (a) The triangular region in w_1 and w_2. (b) Response values at the design points.

of parameter estimates is

$$\mathbf{g} = (\mathbf{W}'_A\mathbf{W}_A)^{-1}\mathbf{W}'_A\mathbf{y} \tag{3.6}$$

and the fitted model in the w_i's, $i = 1, 2, \ldots, q-1$ is

$$\hat{y}(\mathbf{w}) = \mathbf{w}'_p\mathbf{g} \tag{3.7}$$

where $\mathbf{w}'_p$ is a row of the matrix $\mathbf{W}_A$ in Eq. (3.5). Standard procedures for testing the goodness of the fitted model can be used. (See, e.g., Draper and Smith, 1966, Chapter 2.)

3.2. A NUMERICAL EXAMPLE: SENSORY FLAVOR RATING OF FISH PATTIES

In Figure 3.2*b*, average flavor scores are listed at the points of a three-component (three-fish) simplex-centroid design (or a {3, 2} simplex-lattice with an added center point). The scores have been rounded-off to tenths. For purposes of illustration, each of the two-fish blends ($x_i = x_j = \frac{1}{2}$) is replicated twice and the three-fish blend at the centroid is replicated three times.

The design settings in the independent variates w_1 and w_2 corresponding to the mixture component proportions are determined by the two-step transformation in Eqs. (3.1) and (3.2) and are shown in Eq. (3.4). To the observed values a second-degree polynomial model in w_1 and w_2 is to be fitted where the model is of the standard form

$$y = \gamma_0 + \gamma_1 w_1 + \gamma_2 w_2 + \gamma_{11} w_1^2 + \gamma_{22} w_2^2 + \gamma_{12} w_1 w_2 + \epsilon \tag{3.8}$$

In matrix notation, the model represented by Eq. (3.8) over the 12 observations is

$$\underset{\mathbf{y}}{\begin{bmatrix} 4.2 \\ 3.8 \\ 3.6 \\ 5.1 \\ 4.5 \\ 4.7 \\ 3.9 \\ 4.4 \\ 4.5 \\ 4.4 \\ 4.2 \\ 4.3 \end{bmatrix}} = \underset{\mathbf{W}_A}{\begin{bmatrix} 1 & 2\sqrt{6} & 0 & 24 & 0 & 0 \\ 1 & \sqrt{6}/2 & 3/\sqrt{2} & \frac{3}{2} & \frac{9}{2} & 3\sqrt{3}/2 \\ 1 & \sqrt{6}/2 & 3/\sqrt{2} & \frac{3}{2} & \frac{9}{2} & 3\sqrt{3}/2 \\ 1 & -\sqrt{6} & 3\sqrt{2} & 6 & 18 & -6\sqrt{3} \\ 1 & -\sqrt{6} & 0 & 6 & 0 & 0 \\ 1 & -\sqrt{6} & 0 & 6 & 0 & 0 \\ 1 & -\sqrt{6} & -3\sqrt{2} & 6 & 18 & 6\sqrt{3} \\ 1 & \sqrt{6}/2 & -3/\sqrt{2} & \frac{3}{2} & \frac{9}{2} & -3\sqrt{3}/2 \\ 1 & \sqrt{6}/2 & -3/\sqrt{2} & \frac{3}{2} & \frac{9}{2} & -3\sqrt{3}/2 \\ 1 & 0 & 0 & 0 & 0 & 0 \\ 1 & 0 & 0 & 0 & 0 & 0 \\ 1 & 0 & 0 & 0 & 0 & 0 \end{bmatrix}} \underset{\boldsymbol{\gamma}}{\begin{bmatrix} \gamma_0 \\ \gamma_1 \\ \gamma_2 \\ \gamma_{11} \\ \gamma_{22} \\ \gamma_{12} \end{bmatrix}} + \underset{\boldsymbol{\epsilon}}{\begin{bmatrix} \epsilon_1 \\ \epsilon_2 \\ \epsilon_3 \\ \epsilon_4 \\ \epsilon_5 \\ \epsilon_6 \\ \epsilon_7 \\ \epsilon_8 \\ \epsilon_9 \\ \epsilon_{10} \\ \epsilon_{11} \\ \epsilon_{12} \end{bmatrix}}$$

The solution to the normal equations $\mathbf{W}'_A\mathbf{W}_A\mathbf{g} = \mathbf{W}'_A\mathbf{y}$ is the vector $\mathbf{g}$ of parameter estimates which is obtained as in Eq. (3.6)

$$\underset{\mathbf{g}}{\begin{bmatrix} 4.25 \\ -0.11 \\ -0.07 \\ 0.02 \\ -0.01 \\ -0.09 \end{bmatrix}} = \underset{(\mathbf{W}'_A\mathbf{W}_A)^{-1}}{\begin{bmatrix} 0.167 & 0 & 0 & -0.009 & -0.009 & 0 \\ & 0.028 & 0 & -0.004 & 0.004 & 0 \\ & & 0.028 & 0 & 0 & 0.008 \\ & & & 0.003 & -0.001 & 0 \\ & \text{(symmetric)} & & & 0.003 & 0 \\ & & & & & 0.006 \end{bmatrix}} \underset{\mathbf{W}'_A\mathbf{y}}{\begin{bmatrix} 51.60 \\ -4.04 \\ 1.91 \\ 234.45 \\ 235.35 \\ -16.37 \end{bmatrix}}$$

The fitted model in w_1 and w_2 which can be used to obtain an estimate of the response at any point $\mathbf{w} = (w_1, w_2)'$ inside of the triangle is

$$\begin{aligned} \hat{y}(\mathbf{w}) &= \underset{(0.05)}{4.25} - \underset{(0.02)}{0.11}w_1 - 0.07w_2 + \underset{(0.01)}{0.02}w_1^2 - 0.01w_2^2 - \underset{(0.01)}{0.09}w_1w_2 \\ &= \mathbf{w}'_p\mathbf{g} \end{aligned} \tag{3.9}$$

where the elements in $\mathbf{w}'_p$ are the same as a row of the matrix $\mathbf{W}_A$ evaluated at $\mathbf{w}$. The variance of the estimate of the response at $\mathbf{w}$ is

$$\text{var}[\hat{y}(\mathbf{w})] = \mathbf{w}'_p(\mathbf{W}'_A\mathbf{W}_A)^{-1}\mathbf{w}_p\sigma^2 \tag{3.10}$$

where σ^2 is the variance of a single observation and is estimated by the residual mean square in the analysis of variance table, Table 3.1.

The analysis of variance sums of squares formulas for the sources, total, regression, and residual are identical to the corresponding sums of squares formulas presented in Section 2.7. In matrix notation, these formulas are

$$\begin{aligned} \text{SST} &= \mathbf{y}'\mathbf{y} - (\mathbf{1}'\mathbf{y})^2/N \\ \text{SSR} &= \mathbf{g}'\mathbf{W}'_A\mathbf{y} - (\mathbf{1}'\mathbf{y})^2/N \\ \text{SSE} &= \mathbf{y}'\mathbf{y} - \mathbf{g}'\mathbf{W}'_A\mathbf{y} \end{aligned} \tag{3.11}$$

TABLE 3.1. Analysis of variance table for numerical example

Source of Variation	Degrees of Freedom	Sums of Squares	Mean Square	F
Regression	$p-1=5$	SSR = 1.74	0.35	26.9[a]
Residual	$N-p=6$	SSE = 0.08	0.013	

[a]F-test value exceeds tabled $F_{(5,6,0.01)} = 8.75$ at 0.01.

where $\mathbf{1}'$ is $1 \times N$ vector of ones and $(\mathbf{1}'\mathbf{y})^2/N$ is called the "correction for the mean." The degrees of freedom for SST, SSR, and SSE are $N-1$, $p-1$, and $N-p$ respectively where p is the number of parameters (elements of $\boldsymbol{\gamma}$) which are estimated to get the fitted model. The analysis of variance table for the example data is presented in Table 3.1. The value of the coefficient of determination $R^2 = \text{SSR/SST}$ is $R^2 = 0.956$, while the value of R_A^2 is $R_A^2 = 0.921$.

Before we continue with the example, we remark on our feelings concerning the statistics R^2 and R_A^2. Throughout this book we present, with most of the data computations or model-fitting exercises, either the value of R^2 or R_A^2 or both. We do not wish to suggest, however, that these quantities are the only criteria used in determining how well the model fits the data. In addition to the value of R^2 or R_A^2, one should look at the size of the estimate of the observation standard deviation (the square root of the residual mean square in the analysis of variance table) and as a rule compare this measure to some known quantity if possible. For example, if it is known that the standard deviation of the sensory flavor scores is approximately 0.1 units, which is nearly equal to the (residual mean square)$^{1/2} = (0.013)^{1/2}$, there does not seem to be any reason to suspect that an excessive amount of unexplained variation is present after having fitted the model of Eq. (3.9). Also, an inspection of the individual residuals $y_u - \hat{y}_u$, $u = 1, 2, \ldots, 12$, would disclose that the size of the largest residual is 0.2 units, which is only twice the size of the estimate of the observation standard deviation and is therefore well within acceptable limits.

We mention again the use of the C_p statistic for aiding us in deciding on the form of the fitted model (Mallows, 1973). Once an estimate of the error variance is obtained, a value of the C_p statistic can be calculated using $C_p = [\text{RSS}_p/s^2] - (N - 2p)$ for each model fitted. [The RSS_p here is denoted by SSE in Eq. (3.11).] As a rule of thumb, we shall say the form of the fitted model is adequate if the value of $C_p \approx p$, when the estimate of the error variance is assumed to be satisfactory. With the model of Eq. (3.9), $C_6 = 6$.

One final comment on the use of the value of R^2 or R_A^2 concerns the vagueness of these measured quantities. In the formula for SSR shown in Eq. (3.11), the sum of squares corrects for the overall mean because the model in Eq. (3.8) to be fitted contains a constant parameter denoted by γ_0. This correction was not made previously with the mixture component model in Section 2.7. Therefore one might ask, "Since SSR was corrected for the mean, can we use the standard coefficient of determination R^2 with the model in the w's or is this value of R^2 inflated as in the case of the mixture model?" See Question 3.2 at the end of this chapter.

Surface representations in the x_i's and w_i's are the same. To show this we shall fit the model in the x_i's and obtain an estimate of the response at some point in the triangle using the fitted model in the x_i's. Then we shall compare the estimates obtained with the fitted models in the x_i's and in the w_i's.

In matrix notation, the model in the mixture components is

$$\underset{\mathbf{y}}{\begin{bmatrix}4.2\\3.8\\3.6\\5.1\\4.5\\4.7\\3.9\\4.4\\4.5\\4.4\\4.2\\4.3\end{bmatrix}} = \underset{\mathbf{X}}{\begin{bmatrix}1&0&0&0&0&0\\ \frac{1}{2}&\frac{1}{2}&0&\frac{1}{4}&0&0\\ \frac{1}{2}&\frac{1}{2}&0&\frac{1}{4}&0&0\\ 0&1&0&0&0&0\\ 0&\frac{1}{2}&\frac{1}{2}&0&0&\frac{1}{4}\\ 0&\frac{1}{2}&\frac{1}{2}&0&0&\frac{1}{4}\\ 0&0&1&0&0&0\\ \frac{1}{2}&0&\frac{1}{2}&0&\frac{1}{4}&0\\ \frac{1}{2}&0&\frac{1}{2}&0&\frac{1}{4}&0\\ \frac{1}{3}&\frac{1}{3}&\frac{1}{3}&\frac{1}{9}&\frac{1}{9}&\frac{1}{9}\\ \frac{1}{3}&\frac{1}{3}&\frac{1}{3}&\frac{1}{9}&\frac{1}{9}&\frac{1}{9}\\ \frac{1}{3}&\frac{1}{3}&\frac{1}{3}&\frac{1}{9}&\frac{1}{9}&\frac{1}{9}\end{bmatrix}} \underset{\boldsymbol{\beta}}{\begin{bmatrix}\beta_1\\ \beta_2\\ \beta_3\\ \beta_{12}\\ \beta_{13}\\ \beta_{23}\end{bmatrix}} + \underset{\boldsymbol{\epsilon}}{\begin{bmatrix}\epsilon_1\\ \epsilon_2\\ \epsilon_3\\ \epsilon_4\\ \epsilon_5\\ \epsilon_6\\ \epsilon_7\\ \epsilon_8\\ \epsilon_9\\ \epsilon_{10}\\ \epsilon_{11}\\ \epsilon_{12}\end{bmatrix}}$$

and the vector of estimates is

$$\underset{\mathbf{b}}{\begin{bmatrix}4.2\\5.1\\3.9\\-3.6\\1.8\\0.6\end{bmatrix}} = \underset{(\mathbf{X}'\mathbf{X})^{-1}}{\begin{bmatrix}0.981&-0.019&-0.019&-1.778&-1.778&0.222\\ &0.981&-0.019&-1.778&0.222&-1.778\\ &&0.981&0.222&-1.778&-1.778\\ &&&13.333&1.333&1.333\\ &\text{(symmetric)}&&&13.333&1.333\\ &&&&&13.333\end{bmatrix}} \underset{\mathbf{X}'\mathbf{y}}{\begin{bmatrix}16.65\\17.70\\17.25\\3.28\\3.66\\3.73\end{bmatrix}}$$

The fitted model in the mixture components, which can be used to predict the value of the response at any point $\mathbf{x}$, is

$$\begin{aligned}\hat{y}(\mathbf{x}) &= 4.2x_1 + 5.1x_2 + 3.9x_3 - 3.6x_1x_2 + 1.8x_1x_3 + 0.6x_2x_3\\ &\quad (0.981s^2)^{1/2} = (0.11) \qquad (0.42)\\ &= \mathbf{x}_p'\mathbf{b} \qquad (3.12)\end{aligned}$$

where $\mathbf{x}_p = (x_1, x_2, x_3, x_1x_2, x_1x_3, x_2x_3)'$. The numbers in parentheses below the parameter estimates are the estimated standard errors of the estimates where est. s.e.$(b_i) = (0.981s^2)^{1/2} = 0.11$ and est. s.e.$(b_{ij}) = (13.333s^2)^{1/2} = 0.42$ and $s^2 = 0.013$ is the residual mean square obtained from Table 3.1. The same analysis of variance table is used when working in the w_i's as in the x_i's. The variance of the estimate $\hat{y}(\mathbf{x})$ of the response at **x** is

$$\begin{aligned}\text{var}[\hat{y}(\mathbf{x})] &= \mathbf{x}_p'[\text{var}(\mathbf{b})]\mathbf{x}_p \\ &= \mathbf{x}_p'(\mathbf{X}'\mathbf{X})^{-1}\mathbf{x}_p\sigma^2 \end{aligned} \tag{3.13}$$

An estimate of the height of the surface above the triangle can be found for any mixture composition, for example, $x_1 = 0.40$, $x_2 = 0.25$, and $x_3 = 0.35$. The estimate is obtained by substituting the component proportions directly into Eq. (3.12), which is $\hat{y}(\mathbf{x}) = \mathbf{x}_p'\mathbf{b}$, to give

$$\begin{aligned}\hat{y}(0.40, 0.25, 0.35) &= (0.40, 0.25, 0.35, 0.10, 0.14, 0.0875)\begin{bmatrix} 4.2 \\ 5.1 \\ 3.9 \\ -3.6 \\ 1.8 \\ 0.6 \end{bmatrix} \\ &= 4.25 \end{aligned} \tag{3.14}$$

The variance of the estimate $\hat{y}(\mathbf{x})$ at the point $\mathbf{x} = (0.40, 0.25, 0.35)'$ is, from Eq. (3.13)

$$\begin{aligned}\text{var}[\hat{y}(\mathbf{x})] &= \mathbf{x}_p'(\mathbf{X}'\mathbf{X})^{-1}\mathbf{x}_p\sigma^2 \\ &= 0.172\sigma^2 \end{aligned} \tag{3.15}$$

and an estimate of $\text{var}[\hat{y}(\mathbf{x})]$ is obtained by substituting $s^2 = 0.013$ for σ^2 to give $\widehat{\text{var}}[\hat{y}(\mathbf{x})] = 0.002$.

An estimate of the response in the w_1 and w_2 system is obtained as follows. Corresponding to the component proportions $\mathbf{x} = (0.40, 0.25, 0.35)'$, the coordinates of the intermediate variables in Eq. (3.1) are $(\tilde{x}_1, \tilde{x}_2, \tilde{x}_3) = 3(0.40, 0.25, 0.35) - (1, 1, 1) = (0.20, -0.25, 0.05)$. From the intermediate variables the coordinates (w_1, w_2) are

$$\begin{aligned}(w_1, w_2, w_3 = 0) &= \frac{2}{\sqrt{6}}(\tilde{x}_1, \tilde{x}_2, \tilde{x}_3)\begin{bmatrix} 2 & 0 & \sqrt{2} \\ -1 & \sqrt{3} & \sqrt{2} \\ -1 & -\sqrt{3} & \sqrt{2} \end{bmatrix} \\ &= (0.49, -0.42, 0) \end{aligned}$$

Substituting these values of w_1 and w_2 into Eq. (3.9), which is $\hat{y}(\mathbf{w}) = \mathbf{w}_p'\mathbf{g}$, the estimate of the response (or the predicted value of the response for this blend) is

$$\begin{aligned}\hat{y}(0.49, -0.42) &= (1, 0.49, -0.42, 0.49^2, (-0.42)^2, -0.2058)\begin{bmatrix} 4.25 \\ -0.11 \\ -0.07 \\ 0.02 \\ -0.01 \\ -0.09 \end{bmatrix} \\ &= 4.25 \end{aligned} \tag{3.16}$$

and the variance of the estimate of the response at the point $\mathbf{w} = (0.49, -0.42)'$ is

$$\begin{aligned} \mathrm{var}[\hat{y}(0.49, -0.42)] &= \mathbf{w}_p' \,\mathrm{var}(\mathbf{g})\mathbf{w}_p \\ &= \mathbf{w}_p'(\mathbf{W}_A'\mathbf{W}_A)^{-1}\mathbf{w}_p\sigma^2 \\ &= 0.172\sigma^2 \end{aligned} \tag{3.17}$$

The values of the estimates of the response in Eqs. (3.14) and (3.16) are the same and the respective variances in Eqs. (3.15) and (3.17) are the same. This is because we have not changed the shape of the surface representation in writing the models in the x's and in the w's. A plot of the estimated surface contours is presented in Figure 3.3.

Standard response surface techniques that are employed with independent variables can be used now in the (w_1, w_2)-system and on the model $\hat{y}(\mathbf{w}) = \mathbf{w}_p'\mathbf{g}$ in Eq. (3.9). For example, if a more detailed study of the flavor surface is of interest, one can perform such standard techniques as the method of steepest ascent to seek out a maximum point of the surface or the performing of a canonical analysis to determine the nature of the stationary point. These techniques are described in Myers, 1971, Chapter 5. In Section 5.9 we present a method for studying mixture response surfaces that uses gradients along the mixture component axes.

Another transformation that can be used to generate a system of $q-1$ independent variables from the q-component system is presented in

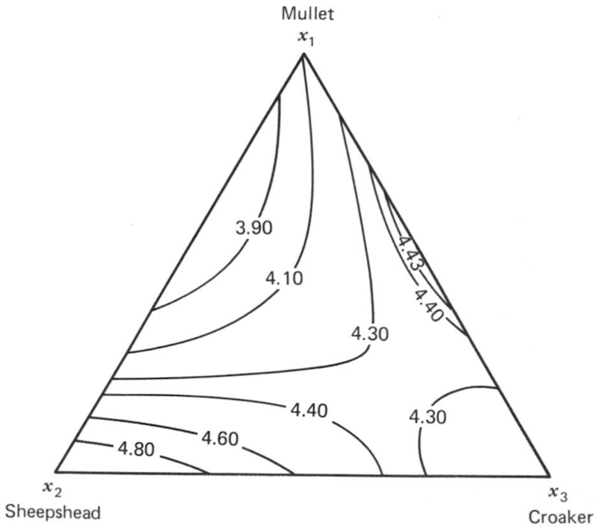

FIGURE 3.3. Surface contours of the estimated flavor of the three-fish patties.

Appendix 3A. (See also Draper and Lawrence, 1965, for $q = 3$ and $q = 4$.) The transformation in Appendix 3A, like the transformation just presented, is extremely simple to perform, but unlike this one, the transformation in the appendix produces a design region that is not centered at the centroid of the composition space. Additional regression techniques that can be used with proportions can be found in Becker (1969).

Let us now consider the experimental situation where the experimenter's interest is not with all possible mixture combinations in the simplex, but rather the interest is directed to a smaller subregion inside the simplex. We call the subregion the *region of interest* and assume initially that the region is to be defined as ellipsoidal in shape. Cuboidal regions are considered briefly later.

3.3. DEFINING A REGION OF INTEREST INSIDE THE SIMPLEX: AN ELLIPSOIDAL REGION

Occasionally experiments are planned to be performed in some reasonably well-defined neighborhood, or *region of interest*, centered about current operating conditions. The reason for exploring the neighborhood is to see if there are other blends in the vicinity of the current conditions that can produce a yield similar or better than is currently being produced. Another reason for restricting the experimentation to a subregion of the simplex is that some blends not inside the subregion either are not feasible or simply not desirable and therefore we do not wish to experiment with them.

To exemplify experimentation concentrated in some subregion, we recall the three-component yarn experiment of Section 2.6 where the blends consisted of proportions of polyethylene (x_1), polystyrene (x_2), and polypropylene (x_3). Now, let us assume the current operating condition is the blend consisting of $x_1 = 0.43$, $x_2 = 0.30$, and $x_3 = 0.27$, but that we are interested in the neighborhood defined *approximately* as

$$0.27 \leq x_1 \leq 0.59, \qquad 0.15 \leq x_2 \leq 0.45, \qquad 0.20 \leq x_3 \leq 0.34 \tag{3.18}$$

The neighborhood and the point (0.43, 0.30, 0.27) is pictured in Figure 3.4.

The region specified by the constraints on x_1, x_2, and x_3 and defined in Eq. (3.18) consists of six vertices and six sides, as seen in Figure 3.4. However, rather than work with the neighborhood region as described exactly by the lower and upper bound constraints in Eq. (3.18), let us define the region of interest to be ellipsoidal and of the form

$$\left(\frac{x_1 - 0.43}{0.16}\right)^2 + \left(\frac{x_2 - 0.30}{0.15}\right)^2 + \left(\frac{x_3 - 0.27}{0.07}\right)^2 \leq 1 \tag{3.19}$$

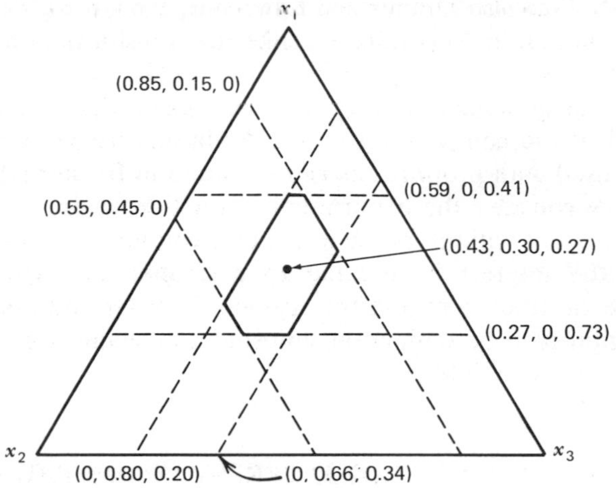

FIGURE 3.4. The region of primary interest as a subset of the triangle.

The ellipsoidal region is centered at the current operating blend $\mathbf{x}_0 = (0.43, 0.30, 0.27)'$ and the constants 0.16, 0.15, and 0.07 represent one half the range of interest for x_1, x_2, and x_3. respectively. By defining the ellipsoidal region as in Eq. (3.19), we are able to exercise considerable flexibility in the size of the region since the intervals of interest for the different components can be different lengths.

In the general q-component case, an ellipsoidal region is expressible as

$$\sum_{i=1}^{q} \left(\frac{x_i - x_{0i}}{h_i}\right)^2 \leq 1 \tag{3.20}$$

where x_{0i} defines the center of the interval of interest for component i and $2h_i$ represents the spread or range of the symmetric interval of interest for component i. Defined in vector notation $\mathbf{x}_0 = (x_{01}, x_{02}, \ldots, x_{0q})'$, the centroid of the ellipsoidal region in Eq. (3.20) might happen to coincide with the centroid of the simplex, that is, $(1/q, 1/q, \ldots, 1/q)'$. The point $(x_{01}, x_{02}, \ldots, x_{0q})$ naturally lies strictly inside the simplex and therefore the condition $x_1 + x_2 + \cdots + x_q = 1$ is also forced on the x_{0i} that is, $x_{01} + x_{02} + \cdots + x_{0q} = 1$.

Having specified the region as ellipsoidal, the methodology can be simplified further by redefining the ellipsoidal region as a unit spherical region in another system of variables. Simplifying that is in the sense that most people feel a *unit spherical region* is easier to work with than an ellipsoidal region. To obtain the unit sphere, we introduce the *intermediate variables* v_i,

$$v_i = \frac{x_i - x_{0i}}{h_i}, \qquad i = 1, 2, \ldots, q \tag{3.21}$$

so that the ellipsoidal region in Eq. (3.20) is now defined in the v_i to be a unit spherical region centered at $v_i = 0$, $1 \leq i \leq q$. In matrix notation, Eq. (3.21) can be written as

$$\mathbf{v} = \mathbf{H}^{-1}(\mathbf{x} - \mathbf{x}_0) \tag{3.22}$$

where $\mathbf{v} = (v_1, v_2, \ldots, v_q)'$, $\mathbf{x}_0 = (x_{01}, x_{02}, \ldots, x_{0q})'$ and $\mathbf{H}$ is the diagonal matrix $\mathbf{H} = \text{diagonal}(h_1, h_2, \ldots, h_q)$.

Let us express N observations of the response by means of the linear first-degree model in the mixture components and then in the intermediate variables. In the mixture components we have

$$\mathbf{y} = \mathbf{X}_c\boldsymbol{\beta} + \boldsymbol{\epsilon} \tag{3.23}$$

where $\mathbf{y}$ is an $N \times 1$ vector of observations, $\mathbf{X}_c$ is an $N \times q$ matrix of rank $q - 1$ with elements $x_{ui} - x_{0i}$ $(1 \leq u \leq N,\ 1 \leq i \leq q)$, $\boldsymbol{\beta}$ is a $q \times 1$ vector of unknown parameters and $\boldsymbol{\epsilon}$ is an $N \times 1$ vector of random errors. Over N observations, Eq. (3.23) is

$$\begin{bmatrix} y_1 \\ y_2 \\ \vdots \\ y_N \end{bmatrix} = \begin{bmatrix} x_{11} - x_{01} & \cdots & x_{1q} - x_{0q} \\ x_{21} - x_{01} & \cdots & x_{2q} - x_{0q} \\ \vdots & & \vdots \\ x_{N1} - x_{01} & \cdots & x_{Nq} - x_{0q} \end{bmatrix} \begin{bmatrix} \beta_1 \\ \beta_2 \\ \vdots \\ \beta_q \end{bmatrix} + \begin{bmatrix} \epsilon_1 \\ \epsilon_2 \\ \vdots \\ \epsilon_N \end{bmatrix}$$

To write the model in Eq. (3.23) in terms of the *intermediate variables*, let $\mathbf{V}$ be the $N \times q$ matrix whose uth row is $\mathbf{v}'_u$ where

$$\mathbf{v}'_u = (\mathbf{x}_u - \mathbf{x}_0)'\mathbf{H}^{-1}$$

Then

$$\mathbf{V} = \mathbf{X}_c\mathbf{H}^{-1} \quad \text{or} \quad \mathbf{X}_c = \mathbf{V}\mathbf{H} \tag{3.24}$$

The model becomes

$$\begin{aligned} \mathbf{y} &= \mathbf{X}_c\boldsymbol{\beta} + \boldsymbol{\epsilon} \\ &= \mathbf{V}\mathbf{H}\boldsymbol{\beta} + \boldsymbol{\epsilon} \\ &= \mathbf{V}\boldsymbol{\gamma} + \boldsymbol{\epsilon} \end{aligned} \tag{3.25}$$

where $\boldsymbol{\gamma} = \mathbf{H}\boldsymbol{\beta}$.

The transformation in Eq. (3.21) changes only the *scale* of the variables and therefore the rank of the coefficient matrix $\mathbf{V}$ in Eq. (3.25) is also $q - 1$, which is the rank of $\mathbf{X}_c$. Since it is often more convenient to use a model of full rank particularly when constructing standard response surface designs and when using design criteria, we shall reparameterize the model in Eq. (3.25) to one of full rank.

In order to define the new variables, we know that because the rank of the matrix $\mathbf{V}$ is $q-1$ there exists the single linear relation among the q columns of $\mathbf{V}$ of the form $\mathbf{VH1}_q = \mathbf{0}_N$. Let us choose a $q \times q$ orthogonal matrix $\mathbf{T}$ (See Appendix 3B for a derivation of the form of $\mathbf{T}$) so that

$$\mathbf{VT} = [\mathbf{W} \quad \mathbf{0}] \tag{3.26}$$

where $\mathbf{W}$ is an $N \times (q-1)$ matrix of rank $q-1$, and $\mathbf{0}$ is an $N \times 1$ vector of zeros. To see that this is possible, we note first that

$$\sum_{i=1}^{q} (x_i - x_{0i}) = \sum_{i=1}^{q} h_i v_i = 0$$

defines a $(q-1)$-dimensional linear manifold in q-space. The transformation in Eq. (3.26) is a rotation of the axes of the intermediate variables about the origin $\mathbf{v} = \mathbf{0}$. The rotation by the matrix $\mathbf{T}$ is chosen so that the constraint on the v's is expressed in the form $w_{uq} = 0$, $(1 \le u \le N)$ and by ignoring this zero coordinate, we are in effect projecting the q-dimensional unit sphere onto the $(q-1)$-dimensional manifold producing once again a unit sphere which is centered at $\mathbf{w} = \mathbf{0}$ where $\mathbf{w} = (w_1, w_2, \ldots, w_{q-1})'$. Hence the region of interest is now a $(q-1)$-dimensional unit spherical region and the variables w_i are mathematically independent. The w_i will be called *design variables* since they will be used to construct designs.

The model in Eq. (3.25) can now be expressed as

$$\begin{aligned} \mathbf{y} &= \mathbf{VTT}'\boldsymbol{\gamma} + \boldsymbol{\epsilon} \\ &= \mathbf{VT}_1\mathbf{T}_1'\boldsymbol{\gamma} + \boldsymbol{\epsilon} \end{aligned} \tag{3.27}$$

where the matrix $\mathbf{T}$ is partitioned as $\mathbf{T} = [\mathbf{T}_1, \mathbf{T}_2]$. The matrix $\mathbf{T}_1$ is $q \times (q-1)$ and $\mathbf{T}_2$ is $q \times 1$ so that in Eq. (3.26), $\mathbf{VT}_1 = \mathbf{W}$ and $\mathbf{VT}_2 = \mathbf{0}$. Now in Eq. (3.27), if we let $\mathbf{T}_1'\boldsymbol{\gamma} = \boldsymbol{\alpha}$, then the model in Eq. (3.27) becomes

$$\mathbf{y} = \mathbf{W}\boldsymbol{\alpha} + \boldsymbol{\epsilon} \tag{3.28}$$

where $\mathbf{W}$ is an $N \times (q-1)$ matrix which contains the levels of the $q-1$ independent variables over all N experiments and $\boldsymbol{\alpha}$ is a $(q-1) \times 1$ vector of unknown parameters.

We now have a model containing $q-1$ independent variables which to work with. Once the estimates $\mathbf{a}$ of $\boldsymbol{\alpha}$ are obtained, if the estimates $\mathbf{g}$ of the parameters in Eq. (3.25) are wanted, they can be obtained by

$$\mathbf{T}'\mathbf{g} = [\mathbf{T}_1'\mathbf{g}, \mathbf{T}_2'\mathbf{g}] = [\mathbf{T}_1'\mathbf{g}, 0]$$
$$= [\mathbf{a}, 0]$$

so that $\mathbf{g} = \mathbf{T}[\mathbf{a}, 0] = \mathbf{T}_1\mathbf{a}$. Thus, by insisting on $\mathbf{T}_2'\mathbf{g} = 0$, we force $\mathbf{g} = \mathbf{T}_1\mathbf{a}$. Finally, the estimates $\mathbf{b}$ of the parameters in the mixture component model in Eq. (3.23) can be obtained from the equation $\mathbf{b} = \mathbf{H}^{-1}\mathbf{g}$ and the prediction equation is $\hat{y}(\mathbf{x}) = \mathbf{x}_c'\mathbf{b}$ where $\mathbf{x}_c'$ resembles a row of $\mathbf{X}_c$.

The variables $w_1, w_2, \ldots, w_{q-1}$ are to be used in the construction of designs as well as in other mathematical calculations such as in the plotting of response contours. [The fitted model will be defined with a constant term in it so that the Eq. (3.28) is altered to read as $\mathbf{y} = \alpha_0\mathbf{1} + \mathbf{W}\boldsymbol{\alpha} + \boldsymbol{\epsilon}$ and the fitted model is $\hat{y}(\mathbf{w}) = a_0 + \mathbf{w}_p'\mathbf{a}$ where $\mathbf{w}_p'$ is a row of $\mathbf{W}$ and $\mathbf{a}$ is the vector of estimates of $\boldsymbol{\alpha}$.] We now show how to determine the experimental levels of the mixture components to be used in the N experiments by expressing the x's in terms of the w's. We have from Eq. (3.26) and $\mathbf{VT}_2 = \mathbf{0}$ and Eq. (3.24)

$$[\mathbf{W}, \mathbf{0}] = \mathbf{VT} = [\mathbf{VT}_1, \mathbf{VT}_2]$$
$$= \mathbf{X}_c\mathbf{H}^{-1}\mathbf{T}$$

where $\mathbf{0}$ is an $N \times 1$ vector of zeros and the elements of the matrix $\mathbf{X}_c$ are $x_{ui} - x_{0i}$, $(1 \leq u \leq N,\ 1 \leq i \leq q)$. Therefore

$$\mathbf{X}_c = [\mathbf{W}, \mathbf{0}]\mathbf{T}'\mathbf{H} = \mathbf{W}\mathbf{T}_1'\mathbf{H} \tag{3.29}$$

If the elements of the matrix $\mathbf{W}$ are denoted by w_{uj} $(1 \leq u \leq N, 1 \leq j \leq q-1)$ and the elements of $\mathbf{T}_1$ by t_{ij} $(1 \leq i \leq q,\ 1 \leq j \leq q-1)$, then

$$x_{ui} = \left(\sum_{l=1}^{q-1} w_{ul}t_{il}\right)h_i + x_{0i} \tag{3.30}$$

Hence, given the values of w_{ul}, we can use Eq. (3.30) to determine the level of the mixture component x_i to be run in the uth experiment, $u = 1, 2, \ldots, N$.

3.4. A NUMERICAL ILLUSTRATION OF THE INVERSE TRANSFORMATION FROM THE DESIGN VARIABLES TO THE MIXTURE COMPONENTS

To illustrate the method of setting up an arbitrary design in the w_i's and then to obtain the corresponding settings in the mixture components, we refer to the beginning of Section 3.3 where in the three-component system the neighborhood region was described in Eq. (3.21) as ellipsoidal of the form

$$\left(\frac{x_1-0.43}{0.16}\right)^2+\left(\frac{x_2-0.30}{0.15}\right)^2+\left(\frac{x_3-0.27}{0.07}\right)^2\leq 1$$

The vector containing the coordinates of the center of the region and the matrix **H** of scale constants are

$$\mathbf{x}_0=(0.43, 0.30, 0.27)', \qquad \mathbf{H}=\text{diag}(0.16, 0.15, 0.07)$$

The elements of the transformation matrix **T** are, using Eqs. (3B.1) and (3B.4) of Appendix 3B,

$$\mathbf{T}=\begin{bmatrix} -0.15/m & -0.16(0.07)/n & 0.16/s \\ 0.16/m & -0.15(0.07)/n & 0.15/s \\ 0 & (0.16^2+0.15^2)/n & 0.07/s \end{bmatrix}=\begin{bmatrix} -0.6839 & -0.2218 & 0.6950 \\ 0.7295 & -0.2079 & 0.6516 \\ 0 & 0.9527 & 0.3041 \end{bmatrix} \tag{3.31}$$

where $m=(0.15^2+0.16^2)^{1/2}$, $n=[0.16^2(0.07)^2+0.15^2(0.07)^2+(0.16^2+0.15^2)^2]^{1/2}$ and $s=(0.16^2+0.15^2+0.07^2)^{1/2}$.

For fitting the first-degree model $E[y(\mathbf{w})]=\alpha_0+\alpha_1 w_1+\alpha_2 w_2$, let us assume that four design points are to be used and the points are positioned on the boundary of a unit spherical region in the w_i's. (We show how to determine the settings in the w_i's shortly.) Then a matrix **W** for Eq. (3.29) is

$$\mathbf{W}=\frac{1}{\sqrt{2}}\begin{bmatrix} 1 & 1 \\ -1 & 1 \\ 1 & -1 \\ -1 & -1 \end{bmatrix}$$

The settings of the mixture components corresponding to the $N=4$ design settings in **W** can be found using Eq. (3.30) or Eq. (3.29) where

TABLE 3.2. Values of the component proportions at the four design points

x_1	x_2	x_3	Design Point
0.328	0.355	0.317	(1)
0.482	0.201	0.317	(2)
0.378	0.399	0.223	(3)
0.532	0.245	0.223	(4)

$$\mathbf{X}_c = [\mathbf{W} \quad \mathbf{0}]\mathbf{T}'\mathbf{H}$$

$$\begin{bmatrix} x_{11}-0.43 & x_{12}-0.30 & x_{13}-0.27 \\ x_{21}-0.43 & x_{22}-0.30 & x_{23}-0.27 \\ x_{31}-0.43 & x_{32}-0.30 & x_{33}-0.27 \\ x_{41}-0.43 & x_{42}-0.30 & x_{43}-0.27 \end{bmatrix} = \frac{1}{\sqrt{2}} \begin{bmatrix} 1 & 1 & 0 \\ -1 & 1 & 0 \\ 1 & -1 & 0 \\ -1 & -1 & 0 \end{bmatrix}$$

$$\times \mathbf{T}' \begin{bmatrix} 0.16 & 0 & 0 \\ 0 & 0.15 & 0 \\ 0 & 0 & 0.07 \end{bmatrix} \quad (3.32)$$

Substituting the matrix **T** in Eq. (3.31) into Eq. (3.32), the design point designations in the mixture components corresponding to the settings in **W** are presented in Table 3.2. The ellipsoidal region is shown in Figure 3.5.

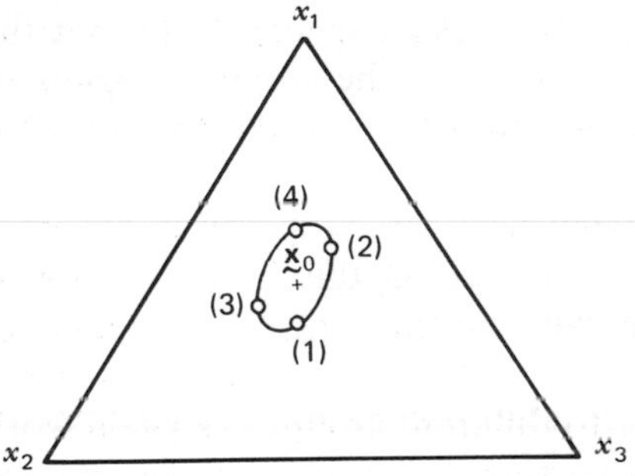

FIGURE 3.5. Design points on the boundary of the ellipsoidal region centered at $\mathbf{x}_0$.

3.5. ENLARGING THE UNIT SPHERICAL REGION OF INTEREST

We have assumed an ellipsoidal region of interest in the space of the mixture components. The centroid and size of this region is to be specified by the experimenter for a particular application. It might be of value to determine how much flexibility one has in enlarging the region of interest within the boundaries of the simplex, keeping of course the region centered at the point of main interest $\mathbf{w} = \mathbf{0}$. This flexibility can be measured by calculating the radius of the largest sphere centered at $\mathbf{w} = \mathbf{0}$ (which is also $\mathbf{v} = \mathbf{0}$) that will fit inside the simplex in the design space and by comparing it with $\rho = 1$, where $\rho = 1$ is the radius of the unit spherical region of interest.

To determine the radius of the largest $(q - 1)$-dimensional sphere, let us define ρ_i as the distance, as measured in the w_i metric, from $\mathbf{w} = \mathbf{0}$ or $\mathbf{x} = \mathbf{x}_0$ to the closest $(q - 2)$-dimensional face opposite the vertex where $x_i = 1$. Then when $h_i = x_{0i}$, for all i,

$$\rho_i = x_{0i}\left\{\frac{1}{x_{0i}^2} + \frac{1}{a - x_{0i}^2}\right\}^{1/2} = \left\{1 + \frac{x_{0i}^2}{a - x_{0i}^2}\right\}^{1/2}, \qquad i = 1, 2, \ldots, q$$

where $a = \Sigma_{l=1}^{q} x_{0l}^2$. On the other hand, if $h_i < x_{0i}$, then $\rho_i = x_{0i}$ $\left\{\frac{1}{h_i^2} + \frac{1}{a - h_i^2}\right\}^{1/2}$ where $a = \Sigma_{i=1}^{\alpha} h_i^2$. If we take

$$\rho^* = \min_{1 \le i \le q} \rho_i \tag{3.33}$$

then ρ^* is the radius of the largest sphere. The largest sphere may be called the *extended region of interest* (in the design space). Note that the extended region of interest can be reduced to the original region of interest, the unit spherical region, by putting $\rho^* = 1$ in the formulas for the designs.

We now discuss a program of design strategy as pertains to using standard response surface designs. We are doing this here because at present we are interested in working with mathematically independent variables in the construction of designs. In Section 4.10, where we discuss categorizing the components according to different groups, we again work with independent variables when constructing designs. The reader who is familiar with first-order and second-order (e.g., central composite design) response surface designs may go on to the next section on cuboidal regions.

3.6. SOME DISCUSSION ON DESIGN STRATEGY WHEN FITTING RESPONSE SURFACES

At the beginning of this chapter, we mentioned that one of the reasons for using independent variables is to enable us to rely or fall back on known response surface techniques, such as the construction of standard designs and the use of known design criteria. Thus familiarity with experimental procedures is the reason why many practitioners choose to work with independent variables.

In an attempt to obtain an efficient distribution of experimental points for the purpose of estimating the elements of the parameter vector $\boldsymbol{\alpha}$ in the model $\mathbf{y} = \mathbf{W}\boldsymbol{\alpha} + \boldsymbol{\epsilon}$, we must keep in mind three considerations. The arrangement should be such that the following conditions are satisfied, provided that they are mutually consistent:

1. The variances of the estimated coefficients should be as small as possible.
2. When fitting a first-degree model, the biases in the estimated coefficients that might occur, if the first-degree model is an inadequate representation, should be as small as possible.
3. Since first-degree designs are the first step in any type of sequential experimentation, we suggest using first-degree designs that can be augmented easily to second-degree designs by the addition of points.

For fitting first- and second-degree models where the objective is to model the surface so that predictions of the response can be made, we suggest the use of *rotatable designs*. A detailed discussion of rotatable designs is contained in Myers (1971), Chapter 7. With rotatable designs, the variance of the estimate of the response $\hat{y}(\mathbf{w})$ is constant at a constant distance from the center of the design. Furthermore, since the magnitude of the var $\hat{y}(\mathbf{w})$ is a function *only* of the distance from the center of the design that the point $\mathbf{w}$ is located and is not a function of the direction, the size of var $\hat{y}(\mathbf{w})$ is independent of the choice of the matrix $\mathbf{T}_1$ in Eq. (3.29). Thus if the design in $\mathbf{W}$ is rotatable, any appropriate $\mathbf{T}_1$ so that $\mathbf{T}$ is orthogonal can be used without affecting the variance of $\hat{y}(\mathbf{w})$. (Note that in using rotatable designs, provided that the design points are within a distance ρ from the point of main interest, a rotation will not take any of them outside the design space where a rotation means the use of an arbitrary transformation matrix $\mathbf{T}_1$.)

3.7. ROTATABLE DESIGNS

Let us define the first-degree model in Eq. (3.28) as

$$\mathbf{y} = \mathbf{W}_1\boldsymbol{\alpha}_1 + \boldsymbol{\epsilon} \tag{3.34}$$

where

$$\mathbf{W}_1 = \begin{bmatrix} 1 & \\ 1 & \\ \vdots & \mathbf{D}_w \\ 1 & \end{bmatrix}$$

$\boldsymbol{\alpha}_1' = (\alpha_0, \boldsymbol{\alpha}')$, and the elements of the design matrix $\mathbf{D}_w$ are the levels of the $q-1$ independent design variables to be used in the N experiments. We include a constant term in Eq. (3.34) and α_0 is the parameter associated with the constant term. Let us define the matrix $\mathbf{D}_w$ to be an $N \times (q-1)$ matrix of the type

$$\mathbf{D}_w = c\begin{bmatrix} \pm 1 & \pm 1 & \cdots & \pm 1 \\ \pm 1 & \pm 1 & \cdots & \pm 1 \\ \vdots & \vdots & & \vdots \\ \pm 1 & \pm 1 & \cdots & \pm 1 \end{bmatrix} \tag{3.35}$$

where c is a scalar quantity that we call the radius multiplier. The design matrix is $\mathbf{D}_w = c\mathbf{W}$ where the matrix $\mathbf{W}$ was introduced in Eq. (3.28). The design points according to the design matrix $\mathbf{D}_w$ in (3.35) lie on a sphere of radius $c\sqrt{q-1}$ where $q-1$ is the number of columns in $\mathbf{D}_w$. Furthermore, the points are positioned symmetrically about the center $(\mathbf{w} = \mathbf{0})$ of the design and therefore the design is known as a first-order, or first-degree, *rotatable design*.

The size of the design in terms of the spread of the points is controlled by fixing the value of the radius multiplier c. The size is important when considering how each of the first two design criteria mentioned previously (namely keeping the variances of the coefficients small and minimizing the biases in the estimated coefficients) can affect the properties of the predictor $\hat{y}(\mathbf{w})$. For example, when the variance of $\hat{y}(\mathbf{w})$ only is to be minimized because the bias of $\hat{y}(\mathbf{w})$ is assumed to be zero, Box (1952) shows that the largest possible design radius is best. In other words, when using a design matrix of the form shown in Eq. (3.35), the "all-variance" design, which is the design with the points spread out as far as possible, should have the value of the radius multiplier

$$c = \frac{\rho^*}{\sqrt{q-1}} \tag{3.36}$$

where ρ^* is the radius of the largest sphere centered at $\mathbf{w} = \mathbf{0}$ that will fit inside the simplex. Thus placing the design points on the perimeter of the largest sphere will result in the minimization of the variance of $\hat{y}(\mathbf{w})$ with the first-degree model.

When the biases in the estimated coefficients of $\hat{y}(\mathbf{w})$ are to be minimized (and the variance of $\hat{y}(\mathbf{w})$ is assumed to be negligible) the spread of the design is reduced by setting the value of c for the design matrix in Eq. (3.35) to be

$$c = \frac{\rho}{\sqrt{q+1}} \tag{3.37}$$

where ρ is the radius of the region of interest. Reducing the spread of the design minimizes the likelihood of picking up curvature of the surface when fitting only the first-degree model.

Before continuing on to second-degree designs, let us say more about the form of the design matrix $\mathbf{D}_w$ in Eq. (3.35). It has been shown in the literature (see, e.g., Box and Draper, 1959) that, when fitting a first-degree model and the true response surface is possibly quadratic, the average mean square error of the estimate $\hat{y}(\mathbf{w})$ is minimized by employing a first-degree orthogonal design that has third-degree moments equal to zero. To ensure that the third-degree moments of the design are zero, one chooses the signs ($\pm$) in $\mathbf{D}_w$ in Eq. (3.35) in such a way that the bottom half of the matrix is the negative replicate of the top half (i.e., the design $\mathbf{D}_w$ is a foldover design). For example, when $q-1=3$ one could use a 2^{3-1} factorial pattern in the top half of $\mathbf{D}_w$ and the bottom half of $\mathbf{D}_w$ would be another 2^{3-1} consisting of elements whose signs are the opposite of the signs of the elements in the top half.

The third design consideration mentioned previously, which is the ease of augmentation of the first-degree design to a second-degree design, is taken care of by the design matrix considered in Eq. (3.35). For example, to form the commonly used central composite design, we simply add $2(q-1)$ axial points and some center point replicates to the matrix $\mathbf{D}_w$.

The design matrix $\mathbf{D}_w$ in Eq. (3.35) consisted of two levels only for each factor w_i and would permit the estimation of the parameters α_0, α_i, and α_{ij}

in the model

$$y = \alpha_0 + \sum_{i=1}^{q-1} \alpha_i w_i + \sum_{i=1}^{q-1} \alpha_{ii} w_i^2 + \sum_{i<j}^{q-1}\sum \alpha_{ij} w_i w_j + \epsilon \tag{3.38}$$

With the addition of center point replicates, however, an estimate of the sum $\alpha_{11} + \alpha_{22} + \cdots + \alpha_{q-1,q-1}$ can be obtained but the experimenter is still without information on the individual α_{ii} $(1 \leq i \leq q-1)$. The α_{ii} are measures of curvature of the surface along the w_i axes.

One of the most useful designs for estimating all the individual parameters in Eq. (3.38) is the central composite design introduced by Box and Wilson (1951). These well-known designs are formed by adding points to the factorial settings where the added points are located on the axes of the w_i $(1 \leq i \leq q-1)$ as well as at the center of the design. For example, with $q-1$ variables, the central composite design which consists of the hypercube, the crosspolytope, and n_0 center point replicates will have a design matrix of the form

$$\mathbf{D}_w = \left[\begin{array}{cccc} \pm c & \pm c & \cdots & \pm c \\ \pm c & \pm c & \cdots & \pm c \\ \vdots & \vdots & & \vdots \\ \pm c & \pm c & \cdots & \pm c \\ \hline -g & 0 & \cdots & 0 \\ g & 0 & \cdots & 0 \\ 0 & -g & & 0 \\ 0 & g & & 0 \\ \vdots & \vdots & & \vdots \\ 0 & 0 & & -g \\ 0 & 0 & & g \\ \hline 0 & 0 & & 0 \\ \vdots & \vdots & & \vdots \\ 0 & 0 & & 0 \end{array}\right] \begin{array}{c} \mathbf{D}_1 \\ \\ \mathbf{D}_2 \\ \\ \mathbf{D}_3 \end{array} \tag{3.39}$$

In $\mathbf{D}_w$, which is horizontally partitioned into the matrices $\mathbf{D}_1$, $\mathbf{D}_2$, and $\mathbf{D}_3$, the elements of the $M \times (q-1)$ matrix $\mathbf{D}_1$ consist of the scalar product of the radius multiplier c and the plus and minus ones in the design matrix of a two-level factorial design. The matrix $\mathbf{D}_2$ is a $2(q-1) \times (q-1)$ matrix whose elements are the axial or star points. The $n_0 \times (q-1)$ matrix $\mathbf{D}_3$ contains n_0 center point replicates. Hence, the total number of experiments to be performed with the central composite design is $N = M + 2(q-1) + n_0$ where $M = 2^{q-1}$ or some fraction of the complete factorial in the $q-1$ variables.

For the second-degree design to be rotatable, we require only that

$$c \leq \frac{\rho_{ccd}}{\sqrt{q-1}} \quad (\rho_{ccc} \leq \rho^*)$$

and

$$g = \sqrt[4]{M}\,c \tag{3.40}$$

where M is the number of design points in the factorial portion $\mathbf{D}_1$ of $\mathbf{D}_w$ and ρ_{ccd} is the radius of the composite design centered at $\mathbf{w} = \mathbf{0}$. As a reminder if the complete factorial in the w_i's is to be run, $M = 2^{q-1}$, but if a fractional replicate of the complete design is run, then $M = 2^{q-p}$ where $p > 1$.

3.8. A SECOND-ORDER ROTATABLE DESIGN FOR A FOUR-COMPONENT SYSTEM

A second-degree model is to be fitted to data collected on a four-component system. For this example the point of main interest is chosen to be $\mathbf{x}_0 = (0.30, 0.25, 0.33, 0.12)'$ and the matrix $\mathbf{H}$ of half-intervals of interest is $\mathbf{H} = \text{diag}(0.20, 0.20, 0.30, 0.10)$.

The design in the $q-1=3$ independent variables w_1, w_2, and w_3 is to be a central composite rotatable design with a design matrix $\mathbf{D}_w$ as in Eq. (3.39). The design points are to be placed on the surface of the largest sphere centered at $\mathbf{w} = \mathbf{0}$ (or $\mathbf{x} = \mathbf{x}_0$) and therefore the value of the radius multiplier c is from Eq. (3.36),

$$c = \frac{\rho^*}{\sqrt{3}} = \frac{1.2348}{\sqrt{3}} = 0.7129$$

where $\rho^* = 0.12\{(1/0.04) + (1/0.14)\}^{1/2} = 1.2348$ from Eq. (3.33).

For the design to be rotatable, we set $g = \sqrt[4]{8}c = 1.682(0.7129) = 1.1990$ in $\mathbf{D}_w$ in Eq. (3.39). Selecting the form of the matrix $\mathbf{T}$ by the method presented in Appendix 3B, from Eq. (3.29) we have, for the settings of the mixture components, corresponding to the settings of the design variables

w_1, w_2, and w_3, $\mathbf{X}_c = \mathbf{D}_w\mathbf{T}_1'\mathbf{H}$ or $\mathbf{X} = \mathbf{D}_w\mathbf{T}_1'\mathbf{H} + [\mathbf{x}_0']$

$$\underset{\mathbf{X}}{\begin{bmatrix} 0.458 & 0.241 & 0.239 & 0.062 \\ 0.291 & 0.408 & 0.239 & 0.062 \\ 0.336 & 0.119 & 0.483 & 0.062 \\ 0.170 & 0.285 & 0.483 & 0.062 \\ 0.431 & 0.214 & 0.177 & 0.178 \\ 0.264 & 0.381 & 0.177 & 0.178 \\ 0.309 & 0.091 & 0.422 & 0.178 \\ 0.142 & 0.259 & 0.422 & 0.178 \\ 0.384 & 0.166 & 0.330 & 0.120 \\ 0.216 & 0.334 & 0.330 & 0.120 \\ 0.361 & 0.311 & 0.208 & 0.120 \\ 0.239 & 0.189 & 0.452 & 0.120 \\ 0.314 & 0.264 & 0.360 & 0.063 \\ 0.287 & 0.237 & 0.299 & 0.177 \\ 0.300 & 0.250 & 0.330 & 0.120 \end{bmatrix}} \underset{=(c)}{=(0.7129)} \underset{\mathbf{W}}{\begin{bmatrix} -1 & -1 & -1 \\ 1 & -1 & -1 \\ -1 & 1 & -1 \\ 1 & 1 & -1 \\ -1 & -1 & 1 \\ 1 & -1 & 1 \\ -1 & 1 & 1 \\ 1 & 1 & 1 \\ -1.682 & 0 & 0 \\ 1.682 & 0 & 0 \\ 0 & -1.682 & 0 \\ 0 & 1.682 & 0 \\ 0 & 0 & -1.682 \\ 0 & 0 & 1.682 \\ 0 & 0 & 0 \end{bmatrix}}$$

$$\times \underset{\mathbf{T}_1'}{\begin{bmatrix} -0.707 & 0.707 & 0 & 0 \\ -0.515 & -0.515 & 0.686 & 0 \\ -0.114 & -0.114 & -0.171 & 0.971 \end{bmatrix}} \underset{\mathbf{H}}{\begin{bmatrix} 0.20 & & & \mathbf{0} \\ & 0.20 & & \\ & & 0.30 & \\ \mathbf{0} & & & 0.10 \end{bmatrix}}$$

$$+ \begin{bmatrix} \mathbf{x}_0' \\ \vdots \\ \mathbf{x}_0' \end{bmatrix}$$

$$+ \begin{bmatrix} 0.30 & 0.25 & 0.33 & 0.12 \\ \vdots & \vdots & \vdots & \vdots \\ 0.30 & 0.25 & 0.33 & 0.12 \end{bmatrix}$$

In Table 3.3 are the design settings in the mixture components for the second-order central composite rotatable designs for $q = 3$ and $q = 4$ where the points are positioned on the largest sphere and also on the unit sphere.

TABLE 3.3. Component proportions corresponding to a central composite rotatable design with design center at centroid of simplex

$q = 3$	Minimum Variance Design $c = \rho^*/\sqrt{2} = 0.866$		Points on Unit Sphere $c = 0.7071$		
x_1	x_2	x_3	x_1	x_2	x_3
0.6553	0.2470	0.0976	0.5962	0.2629	0.1408
0.2470	0.6553	0.0976	0.2629	0.5962	0.1408
0.4196	0.0113	0.5690	0.4037	0.0704	0.5258
0.0113	0.4196	0.5690	0.0704	0.4037	0.5258
0.6219	0.0447	0.3333	0.5690	0.0976	0.3333
0.0447	0.6219	0.3333	0.0976	0.5690	0.3333
0.5000	0.5000	0.0000	0.4694	0.4694	0.0611
0.1667	0.1667	0.6666	0.1973	0.1973	0.6055
0.3333	0.3333	0.3333	0.3333	0.3333	0.3333

$q = 4$	$c = \rho^*/\sqrt{3} = 0.6667$			$c = 0.5775$			
x_1	x_2	x_3	x_4	x_1	x_2	x_3	x_4
0.4840	0.2483	0.1620	0.1057	0.4527	0.2485	0.1738	0.1250
0.2483	0.4840	0.1620	0.1057	0.2485	0.4527	0.1738	0.1250
0.3479	0.1122	0.4342	0.1057	0.3348	0.1307	0.4095	0.1250
0.1122	0.3479	0.4342	0.1057	0.1307	0.3348	0.4095	0.1250
0.3878	0.1521	0.0658	0.3943	0.3693	0.1652	0.0905	0.3750
0.1521	0.3878	0.0658	0.3943	0.1652	0.3693	0.0905	0.3750
0.2517	0.0160	0.3380	0.3943	0.2515	0.0473	0.3262	0.3750
0.0160	0.2517	0.3380	0.3943	0.0473	0.2515	0.3262	0.3750
0.4482	0.0518	0.2500	0.2500	0.4217	0.0783	0.2500	0.2500
0.0518	0.4482	0.2500	0.2500	0.0783	0.4217	0.2500	0.2500
0.3644	0.3644	0.0211	0.2500	0.3491	0.2491	0.0518	0.2500
0.1356	0.1356	0.4789	0.2500	0.1059	0.1509	0.4482	0.2500
0.3309	0.3309	0.3309	0.0073	0.3201	0.3201	0.3201	0.0397
0.1691	0.1691	0.1691	0.4927	0.1799	0.1799	0.1799	0.4603
0.2500	0.2500	0.2500	0.2500	0.2500	0.2500	0.2500	0.2500

We now discuss briefly inscribing a cuboidal region of interest completely inside the simplex as an alternative to the ellipsoidal region of interest. In some applications the cuboidal region is a more realistic region than an ellipsoidal region, particularly when the sides of the cuboidal region are completely specified by imposing additional constraints on the component proportions.

3.9. DEFINING A CUBOIDAL REGION OF INTEREST IN THE MIXTURE SYSTEM

A cuboidal region inside the simplex consists of all points simultaneously belonging to the intervals

$$x_{0i} \pm h_i, \qquad i = 1, 2, \ldots, q \tag{3.41}$$

where x_{0i} denotes the center of the interval of interest for component i and $2h_i$ is the range of interest of component i. We could also express the q simultaneous intervals as $x_{0i} - h_i \leq x_i \leq x_{0i} + h_i$, $i = 1, 2, \ldots, q$ assuming of course that all the points or proportions assigned to the components satisfy the property $x_1 + x_2 + \cdots + x_q = 1$.

As with the ellipsoidal region it would be helpful to us when constructing designs in the w_i's if we had a distance formula to aid us in determining the largest cube centered at $\mathbf{x} = \mathbf{x}_0$ that will fit inside the simplex. Such a formula is arrived at by recalling that the distance from $\mathbf{x} = \mathbf{x}_0$ to the closest boundary face opposite the vertex $x_i = 1$ is

$$\rho_i = \left\{ 1 + \frac{x_{0i}^2}{\sum_{\substack{j=1 \\ j \neq i}}^{q} x_{0j}^2} \right\}^{1/2}, \qquad i = 1, 2, \ldots, q \tag{3.42}$$

Since the shortest distance or smallest valued ρ_i will be the greatest possible distance from $\mathbf{x} = \mathbf{x}_0$ to the vertices of the cube, the largest cube in the simplex centered at $\mathbf{x}_0$ is found by calculating

$$\rho^* = \min_{1 \leq i \leq q} \rho_i \tag{3.43}$$

Let us write the range of values for the w_j as

$$-c \leq w_j \leq c, \qquad j = 1, 2, \ldots, q-1 \tag{3.44}$$

where c is at most equal to ρ^* in Eq. (3.43). If a design in the w_j's is chosen by selecting a design matrix $\mathbf{D}_w$ of the form in Eq. (3.35), then

$$c = \frac{\rho}{\sqrt{q-1}} \tag{3.45}$$

which guarantees that the points are positioned on the vertices of the inscribed cube.

Let us use the three-component example already discussed to illustrate how to define a cuboidal region. Earlier the design point settings were placed on the perimeter of the unit spherical region and the mixture component settings are listed in Table 3.2. Let us position the design points at the vertices of the largest cube centered at $\mathbf{x}_0 = (0.43, 0.30, 0.27)'$ that will fit inside the triangle assuming of course that the region of interest extends to the perimeter of the cube. In this case we alter the elements of the matrix of scale constants to be $\mathbf{H} = \text{diag}(0.43, 0.30, 0.27)$. The distances ρ_1, ρ_2, and ρ_3 to the respective edges are, from Eq. (3.42),

$$\rho_1 = 0.43\left\{\frac{1}{0.43^2} + \frac{1}{0.30^2 + 0.27^2}\right\}^{1/2} = 1.46, \qquad \rho_2 = 1.16, \qquad \rho_3 = 1.12$$

From Eq. (3.43), the largest cube has radius, from the center to the nearest vertex, $\rho^* = \rho_3 = 1.12$ and therefore the settings for w_1 and w_2, according to Eqs. (3.44) and (3.45), are

$$-\frac{1.12}{\sqrt{2}} \leq w_j \leq \frac{1.12}{\sqrt{2}}$$

$$-0.79 \leq w_j \leq 0.79, \qquad i = 1, 2.$$

Upon substituting the values of w_1 and w_2 into the matrix $\mathbf{W}$ as in Eq. (3.32), the corresponding settings in the mixture components are

$$\underset{\mathbf{X}}{\begin{bmatrix} 0.107 & 0.433 & 0.460 \\ 0.497 & 0.043 & 0.460 \\ 0.363 & 0.557 & 0.080 \\ 0.753 & 0.168 & 0.080 \end{bmatrix}} = \underset{[\mathbf{W} \quad \mathbf{0}]}{\begin{bmatrix} 0.79 & 0.79 & 0 \\ -0.79 & 0.79 & 0 \\ 0.79 & -0.79 & 0 \\ -0.79 & -0.79 & 0 \end{bmatrix}} \underset{\mathbf{T}'}{\begin{bmatrix} -0.572 & 0.820 & 0 \\ -0.375 & -0.262 & 0.889 \\ 0.729 & 0.509 & 0.458 \end{bmatrix}}$$

$$\times \mathbf{H} + \begin{bmatrix} \mathbf{x}_0' \\ \mathbf{x}_0' \\ \mathbf{x}_0' \\ \mathbf{x}_0' \end{bmatrix}$$

$$\times \mathbf{H} + \begin{bmatrix} 0.43 & 0.30 & 0.27 \\ 0.43 & 0.30 & 0.27 \\ 0.43 & 0.30 & 0.27 \\ 0.43 & 0.30 & 0.27 \end{bmatrix}$$

and the design configuration is shown in Figure 3.6. The use of a different

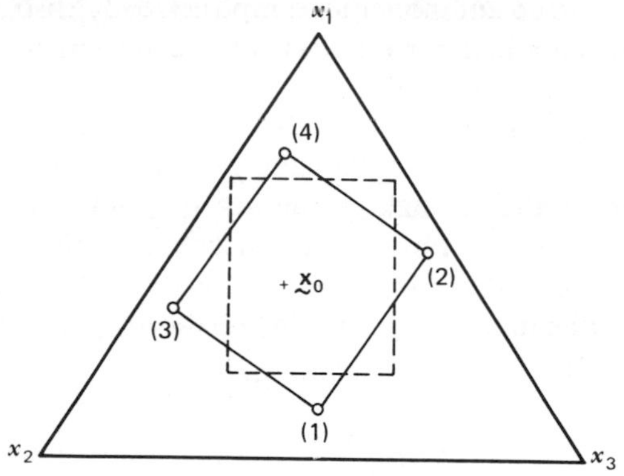

FIGURE 3.6. Design points located at the vertices of the largest inscribed cube centered at $\mathbf{x}_0$. The dotted cube is the solid cube rotated.

orthogonal matrix $\tilde{\mathbf{T}}$ produced the following mixture component proportions

$$
\begin{array}{ccc}
\mathbf{X} & = & [\mathbf{W}\ \ \mathbf{0}] \qquad\qquad\qquad \tilde{\mathbf{T}}'
\end{array}
$$

$$
\begin{bmatrix} 0.662 & 0.330 & 0.008 \\ 0.198 & 0.587 & 0.215 \\ 0.662 & 0.013 & 0.325 \\ 0.198 & 0.270 & 0.532 \end{bmatrix} = \begin{bmatrix} 0.79 & 0.79 & 0 \\ -0.79 & 0.79 & 0 \\ 0.79 & -0.79 & 0 \\ -0.79 & -0.79 & 0 \end{bmatrix} \begin{bmatrix} 0.684 & -0.542 & -0.488 \\ 0 & 0.670 & -0.743 \\ 0.729 & 0.509 & 0.455 \end{bmatrix}
$$

$$
\times \mathbf{H} + \begin{bmatrix} \mathbf{x}_0' \\ \mathbf{x}_0' \\ \mathbf{x}_0' \\ \mathbf{x}_0' \end{bmatrix}
$$

$$
\times \mathbf{H} + \begin{bmatrix} 0.43 & 0.30 & 0.27 \\ 0.43 & 0.30 & 0.27 \\ 0.43 & 0.30 & 0.27 \\ 0.43 & 0.30 & 0.27 \end{bmatrix}
$$

The design configuration caused by the rotation $\tilde{\mathbf{T}}$ is shown as the dotted cuboidal region in Figure 3.6.

3.10. THE INCLUSION OF PROCESS VARIABLES

In the earlier example, in which the automobile fuels A and B were blended together to see if an increase in miles per gallon could be achieved when compared to the simple average of the mileage of the individual fuels, all the experimental trials were performed using the same automobile, which was driven at a constant speed. Now, in addition to the joint effects of the fuels, it seems reasonable to want to study the effect of varying the driving speed or of the size of the automobile, particularly as each influences the blending behavior of the fuels. These latter two experimental factors (driving speed and automobile size) do not form any part of the mixtures and therefore are not called components. Instead they are called *process variables* of the experiment. (Another example of process variables is in cattle feeding trials in which factors like age of cattle, breed, lactation, etc., are varied in addition to the proportions of different feeds.)

In order to include process variables in the design and analysis of mixing experiments, we consider the case where the mixture components are transformed to $q-1$ mathematically independent variables $w_1, w_2, \ldots, w_{q-1}$. This will facilitate the combining of the two types of variables, because the process variables are independent variables and combining two sets of independent variables (the ingredients and the process) into one set is relatively easy. For convenience, we assume that in the space of the mixture components the experimenter's interest is concentrated in an ellipsoidal region within the simplex. The combining of process variables with the original mixture components where lattice designs are used in the mixture components is discussed in Chapter 5.

The process variables are not linear functions of the chemical components and will be denoted by z_j $(1 \leq j \leq n)$. The number of process variables we can consider can be any positive integer n such that $q-1+n<N$, where N is the total number of observations. In most practical situations, however, it would seem reasonable to assume that n is small (say $n \leq 6$) since the effects of q mixture components are also being investigated.

In defining the region of interest, which now includes both the mixture components and the process variables, let us first consider the case where the process variables are each performed at only two levels. If we keep the ellipsoidal region of interest for the mixture components as in Eq. (3.20) and define a separate region for the process variables, the inclusion of the process variables does not affect the design settings in the composition space whose coordinates are at $w_1, w_2, \ldots, w_{q-1}$. For the process variables z_j which are each at two levels, the region of interest is an n-dimensional hypercube. The two regions are then combined to form a

$(q-1+n)$-dimensional region of interest. An example of combining the separate regions of interest into one region (the cylinder) is shown in the Figure 3.7, where we have three components x_1, x_2, and x_3 and the single process variable z_1, that is, $q=3$ and $n=1$.

For the purpose of fitting a low-order polynomial in the w_i's and the z_j's, let us consider one of two possible model forms, keeping in mind that only the w_i are linear functions of the mixture components. In fact, for the remainder of this section, we shall refer to the w_i as MRVs (mixture-related variables). The first form presented is for the case where the response is assumed to be planar in both the MRVs and the process variables and therefore the polynomial model is written

$$y(\mathbf{w},\mathbf{z}) = \alpha_0 + \sum_{i=1}^{q-1} \alpha_i w_i + \sum_{j=1}^{n} \gamma_j z_j + \epsilon \tag{3.46}$$

The α_i, $i=1,2,\ldots,q-1$ are parameters associated with the MRVs, and the γ_j are parameters associated with the process variables. The use of the model in Eq. (3.46) assumes that the w_i's and the z_j's are independent, in the sense that the effects of the proccss variables do not influence the blending characteristics of the mixture components. This assumption is

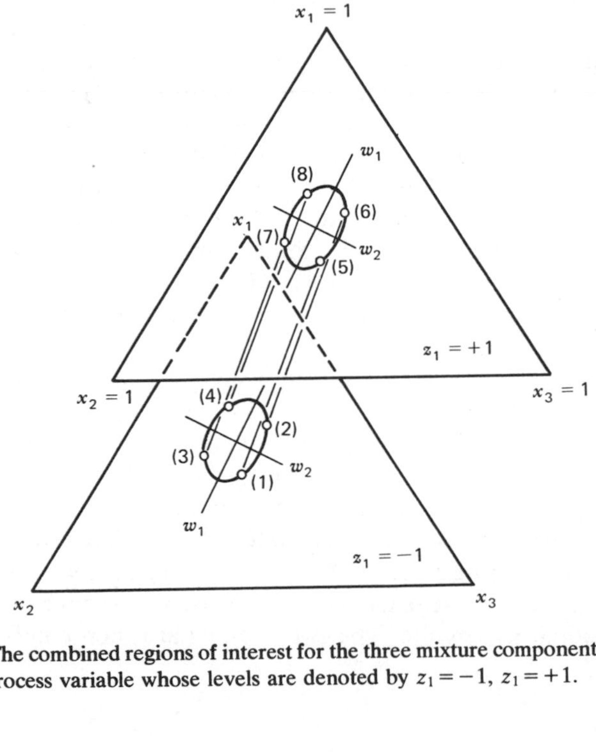

FIGURE 3.7. The combined regions of interest for the three mixture components 1, 2, and 3, and the single process variable whose levels are denoted by $z_1=-1$, $z_1=+1$.

questionable in many experiments and therefore one would probably prefer to consider a different polynomial model form.

A second model is one which contains crossproduct terms between the MRVs and the process variables:

$$y(\mathbf{w}, \mathbf{z}) = \alpha_0 + \sum_{i=1}^{q-1} \alpha_i w_i + \sum_{j=1}^{n} \gamma_j z_j + \sum_{i=1}^{q-1} \sum_{j=1}^{n} \delta_{ij} w_i z_j + \epsilon \tag{3.47}$$

This model enables one to obtain some measure of dependency that is present between the MRVs and the process variables through the parameter estimates $\hat{\delta}_{ij}$ of the δ_{ij}. For example, if the fuels A and B are synergistic to one another at low speeds but at high speeds, the fuels are merely additive in their effects, we would like to discover this dependency of effects at different speeds and this we could do with the $\delta_{ij} w_i z_j$ in Eq. (3.47). We might even add terms such as $\alpha_{ii'} w_i w_{i'}$ $(i < i')$ and $\gamma_{jj'} z_j z_{j'}$ $(j < j')$ to the model if it is desired to obtain some measure of the lack of fit in the separate regions of interest.

The structure of the error term ϵ in the Eqs. (3.46) and (3.47) will depend on the manner (or sequence) in which the experiments are performed. In an actual performance of the experimental runs, the N runs should be performed in a complete random order to insure the independence of the errors. If the runs are not performed at random but rather at each of the design settings of the process variables a set of experiments in the MRVs is performed, then the design would be a split-plot design because of the restricted randomization. The design settings among the process variables would be like the main-plot treatments and the settings among the MRVs would be like subplot treatments, randomized in each main-plot treatment. In this case two error variances would be needed for performing tests of significance; one error variance to test the differences among the treatments in the MRVs and a second error variance to test the differences between the means of the runs at each treatment combination in the process variables. In order to obtain estimates of these error variances, the entire set of N experiments would have to be replicated at least twice.

3.11. A NUMERICAL EXAMPLE INVOLVING THREE MIXTURE COMPONENTS AND ONE PROCESS VARIABLE

To illustrate the setting up of the design region in a three-component triangle and to discuss the analysis of the data, we recall the three-component example of the Section 3.3 where the ellipsoidal region of interest was introduced. The proportions of the three components are denoted by x_1, x_2, and x_3 and now we include one process variable and let z_1 denote the level of the process variable as well as the variable itself. In vector notation, the ellipsoidal region of interest of Eq. (3.19) in the mixture components is defined as

$$(\mathbf{x}-\mathbf{x}_0)'\mathbf{H}^{-1}(\mathbf{x}-\mathbf{x}_0) \le 1 \tag{3.48}$$

where

$$\begin{aligned}(\mathbf{x}-\mathbf{x}_0)' &= (x_1 - x_{01}, x_2 - x_{02}, x_3 - x_{03}) \\ &= (x_1 - 0.43, x_2 - 0.30, x_3 - 0.27) \\ \mathbf{H} &= \text{diag}(0.16, 0.15, 0.07)\end{aligned}$$

A planar surface over the ellipsoidal region in Eq. (3.48) is to be studied at a high level (+1) and at a low level (−1) of the process variable z_1.

The model in Eq. (3.47) is written in matrix notation as

$$\mathbf{y} = \mathbf{W}_{p}\boldsymbol{\alpha}_{p} + \boldsymbol{\epsilon} \tag{3.49}$$

where the subscript p signifies the inclusion of a process variable. At each level of the process variable, a design matrix $\mathbf{D}_w$ in w_1 and w_2 like the $\mathbf{W}$ used in Eq. (3.32) is set up so that the form of the matrix $\mathbf{W}_p$ and the vector $\boldsymbol{\alpha}_p$ of parameters to be estimated are

$$\mathbf{W}_p = \begin{bmatrix} 1 & c\begin{bmatrix} 1 & 1 \\ -1 & 1 \\ 1 & -1 \\ -1 & -1 \\ 1 & 1 \\ -1 & 1 \\ 1 & -1 \\ -1 & -1 \end{bmatrix} & \begin{matrix} -1 \\ -1 \\ -1 \\ -1 \\ 1 \\ 1 \\ 1 \\ 1 \end{matrix} & c\begin{bmatrix} -1 & -1 \\ 1 & -1 \\ -1 & 1 \\ 1 & 1 \\ 1 & 1 \\ -1 & 1 \\ 1 & -1 \\ -1 & -1 \end{bmatrix} \end{bmatrix} \qquad \boldsymbol{\alpha}_p = \begin{bmatrix} \alpha_0 \\ \alpha_1 \\ \alpha_2 \\ \gamma_1 \\ \delta_{11} \\ \delta_{21} \end{bmatrix}$$

where c is the scalar radius multiplier. Setting the design points on the perimeter of the unit sphere centered at $\mathbf{w} = \mathbf{0}$ is equivalent to setting the value of the radius multiplier c equal to $c = 1/\sqrt{2}$. Using the same orthogonal matrix $\mathbf{T}$ as in Eq. (3.31), the settings of the mixture components at each level z_1 are presented in Table 3.4. Also presented in Table 3.4 are the observed values of the response at the eight design point settings in $\mathbf{W}_p$.

The least squares estimates of the elements of $\boldsymbol{\alpha}_p$ in Eq. (3.49) are

$$\underset{\mathbf{a}_p}{\begin{bmatrix} 19.33 \\ -1.77 \\ -0.21 \\ 4.65 \\ -0.04 \\ -1.03 \end{bmatrix}} = \underset{(\mathbf{W}_p'\mathbf{W}_p)^{-1}}{\frac{1}{8c^2}\begin{bmatrix} c^2 & & & & & \mathbf{0} \\ & 1 & & & & \\ & & 1 & & & \\ & & & c^2 & & \\ & & & & 1 & \\ \mathbf{0} & & & & & 1 \end{bmatrix}} \underset{\mathbf{W}_p'\mathbf{y}}{\begin{bmatrix} 154.60 \\ -7.07 \\ -0.85 \\ 37.20 \\ -0.14 \\ -4.10 \end{bmatrix}}$$

TABLE 3.4. Design point settings and observed response values with three mixture components at two levels of one process variable

Design Point	Design Settings		Mixture Components			Process Variable (z_1)	Observed Response (y)
	w_1	w_2	x_1	x_2	x_3		
1	0.707	0.707	0.328	0.355	0.317	−1	14.3
2	−0.707	0.707	0.482	0.201	0.317	−1	16.2
3	0.707	−0.707	0.378	0.399	0.223	−1	12.6
4	−0.707	−0.707	0.532	0.245	0.223	−1	15.6
5	0.707	0.707	0.328	0.355	0.317	1	21.3
6	−0.707	0.707	0.482	0.201	0.317	1	24.9
7	0.707	−0.707	0.378	0.399	0.223	1	24.1
8	−0.707	−0.707	0.532	0.245	0.223	1	25.6

and the analysis of variance table is Table 3.5. In the sums of squares formulas, $\mathbf{1}'$ is a 1×8 vector of ones. The F-test of the fitted model is significant $F = 54.2 > F_{(5,2,\alpha=0.05)} = 19.30$ and the value of $R^2 = (189.87/191.28) = 0.993$ is high enough for us to be confident in using the fitted model for prediction purposes. Also $R_A^2 = 0.974$ supports this confidence.

Predicted values of the response corresponding to blends at either level of the process variable (at $z_1 = -1$ or $z_1 = +1$) can be obtained using the model $\hat{y}(\mathbf{w}, z_1) = \mathbf{w}_p'\mathbf{a}_p$ where the elements of the 1×6 vector $\mathbf{w}_p'$ correspond to the elements in a row of the matrix $\mathbf{W}_p$. For example, let us write the prediction equation

$$\underset{}{\hat{y}(\mathbf{w}, z_1) = \underset{(0.30)}{19.33} - \underset{(0.42)}{1.77}w_1 - \underset{(0.42)}{0.21}w_2 + \underset{(0.30)}{4.65}z_1 - \underset{(0.42)}{0.04}w_1z_1 - \underset{(0.42)}{1.03}w_2z_1} \tag{3.50}$$

and let us find the estimate of the response at the coordinate $(w_1, w_2) = (c = 0.707, 0)$ and at $z_1 = -1$. Then $\mathbf{w}_p = (1, c = 0.707, 0, -1, -0.707, 0)'$

TABLE 3.5. Analysis of variance table for the three-component and one-process-variable data

Source of Variation	Degrees of Freedom	Sums of Squares	Mean Square	F
Regression	5	$\{\mathbf{a}_p'\mathbf{W}_p'\mathbf{y} - (\mathbf{1}'\mathbf{y})^2/8\} = 189.87$	37.97	54.2
Residual	2	$\{\mathbf{y}'\mathbf{y} - \mathbf{a}_p'\mathbf{W}_p'\mathbf{y}\} = 1.41$	0.70	
Total	7	$\{\mathbf{y}'\mathbf{y} - (\mathbf{1}'\mathbf{y})^2/8\} = 191.28$		

and

$$\begin{aligned}\hat{y}(0.707, 0, -1) &= \mathbf{w}'_{\mathscr{p}}\mathbf{a}_{\mathscr{p}}\\ &= 19.33 - 1.77(0.707) + 4.65(-1) - 0.40(-0.707)\\ &= 13.45 \qquad (3.51)\end{aligned}$$

An estimate of the variance of $\hat{y}(0.707, 0, -1)$ is

$$\begin{aligned}\widehat{\text{var}}[\hat{y}(0.707, 0, -1)] &= \mathbf{w}'_{\mathscr{p}}(\mathbf{W}'_{\mathscr{p}}\mathbf{W}_{\mathscr{p}})^{-1}\mathbf{w}_{\mathscr{p}}s^2\\ &= (0.50)0.70 = 0.35 \qquad (3.52)\end{aligned}$$

3.12. SETTING UP THE MODEL IN THE ORIGINAL MIXTURE COMPONENTS

An equally effective model to Eq. (3.47) can be setup in the *mixture components* x_1, x_2, and x_3 with z_1. Borrowing from the material in Section 5.11, such a model would be of the form

$$\begin{aligned}y(\mathbf{x}, z_1) &= \sum_{i=1}^{3} [\gamma_i^0 + \gamma_i^1 z_1]x_i + \epsilon\\ &= \gamma_1^0 x_1 + \gamma_2^0 x_2 + \gamma_3^0 x_3 + \gamma_1^1 z_1 x_1 + \gamma_2^1 z_1 x_2 + \gamma_3^1 z_1 x_3 + \epsilon \qquad (3.53)\end{aligned}$$

where the terms in the model represent the mixture components x_i and products of the x_i with z_1. The coefficients γ_i^0, $i = 1, 2$ and 3, represent the heights of the surface above the vertices where $x_i = 1$, $x_j = 0$, $j \neq i$, averaged over the two levels of z_1 (even though no observations are collected at $x_i = 1$) while the coefficients γ_i^1 represent the effect of z_1 on these heights (the effect of z_1 is measured by taking one-half of the difference between γ_i^0 at $z_1 = +1$ and γ_i^0 at $z_1 = -1$).

Let us take the mixture settings for x_1, x_2, x_3, and the y_u values, from Table 3.4. The least squares estimates of the elements $\boldsymbol{\gamma}_{\mathscr{p}}$ in the model $\mathbf{y} = \mathbf{X}_{\mathscr{p}}\boldsymbol{\gamma}_{\mathscr{p}} + \boldsymbol{\epsilon}$ are

$$\mathbf{g}_{\mathscr{p}} = (\mathbf{X}'_{\mathscr{p}}\mathbf{X}_{\mathscr{p}})^{-1}\mathbf{X}'_{\mathscr{p}}\mathbf{y}$$

$$= (\mathbf{X}'_{\mathscr{p}}\mathbf{X}_{\mathscr{p}})^{-1}\begin{bmatrix} 0.328 & 0.355 & 0.317 & -0.328 & -0.355 & -0.317\\ 0.482 & 0.201 & 0.317 & -0.482 & -0.201 & -0.317\\ 0.378 & 0.399 & 0.223 & -0.378 & -0.399 & -0.223\\ 0.532 & 0.245 & 0.223 & -0.532 & -0.245 & -0.223\\ 0.328 & 0.355 & 0.317 & 0.328 & 0.355 & 0.317\\ 0.482 & 0.201 & 0.317 & 0.482 & 0.201 & 0.317\\ 0.378 & 0.399 & 0.223 & 0.378 & 0.399 & 0.223\\ 0.532 & 0.245 & 0.223 & 0.532 & 0.245 & 0.223 \end{bmatrix}' \begin{bmatrix} 14.3\\ 16.2\\ 12.6\\ 15.6\\ 21.3\\ 24.9\\ 24.1\\ 25.6 \end{bmatrix}$$

$$\mathbf{g}_A = \begin{bmatrix} 27.11 \\ 10.87 \\ 16.32 \\ 8.95 \\ 8.63 \\ -6.62 \end{bmatrix} \tag{3.54}$$

Substituting the parameter estimates into the model in Eq. (3.53), we have

$$\underset{}{\hat{y}(\mathbf{x}, z_1)} = \underset{(2.38)}{27.11x_1} + \underset{(2.80)}{10.87x_2} + \underset{(4.61)}{16.32x_3} + \underset{(2.38)}{8.95z_1x_1} + \underset{(2.80)}{8.63z_1x_2} - \underset{(4.61)}{6.62z_1x_3} \tag{3.55}$$

The settings x_1, x_2, and x_3 corresponding to the coordinate settings $(w_1, w_2, w_3 = 0) = (0.707, 0\,0)$ are found from Eq. (3.29) using $(\mathbf{x} - \mathbf{x}_0)' = (\mathbf{w}', 0)\mathbf{T}'\mathbf{H}$,

$$\begin{aligned}(x_1 - x_{01}, x_2 - x_{02}, x_3 - x_{03}) &= (0.707, 0, 0)\mathbf{T}' \operatorname{diag}(0.16, 0.15, 0.07) \\ &= (-0.077, 0.077, 0)\end{aligned}$$

and therefore,

$$\begin{aligned}(x_1, x_2, x_3) &= (-0.077, 0.077, 0) + (0.43, 0.30, 0.27) \\ &= (0.353, 0.377, 0.270)\end{aligned}$$

Substituting these values of x_1, x_2, and x_3 into Eq. (3.55) along with $z_1 = -1$, the predicted value of the response is

$$\begin{aligned}\hat{y}(0.353, 0.377, 0.270, -1) &= 27.11(0.353) + 10.87(0.377) + 16.32(0.270) \\ &\quad + 8.95(-0.353) + 8.63(-0.377) - 6.62(-0.270) \\ &= 13.45\end{aligned} \tag{3.56}$$

which is the same value predicted previously with Eq. (3.51) using the model in the w_i's and z_1. The estimated variance of the estimate of the response in Eq. (3.56) is $\widehat{\text{var}}[\hat{y}(\mathbf{x}, z_1)] = 0.35$, the same as in Eq. (3.52).

3.13. A DISCUSSION ON THE COMBINED MODEL VERSUS THE SEPARATE MIXTURE MODELS AT THE PROCESS VARIABLE SETTINGS

The combined model in the mixture components and process variable in either Eq. (3.50) or (3.55) can be used to predict values of the response to any mixture contained inside or on the surface of the cylinder in Figure 3.7. However, should one wish to isolate the predictions to mixtures located at the bases of the cylinder (corresponding to $z_1 = -1$ and $z_1 = +1$, respectively) these models are

$$\text{at } z_1 = -1 \qquad \begin{aligned} \hat{y}(\mathbf{w}, -1) &= 14.68 - 1.73w_1 + 0.82w_2 \\ \hat{y}(\mathbf{x}, -1) &= 18.16x_1 + 2.24x_2 + 22.94x_3 \end{aligned} \tag{3.57}$$

$$\text{at } z_1 = +1 \qquad \begin{aligned} \hat{y}(\mathbf{w}, +1) &= 23.98 - 1.81w_1 - 1.24w_2 \\ \hat{y}(\mathbf{x}, +1) &= 36.06x_1 + 19.50x_2 + 9.70x_3 \end{aligned} \tag{3.58}$$

The models in Eq. (3.57) are identical in form to the models that would have been found using the four observations only at $z_1 = -1$ whereas the models in Eq. (3.58) are identical to the models found using the four observations at $z_1 = +1$ only.

The reason for displaying the reduced models in Eqs. (3.57) and (3.58) is to illustrate how the parameter estimates in the complete model in Eq. (3.55) are expressible as functions of the estimates in Eqs. (3.57) and (3.58). In other words, the values of the parameter estimates in the combined mixture components process variable model in Eq. (3.55) are simple functions of the values of the parameter estimates in the separate models at $z_1 = -1$ and at $z_1 = +1$. The formulas for the estimates g_i^0, g_i^1, $i = 1$, 2, and 3 in Eq. (3.55) are

$$\begin{aligned} g_1^0 &= 27.11 = \tfrac{1}{2}[18.16 + 36.06] \qquad & g_1^1 &= 8.95 = \tfrac{1}{2}[36.06 - 18.16] \\ g_2^0 &= 10.87 = \tfrac{1}{2}[2.24 + 19.50] \qquad & g_2^1 &= 8.63 = \tfrac{1}{2}[19.50 - 2.24] \\ g_3^0 &= 16.32 = \tfrac{1}{2}[22.94 + 9.70] \qquad & g_3^1 &= -6.62 = \tfrac{1}{2}[9.70 - 22.94] \end{aligned} \tag{3.59}$$

The g_i^0, $i = 1$, 2 and 3 are the averages of the individual estimated surface heights at both levels of z_1, while the g_i^1, $i = 1$, 2 and 3 are the average differences between the individual estimated heights at the levels of z_1. Since the heights of the surface at $z_1 = -1$ and $z_2 = +1$ are independent, the variances of the estimates g_i^0 and g_i^1 are found to be $\text{var}(g_i^0) = \text{var}(g_i^1) = \frac{1}{4}[2 \times \text{var(height estimate)}] = \frac{1}{2}\text{var(height estimate)}$ at $z_1 = -1$ or at $z_1 = +1$.

According to the size of the parameter estimates obtained in Eq. (3.59), there seems present an effect of changing the level of z_1 on the estimated height of the surface. With components one and two, the estimated heights of the surface at the respective vertices $x_1=1$, and $x_2=1$ at $z_1=+1$ are greater than the estimated heights of the surface at the respective vertices with $z_1=-1$. This is because the estimates g_1^1 and g_2^1 are both positive and, when put in ratio form relative to their standard errors,

$$\frac{g_1^1}{\sqrt{\widehat{\text{var}}(g_1^1)}}=\frac{8.95}{\sqrt{5.63}}=3.77, \qquad \frac{g_2^1}{\sqrt{\widehat{\text{var}}(g_2^1)}}=\frac{8.63}{\sqrt{7.78}}=3.09 \qquad (3.60)$$

each ratio is greater than the tabled $t_{(2,0.05)}=2.92$ value indicating the estimates g_1^1 and g_2^1 are significantly greater than zero. With component three on the other hand, $g_3^1=-6.62$, implying the surface is higher at $z_1=-1$ than at $z_1=+1$, but the ratio $g_3^1/\sqrt{\widehat{\text{var}}(g_3^1)}=-1.44$ is not different from zero at the 0.05 level of significance and so we do not make the statement that z_1 affects the blending of the pure component 3.

An alternative model to Eq. (3.55) is expressed as

$$y(\mathbf{x}, z_1)=\beta_1x_1+\beta_2x_2+\beta_3x_3+\gamma_1z_1+\gamma_{11}z_1x_1+\gamma_{12}z_1x_2+\epsilon \qquad (3.61)$$

where the β_i, $i=1$, 2, and 3 represent the heights of the surface at $x_i=1$, $x_j=0$, $j\neq i$, respectively, and γ_1 represents the effect of z_1 on the height of the surface at $x_3=1$, $x_1=x_2=0$ (this was measured previously by γ_3^1 in the model in Eq. (3.53)). The parameters γ_{1i}, $i=1$ and 2, represent the differences in the effect of z_1 on the heights at $x_1=1$ versus $x_3=1$ and at $x_2=1$ versus $x_3=1$, respectively. In other words, with the parameter estimates in Eq. (3.59), the estimates of the parameters in Eq. (3.61) are

$$\begin{aligned} &b_i=g_i^0, \qquad i=1, 2, \text{ and } 3\\ &g_1=g_3^1\\ &g_{11}=g_1^1-g_3^1, \qquad g_{12}=g_2^1-g_3^1 \end{aligned} \qquad (3.62)$$

Thus the model in Eq. (3.61) is useful when one of the mixture ingredients is a baseline or standard ingredient (such as with component 3 above, for example) and one is interested in investigating the effect of the process variable on comparisons made between the other ingredients and the standard ingredient. These comparisons are measured by the estimates g_{1i}, $i=1,2$ in Eq. (3.62).

Although we have discussed the design and the analysis where the w_i's and the process variable z_1 are all at two levels only, it is not necessary to limit the number of experimental levels of the variables to two. It would be quite easy to use a simplex design consisting of q points in the $q-1$ dimensions of the w_i and use either a simplex or a two-level factorial arrangement in the process variables. In Section 5.12 the analysis of an experimental program is presented in which a simplex-centroid design in x_1, x_2, and x_3 was set up in combination with a two-level factorial arrangement in three process variables z_1, z_2, and z_3.

3.14. SUMMARY OF CHAPTER 3

In this chapter a transformation is made from the dependent q-component system to $q-1$ mathematically independent variables. In the $q-1$ independent variable system, standard regression procedures (i.e., model fitting) and known response surface designs are suggested. In particular, rotatable designs are recommended for fitting polynomial models because of the invariance property of var $\hat{y}(\mathbf{x})$ with respect to the form of the transformation matrix $\mathbf{T}$ that is used in going from the x_i's to the w_i's.

When the entire simplex region is not of interest and the experimenter wishes to concentrate the experimentation to some subregion of the simplex, the subregion might be defined as ellipsoidal or as cuboidal in shape. Both types of regions are considered.

Combining n process variables and the $q-1$ independent transformed variates in mixture experiments is the final topic covered in this chapter. The designs are standard response surface designs (i.e., factorial arrangements) but the polynomial models are written either to contain the w_i's and the process variables or the mixture component proportions, the x_i's, and the process variables. A numerical example helps to illustrate the difference in the various model forms.

3.15. REFERENCES AND RECOMMENDED READING

Becker, N. G. (1969). Regression problems when the predictor variables are proportions. *J. R. Stat. Soc., B*, **31**, 107–112.

Becker, N. G. (1970). Mixture designs for a model linear in the proportions. *Biometrika*, **57**, No. 2, 329–338.

Box, G. E. P. (1952). Multi-factor designs of first order. *Biometrika*, **39**, 49–57.

Box, G. E. P. and N. R. Draper (1959). A basis for the selection of a response surface design. *J. Am. Stat. Assoc.*, **54**, 622–654.

Box, G. E. P. and K. B. Wilson (1951). On the experimental attainment of optimum conditions. *J. R. Stat. Soc., B*, **13**, 1–45.

Cornell, J. A. (1968). A response surface approach to the mixture problem when the mixture components are categorized. Ph.D. thesis, Virginia Polytechnic Institute, Blacksburg, VA.

Cornell, J. A. (1971). Process variables in the mixture problem for categorized components. *J. Am. Stat. Assoc.*, **66**, 42–48.

Draper, N. R. and W. E. Lawrence (1965). Mixture designs for three factors. *J. R. Stat. Soc., B*, **27**, 450–465.

Draper, N. R. and W. E. Lawrence (1965). Mixture designs for four factors. *J. R. Stat. Soc., B*, **27**, 473–478.

Draper, N. R. and H. Smith (1966). *Applied Regression Analysis*. Wiley, New York.

Hare, L. B. (1979). Designs for mixture experiments involving process variables. *Technometrics*, **21**, 159–173.

Myers, R. H. (1971). *Response Surface Methodology*. Allyn and Bacon, Boston.

Scheffé, H. (1963). The simplex-centroid design for experiments with mixtures. *J. R. Stat. Soc., B*, **25**, 235–263.

Thompson, W. O. and R. H. Myers (1968). Response surface designs for experiments with mixtures. *Technometrics*, **10**, 739–756.

QUESTIONS FOR CHAPTER 3

3.1. State some rational for working with an independent variable system instead of with the dependent mixture components. What are some disadvantages to working with transformed mixtures.

3.2. In the system of $q-1$ transformed independent variables $w_1, w_2, \ldots, w_{q-1}$, the adequacy of the fitted model containing p terms can be tested using $F = [(N-p)R^2/(p-1)(1-R^2)]$ where R^2 is the square of the multiple correlation coefficient and N is the total number of observations. Is the value of R^2 inflated with the model in the w_i's as we claim it is with the fitted model in the mixture components? If so why, if not, why not?

3.3. In Section 3.11, the fitted model in w_1, w_2, z_1, presented by Eq. (3.50), is

$$\hat{y}(\mathbf{w}, z_1) = 19.33 - 1.77w_1 - 0.21w_2 + 4.65z_1 - 0.04w_1z_1 - 1.03w_2z_1$$

Explain the meanings attached to the coefficient estimates. Make some recommendations about the settings of w_1 and w_2 or the proportions x_1, x_2, and x_3 that may result in large positive values of $\hat{y}(\mathbf{w}, z_1)$. Set up a 95% confidence interval for the true value of the response at $(w_1, w_2, z_1) = (0.707, 0, -1)$.

3.4. Given the point $\mathbf{x}_0 = (0.39, 0.14, 0.47)'$ and the matrix $\mathbf{H} = \text{diag}(0.25, 0.14, 0.30)$ associated with an ellipsoidal region defined in x_1, x_2, and x_3, construct a first-order rotatable design using four points only and place the points on the boundary of the unit sphere centered at $\mathbf{x}_0$. List the component proportions corresponding to the design settings.

3.5. The current operating condition in a four-component blending system is at $\mathbf{x}_0 = (0.37, 0.15, 0.22, 0.26)'$. Set up a second-order rotatable design with the points positioned on the largest sphere centered at $\mathbf{x}_0$. List the corresponding component proportions.

3.6. The following data set was taken from Table 2.9. The data represent average relative percentages of mite numbers on plants taken seven days after spraying combinations of pesticides on the plants. Only chemicals 1, 2, and 3 are considered here.

	Component proportions						
$x_1 =$	1	0	$\frac{1}{2}$	0	$\frac{1}{2}$	0	$\frac{1}{3}$
$x_2 =$	0	1	$\frac{1}{2}$	0	0	$\frac{1}{2}$	$\frac{1}{3}$
$x_3 =$	0	0	0	1	$\frac{1}{2}$	$\frac{1}{2}$	$\frac{1}{3}$
$y_u =$	1.8	25.4	4.9	28.6	3.1	3.4	22.0

Fit a second-degree (quadratic) model in w_1 and w_2 where the w_i's are defined by the transformation presented in Eq. (3.2). If an estimate of the error variance is $s^2 = 20$, what inferences can you make concerning the shape of the surface over the triangle?

3.7. Refer to Table 2.4 and the average elongation values of the yarn. Set up a first-degree model in w_1 and w_2 and fit the model to the yarn elongation values. Using $s^2 = 0.73$ as an estimate of the error variance, test for significance of the planar model. If there is evidence that the planar model is an inadequate representation, fit a second-order model to the elongation values and retest. Once a model has been decided upon, predict the response at $x_1 = \frac{2}{3}$, $x_2 = \frac{1}{3}$, and $x_3 = 0$. Set up a 95% confidence interval for the true mean elongation of the yarn. Compare this interval length with the interval in Section 2.4.

3.8. You wish to set up a first-order design that is symmetrical about the point $(x_1, x_2, x_3) = (\frac{1}{3}, \frac{1}{3}, \frac{1}{3})$. You have available resources for possibly five or at the very most six experimental runs. Set up an "all-variance" design using as many three-component blends as possible. Hint: Should a spherical region or a cuboidal region be used?

3.9. Beef patties were cooked at three separate time-temperature conditions (c_1, c_2, and c_3). The patties were prepared by blending two types of ground beef (denoted by A and B). The blends were selected according to a {2, 3} lattice arrangement. Three replications of each blend-cooking treatment combination were performed. The data represent texture test readings as measured in grams of force $\times 10^{-3}$.

	Replication 1			Replication 2			Replication 3		
$(x_A : x_B)$	c_1	c_2	c_3	c_1	c_2	c_3	c_1	c_2	c_3
$1:0$	1.5	2.0	3.0	1.4	1.8	2.1	1.4	2.0	3.0
$\frac{2}{3}:\frac{1}{3}$	2.8	3.4	2.6	3.2	3.5	3.2	2.6	3.4	2.6
$\frac{1}{3}:\frac{2}{3}$	2.4	2.6	2.6	2.0	1.8	1.5	2.2	2.1	1.8
$0:1$	1.2	1.2	1.8	1.2	1.4	2.0	0.8	1.8	2.0

Let the cooking combinations represent equally spaced levels of a quantitative factor where c_1, c_2, and c_3 are the low, medium, and high levels respectively.

a. Use the data from replication 1 only and fit the model (estimate the coefficients γ_A^i, γ_B^i, γ_{AB}^i, δ_{AB}^i, $i = 0, 1, 2$)

$$y_u = \sum_{i=0,1,2} [\gamma_A^i x_A + \gamma_B^i x_B + \gamma_{AB}^i x_A x_B + \delta_{AB}^i x_A x_B (x_A - x_B)] c^i + \epsilon$$

where $c^0 = 1$, c^1 represents the linear term for the cooking treatments and c^2 is the curvilinear term for cooking treatments.

b. Use the data from replications 1 and 2 only and fit the model

$$y_u = \sum_{i=0,1} [\gamma_A^i x_A + \gamma_B^i x_B + \gamma_{AB}^i x_A x_B] c^i + \epsilon$$

where ϵ is comprised of two error sources arising from the following restriction on the randomization of the treatment combinations; in replication 1, the four patties $(1, 0)$, $(\frac{2}{3}, \frac{1}{3})$, $(\frac{1}{3}, \frac{2}{3})$ and $(0, 1)$ were prepared at cooking level c_2, then four more patties were prepared at c_1 and finally four patties were prepared at c_3. In replication 2, again the four blends were cooked together only the cooking levels were run in the order c_1, c_3, and c_2. Set up the analysis of variance table for a split-plot experiment where the cooking conditions represent the main-plot treatments and the four mixture blends are the subplot treatments. Calculate the sums of squares quantities and test for significance of the treatment effects.

c. Analyze the data from all three replications and set up the ANOVA table. Describe the blending behavior between beef types A and B. Does raising the cooking level ($c_1 < c_2 < c_3$) increase the texture (harden the texture) of the patties? Are combinations of A and B affected similar to the pure beef patties by the cooking treatments? Fully explain your analyses. Assume the 12 patties in each replication were prepared in a completely random order.

APPENDIX 3A. AN ALTERNATIVE TRANSFORMATION FROM THE MIXTURE COMPONENT SYSTEM TO THE INDEPENDENT VARIABLE SYSTEM

Let us define the curtailed vector $\mathbf{x}_c = (x_1, x_2, \ldots, x_{q-1})'$ by ignoring component q. Using the $(q-1)\times(q-1)$ matrix $\mathbf{A} = a\mathbf{I} + b\mathbf{J}$ where the scalars a and b are $a = 1$, $b = 1/(1+\sqrt{q})$, $\mathbf{I}$ is an identity matrix of order $q-1$ and $\mathbf{J}$ is a square matrix of ones, the $(q-1)\times 1$ curtailed vector $\mathbf{w}_c = (w_1, w_2, \ldots, w_{q-1})'$ is obtained by multiplying

$$\mathbf{w}_c = \mathbf{A}\mathbf{x}_c \tag{3A.1}$$

Going by the form of the matrix $\mathbf{A}$, we project each x_i-axis, $i = 1, 2, \ldots, q-1$, onto the simplex with an appropriate scale change (scale $= \sqrt{q}$) making the transformation, from $\mathbf{x}_c$ to $\mathbf{w}_c$, orthogonal (through $\mathbf{A}$). Also, prior to making the transformation in Eq. (3A.1), the w_q-axis is made orthogonal to the simplex by forcing the vertex $x_q = 1$ to be located at the origin of the new system $\mathbf{w}_c = (0, 0, \ldots, 0)'$.

Unlike the transformation presented in Section 3.1, this transformation forces all points in the composition space, which of course are nonnegative valued, to take on only nonnegative values in the w-system. This is because the elements of the matrix $\mathbf{A}$ in Eq. (3A.1) are positive and thus locate the design in the quadrant where $w_i \geq 0$ for all $i = 1, 2, \ldots, q-1$.

To show how the values of the w_i are nonnegative, let us set $q = 3$ so that the matrix $\mathbf{A}$ becomes $\mathbf{A} = [\mathbf{I} + 1/(1+\sqrt{3})\mathbf{J}]$. With the first two columns [because we are using the curtailed vector $\mathbf{x}_c' = (x_1, x_2)$] of the $N \times 3$ matrix $\mathbf{X}$, the w_i coordinates are

$$\mathbf{w}_c = \begin{bmatrix} w_1 \\ w_2 \end{bmatrix} = \frac{1}{1+\sqrt{3}} \begin{bmatrix} 2+\sqrt{3} & 1 \\ 1 & 2+\sqrt{3} \end{bmatrix} \begin{bmatrix} 1 & \frac{1}{2} & 0 & 0 & 0 & \frac{1}{2} & \frac{1}{3} \\ 0 & \frac{1}{2} & 1 & \frac{1}{2} & 0 & 0 & \frac{1}{3} \end{bmatrix}$$

$$\begin{bmatrix} w_1 \\ w_2 \end{bmatrix} = \frac{1}{1+\sqrt{3}} \begin{bmatrix} 2+\sqrt{3} & \dfrac{3+\sqrt{3}}{2} & 1 & \frac{1}{2} & 0 & \dfrac{2+\sqrt{3}}{2} & \dfrac{3+\sqrt{3}}{3} \\ 1 & \dfrac{3+\sqrt{3}}{2} & 2+\sqrt{3} & \dfrac{2+\sqrt{3}}{2} & 0 & \frac{1}{2} & \dfrac{3+\sqrt{3}}{3} \end{bmatrix}$$

The design is displayed in Figure 3.8. As with the transformation presented in

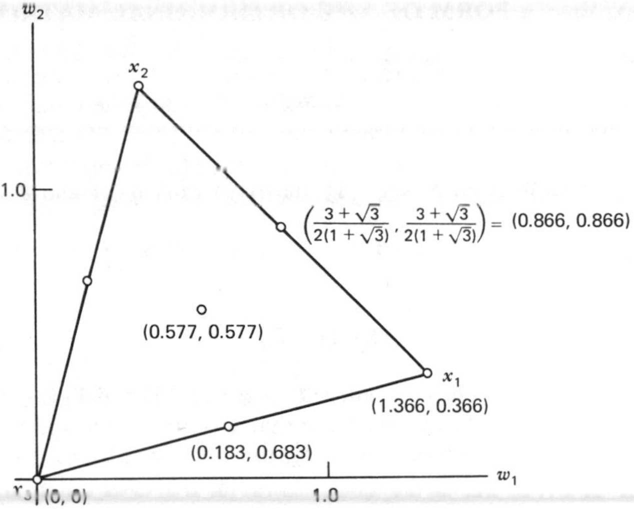

FIGURE 3.8. The simplex centroid design in (w_1, w_2).

Section 3.1, with this transformation the symmetry of the configuration in the **x**-system is preserved in the **w**-system.

Owing to the form of the matrix **A** which is used in Eq. (3A.1) the inverse transformation which expresses the composition variables $\mathbf{x}_c$, in terms of the w_i, is easily employed. First one finds the inverse matrix $\mathbf{A}^{-1}$ of **A**, and then one solves for $\mathbf{x}_c = \mathbf{A}^{-1}\mathbf{w}_c$, or

$$\mathbf{x}_c = \left[\mathbf{I} - \frac{1}{q+\sqrt{q}}\mathbf{J}\right]\mathbf{w}_c \tag{3A.2}$$

The design coordinates or settings in the mixture components corresponding to the design settings in the unconstrained variables over all N observations are easily found using the inverse transformation $\mathbf{x}_c = \mathbf{A}^{-1}\mathbf{w}_c$. To show this let $\mathbf{w}_c'$ be a row of the $N \times (q-1)$ design matrix $\mathbf{W}_c$ where the design matrix $\mathbf{W}_c$ covers the design points from which the N observations are collected. Then the elements of the $N \times (q-1)$ design matrix $\mathbf{X}_c$, where the uth row contains the component settings corresponding to the uth experimental design point in the **w**'s, can be expressed as

$$\mathbf{X}_c = \mathbf{W}_c\left[\mathbf{I} - \frac{1}{q+\sqrt{q}}\mathbf{J}\right] \tag{3A.3}$$

where $\mathbf{W}_c$ is the $N \times (q-1)$ design matrix in the w_i's. Since the $N \times (q-1)$ matrix $\mathbf{W}_c$ is called the design matrix, the $w_1, w_2, \ldots, w_{q-1}$ are the design variables.

APPENDIX 3B. A FORM OF THE ORTHOGONAL MATRIX T

In Section 3.3, the matrix **T** was used to rotate the axes of the intermediate variables, the v_i, to project the $(q-1)$-dimensional unit sphere onto the $(q-1)$-dimensional linear manifold producing a unit sphere which was centered at $\mathbf{w}=\mathbf{0}$ where $\mathbf{w}=(w_1, w_2, \ldots, w_{q-1})'$. The transformation was shown in Eq. (3.26) to be $\mathbf{VT}=[\mathbf{W}, \mathbf{0}]$ where $\mathbf{W}$ is an $N\times(q-1)$ matrix of rank $q-1$ and $\mathbf{0}$ is an $N\times 1$ vector of zeros.

To derive a form for the matrix **T**, let us partition **T** like the partitioning of **VT** in Eq. (3.26), that is,

$$\mathbf{T}=[\mathbf{T}_1, \mathbf{T}_2] \tag{3B.1}$$

where $\mathbf{VT}_1=\mathbf{W}$ and $\mathbf{VT}_2=\mathbf{0}$. The matrix $\mathbf{T}_1$ is $q\times(q-1)$ and $\mathbf{T}_2$ is $q\times 1$.

Sufficient conditions on the matrix **T** are that **T** is orthogonal and $\mathbf{VT}_2=0$. One such matrix $\mathbf{T}_2$ can be described by using the relation $\Sigma_{i=1}^{q} h_i v_i=0$ to find its elements. Let

$$\tilde{h}_i=\frac{h_i}{\left(\sum_{i=1}^{q} h_i^2\right)^{1/2}} \tag{3B.2}$$

so that $\Sigma_{i=1}^{q} \tilde{h}_i v_i=0$. The vector $\mathbf{T}_2$ can be defined simply as

$$\mathbf{T}_2=\tilde{\mathbf{h}}=\begin{bmatrix}\tilde{h}_1\\ \tilde{h}_2\\ \vdots\\ \tilde{h}_q\end{bmatrix} \tag{3B.3}$$

so that $\mathbf{VT}_2=\mathbf{0}$. The elements of the $q-1$ columns of the matrix $\mathbf{T}_1$ can also be constructed using the h_i's $(1\le i\le q)$, the only requirement is that the matrix **T** be orthogonal. One such case is to denote by t_{ij} the element in the ith row and jth column of the matrix $\mathbf{T}_1$. If we let

$$\begin{array}{lll}
t_{11}=-h_2 & t_{12}=-h_1h_3 & t_{13}=-h_1h_4\\
t_{21}=h_1 & t_{22}=-h_2h_3 & t_{23}=-h_2h_4\\
t_{i1}=0\quad i=3,4,\ldots,q & t_{32}=h_1^2+h_2^2 & t_{33}=-h_3h_4\\
 & t_{i2}=0\quad i>3 & t_{43}=h_1^2+h_2^2+h_3^2\\
 & & t_{i3}=0\quad i>4
\end{array} \tag{3B.4}$$

and then normalize the columns in Eq. (3B.4), the first three columns of the matrix $\mathbf{T}_1$ are defined.

To illustrate numerically the calculations used in constructing the matrix $\mathbf{T}=[\mathbf{T}_1, \mathbf{T}_2]$, let us refer to the three-component example of Section 3.4 where the matrix **H** of scale constants is defined as

$$\mathbf{H}=\begin{bmatrix} h_1=0.16 & 0 & 0 \\ 0 & h_2=0.15 & 0 \\ 0 & 0 & h_3=0.07 \end{bmatrix}$$

To obtain the elements of the vector $\mathbf{T}_2$ in (3B.3), the quantity $(h_1^2+h_2^2+h_3^2)^{1/2}$ equals $(0.16^2+0.15^5+0.07^2)^{1/2}=0.2302$ and thus

$$\mathbf{T}_2=\begin{bmatrix} 0.16/0.2302 \\ 0.15/0.2302 \\ 0.07/0.2302 \end{bmatrix}=\begin{bmatrix} 0.6950 \\ 0.6516 \\ 0.3041 \end{bmatrix}$$

From (3B.4), $t_{11}=-0.15$, $t_{21}=0.16$, $t_{12}=-0.16(0.07)$, $t_{22}=-0.15(0.07)$ and $t_{32}=0.16^2+0.15^2$. The matrix $\mathbf{T}_1$ is

$$\mathbf{T}_1=\begin{bmatrix} t_{11}/(t_{11}^2+t_{21}^2)^{1/2} & t_{12}/(t_{12}^2+t_{22}^2+t_{32}^2)^{1/2} \\ t_{21}/(t_{11}^2+t_{21}^2)^{1/2} & t_{22}/(t_{12}^2+t_{22}^2+t_{32}^2)^{1/2} \\ 0 & t_{32}/(t_{12}^2+t_{22}^2+t_{32}^2)^{1/2} \end{bmatrix}=\begin{bmatrix} -0.6839 & -0.2218 \\ 0.7295 & -0.2079 \\ 0 & 0.9527 \end{bmatrix}$$

A final check on the orthogonality of $\mathbf{T}=[\mathbf{T}_1,\mathbf{T}_2]$ is made by verifying that $\mathbf{T}'\mathbf{T}=\mathbf{T}\mathbf{T}'=\mathbf{I}$ where $\mathbf{I}$ is the identity matrix.

CHAPTER 4

Multiple Constraints on the Component Proportions

In Chapter 2, we emphasized the construction of lattice designs and the fitting of model equations which enabled us to cover or explore all or almost all of the simplex region. In Chapter 3, we continued with the freedom of exploring the entire simplex except in the cases where the region of interest was chosen to be an ellipsoidal region or a cuboidal region contained inside of the simplex. For the most part, in both chapters, the only real restrictions on the component proportions were $0 \leq x_i \leq 1$, $i = 1, 2, \ldots, q$ and $x_1 + x_2 + \cdots + x_q = 1$.

Frequently, one is not completely at freedom to explore the entire simplex because of certain additional restrictions that are placed on the component proportions. For example, in the yarn-manufacturing experiment of Section 2.4, one might be interested in the properties of the yarn spun only from mixtures in which the fractional proportions of polystyrene and polypropylene are greater than or equal to $x_2 \geq 0.20$ and $x_3 \geq 0.35$, respectively. These lower-bound restrictions placed on x_2 and x_3 would limit the desired mixtures to a subregion of the simplex. Limiting the experimentation to some subregion of the simplex arises also from the placing of upper bounds on some of the component proportions. Still another case which occurs frequently is where both lower and upper bounds are placed on some or all of the component proportions. In any of these situations, if some subset of the simplex is the region one is confined to, and if one is able to isolate the design and modeling efforts to the subregion, then it is clear that a decrease in experimentation cost and time, as well as an increase in precision of the model estimates, should result by isolating the experimentation to the smaller region. We now discuss the methodology necessary to design and model blending systems where lower bounds only are placed on the component proportions.

4.1. LOWER-BOUND RESTRICTIONS ON SOME OF THE COMPONENT PROPORTIONS

We begin our study of how to design an experiment when the component proportions are restricted by lower bounds by assuming, for reasons of simplicity, that there are only three components in the system. Furthermore, let us assume that the current product is being manufactured within the following blending limits:

$$x_1 = 0.45 \pm 0.01, \qquad x_2 = 0.30 \pm 0.02, \qquad x_3 = 0.25 \pm 0.02$$

However, we would like to study the shape of the surface over the region of the simplex that is defined by the placing of the lower bounds

$$x_1 \geq 0.35, \qquad x_2 \geq 0.20, \qquad x_3 \geq 0.15 \tag{4.1}$$

on the respective component proportions. In such a region, at least a proportion $x_1 = 0.35$ of component 1 is required to be present in each blend, combined with at least a proportion $x_2 = 0.20$ of component 2 combined with at least $x_3 = 0.15$ of component 3. Of course, not all three components can assume these lower bounds simultaneously since in this case the sum $x_1 + x_2 + x_3 = 0.70$ is less than unity and thus would not form a valid mixture.

Corresponding to the restrictions on the x_i listed in Eq. (4.1), the factor space of feasible mixtures involving components 1, 2, and 3, is represented by the interior triangle shown in Figure 4.1. The placing of lower bounds only on the component proportions does not distort the shape of the subregion because the subregion retains the shape of a regular

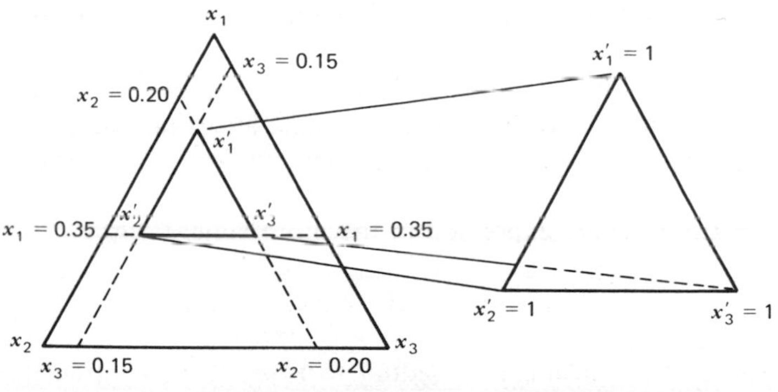

FIGURE 4.1. A subregion of the original simplex redefined as a simplex in the pseudocomponents x'_i, $i = 1, 2, 3$.

simplex or regular triangle in three components. If the sizes of the lower bounds are equal, then the centroid of the subregion remains at the centroid of the simplex in the original components. The figure has sides of equal length and angles of equal magnitudes. (If upper bounds are placed on the component proportions however, or if both upper and lower bounds are placed on the component proportions, the resulting region usually will not assume the shape of a simplex. Only in certain special cases, such as when both upper and lower bounds are placed on the proportions and the ranges of the restricted proportions are equal, can we redefine the system using pseudocomponents whose region will take the form of a regular simplex.)

4.2. INTRODUCING PSEUDOCOMPONENTS

In an attempt to describe the shape of the surface over the smaller subregion with some form of polynomial equation, the model can be written either in the original components whose proportions are constrained in defining the region, or one can introduce superficial components called *pseudocomponents* (or false components) which are generated from the original components. The pseudocomponents are combinations of (or scalar fractions of) the original real components, and the reason for introducing the pseudocomponents is that most of the time both the construction of the designs and the fitting of the models are much simpler when done in the pseudocomponent system than when done in the original component system. One word of caution, however, is that one must keep in mind the pseudocomponents are "pseudo" and if one wishes to make inferences concerning the components that actually comprise the blending system, the inverse transformation from the pseudocomponents back to the original components must be performed prior to making the inferences about the original system.

To show how the pseudocomponents are defined in terms of the original components and their lower bounds, let us talk in general terms by saying the system consists of q components and let $a_i \geq 0$ denote the lower bound for component i, $i = 1, 2, \ldots, q$. The lower-bound constraints in Eq. (4.1) are expressed in the more general form

$$0 \leq a_i \leq x_i, \qquad \text{for } i = 1, 2, \ldots, q \tag{4.2}$$

where some of the a_i might be equal to zero. If we subtract the lower bound a_i from x_i, divide the difference by 1—(sum of the a_i's), then the *pseudocomponent* x'_i is defined, using the linear transformation, as

$$x'_i = \frac{x_i - a_i}{1 - L} \tag{4.3}$$

where $L = \Sigma_{i=1}^{q} a_i < 1$. To illustrate, let us refer to the earlier restrictions in Eq. (4.1), where $L = 0.35 + 0.20 + 0.15 = 0.70$, so that the pseudocomponents are

$$x'_1 = \frac{x_1 - 0.35}{0.30}, \qquad x'_2 = \frac{x_2 - 0.20}{0.30}, \qquad x'_3 = \frac{x_3 - 0.15}{0.30} \tag{4.4}$$

The factor space shown in Figure 4.1 is a regular 2-dimensional simplex in the pseudocomponents x'_i.

The ease in constructing designs in the pseudocomponent system can be illustrated as follows. For simplicity, let us choose a second-degree polynomial to model the surface over the region in the x'_i's, and let us choose a {3, 2} lattice in the x'_i's at which to observe the values of the response. The design settings $x'_i = 0, \frac{1}{2}, 1$ are shown on the left side of Table 4.1. The settings in the original x_i components, corresponding to the lattice settings in the x'_i, are obtained by reversing the equations in Eq. (4.4) and solving. In other words, the settings in the original components are obtained using

$$x_i = a_i + (1 - L)x'_i \tag{4.5}$$

so that for $i = 1$ and $a_1 = 0.35$, $L = 0.70$, we have for the value of x_1 corresponding to $x'_1 = 1.0$, $x_1 = 0.35 + (0.30)1.0 = 0.65$. Similarly, corresponding to $x'_1 = 0.5$, we have $x_1 = 0.35 + (0.30)(0.5) = 0.50$. The remaining settings in the original components are presented in Table 4.1.

Once the mixture blends in the original system are defined from the pseudocomponent settings, the next step is to collect observed values of the response at the design settings so that a model either in terms of the pseudocomponents, or in terms of the original components can be obtained. A second-degree polynomial in the pseudocomponents is

$$\eta = \gamma_1 x'_1 + \gamma_2 x'_2 + \gamma_3 x'_3 + \gamma_{12} x'_1 x'_2 + \gamma_{13} x'_1 x'_3 + \gamma_{23} x'_2 x'_3 \tag{4.6}$$

while the corresponding model in the original components would be of the form

$$\eta = \gamma_1 \frac{(x_1 - a_1)}{1 - L} + \gamma_2 \frac{(x_2 - a_2)}{1 - L} + \gamma_3 \frac{(x_3 - a_3)}{1 - L} + \gamma_{12} \frac{(x_1 - a_1)(x_2 - a_2)}{(1 - L)^2}$$
$$+ \gamma_{13} \frac{(x_1 - a_1)(x_3 - a_3)}{(1 - L)^2} + \gamma_{23} \frac{(x_2 - a_2)(x_3 - a_3)}{(1 - L)^2}$$

TABLE 4.1. Original component settings and pseudocomponent settings

Pseudocomponent Settings			Original Component Settings			
x'_1	x'_2	x'_3	x_1	x_2	x_3	Data Values
1	0	0	0.65	0.20	0.15	28.6
0.5	0.5	0	0.50	0.35	0.15	42.4
0	1	0	0.35	0.50	0.15	20.0
0	0.5	0.5	0.35	0.35	0.30	12.5
0	0	1	0.35	0.20	0.45	15.3
0.5	0	0.5	0.50	0.20	0.30	32.7

or

$$\eta = \beta_1 x_1 + \beta_2 x_2 + \beta_3 x_3 + \beta_{12} x_1 x_2 + \beta_{13} x_1 x_3 + \beta_{23} x_2 x_3 \tag{4.7}$$

where the β's can be expressed in terms of the γ's as

$$\begin{aligned}
\beta_1 &= \frac{\gamma_{12} a_2(a_1 - 1) + \gamma_{13} a_3(a_1 - 1) + \gamma_{23} a_2 a_3}{(1-L)^2} + \frac{\gamma_1 - \sum_{i=1}^{3} \gamma_i a_i}{1-L} \\
\beta_2 &= \frac{\gamma_{12} a_1(a_2 - 1) + \gamma_{13} a_1 a_3 + \gamma_{23} a_3(a_2 - 1)}{(1-L)^2} + \frac{\gamma_2 - \sum_{i=1}^{3} \gamma_i a_i}{1-L} \\
\beta_3 &= \frac{\gamma_{12} a_1 a_2 + \gamma_{13} a_1(a_3 - 1) + \gamma_{23} a_2(a_3 - 1)}{(1-L)^2} + \frac{\gamma_3 - \sum_{i=1}^{3} \gamma_i a_i}{1-L} \\
\beta_{ij} &= \gamma_{ij}/(1-L)^2, \qquad i, j = 1, 2 \text{ and } 3, \qquad i < j
\end{aligned} \tag{4.8}$$

For the pseudocomponent model in Eq. (4.6), the interpretations of the parameters γ_i and γ_{ij}, $i, j = 1$, 2, and 3, $i < j$, in the pseudocomponent system is the same as was previously described in the earlier chapters when discussing the regular component models. The γ_i, $i = 1$, 2, and 3, represent the heights of the surface above the triangle at the vertices $x'_i = 1$, $i = 1$, 2, and 3, respectively, in the pseudocomponent system and the γ_{ij}, $i < j$, represent deviations from the planar surface, that is, nonlinear blending of the $x'_i x'_j$ binaries. The β_i and β_{ij}, $i = 1, 2$, and 3, $i < j$, on the other hand, are not so easily defined. From Eq. (4.8), the β_i are rather complicated linear functions of the γ_i and γ_{ij}. This is illustrated using the data values from Table 4.1.

4.3. A NUMERICAL ILLUSTRATION OF A PSEUDOCOMPONENT MODEL

With the pseudocomponent model in Eq. (4.6), the estimates g_i and g_{ij} of the model parameters γ_i and γ_{ij}, respectively, are found using the same formulas as in Section 2.4, that is, $g_i = y_i$, $g_{ij} = 4y_{ij} - 2(y_i + y_j)$, $i < j$. Using the data values from Table 4.1, the estimates are

$$g_1 = 28.6, \qquad g_2 = 20.0, \qquad g_3 = 15.3$$

$$g_{12} = 4(42.4) - 2(28.6 + 20.0) = 72.4, \qquad g_{13} = 43.0, \qquad g_{23} = -20.6$$

Substituting these estimates into the model form of Eq. (4.6), the prediction equation for the response in the pseudocomponent system, is

$$\hat{y}(\mathbf{x}') = 28.6x_1' + 20.0x_2' + 15.3x_3' + 72.4x_1'x_2' + 43.0x_1'x_3' - 20.6x_2'x_3' \tag{4.9}$$

The corresponding set of parameter estimates for the model in the original components as expressed by Eq. (4.7) is

$$\begin{aligned} b_1 &= \frac{72.4(0.20)(0.35 - 1.0) + 43.0(0.15)(0.35 - 1.0) + (-20.6)(0.20)(0.15)}{(1 - 0.70)^2} \\ &\quad + \frac{28.6 - 28.6(0.35) - 20.0(0.20) - 15.3(0.15)}{1 - 0.70} \\ &= -117.04 \end{aligned}$$

$$b_2 = -160.38, \qquad b_3 = -50.27, \qquad b_{12} = \frac{72.4}{(1 - 0.70)^2} = 804.44,$$

$$b_{13} = 477.78, \qquad b_{23} = -228.89$$

and the prediction equation for the response in the original components, rounding off the estimates to tenths, is

$$\hat{y}(\mathbf{x}) = -117.0x_1 - 160.4x_2 - 50.3x_3 + 804.4x_1x_2 + 477.8x_1x_3 - 228.9x_2x_3 \tag{4.10}$$

for the region where $x_1 \geq 0.35$, $x_2 \geq 0.20$, and $x_3 \geq 0.15$.

Contour plots of estimated response values can be drawn either with the pseudocomponent model in Eq. (4.9) or with the model in the original components in Eq. (4.10). In Figure 4.2 is presented the contour plot of the estimated surface using the pseudocomponent model (4.9).

In summary, when one or more of the component proportions is restricted by lower bounds and the sum of the lower bounds is less than

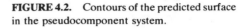

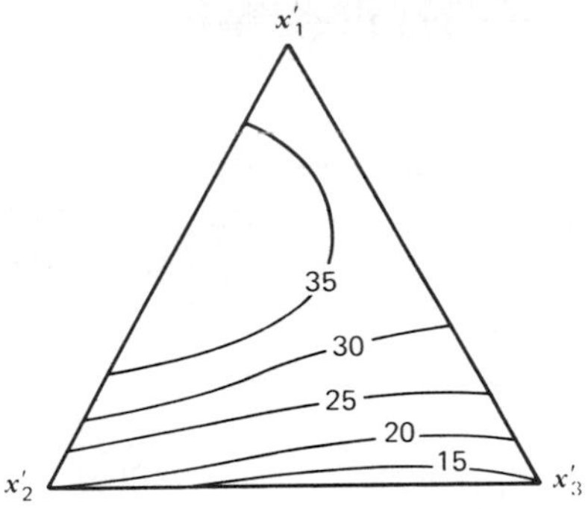

FIGURE 4.2. Contours of the predicted surface in the pseudocomponent system.

unity, $\Sigma_{i=1}^{q} a_i < 1$, the resulting subspace is a simple homomorphic transformation of the original simplex space. The lattice designs may be applied directly to the subspace and the surface can be modelled by a polynomial in the pseudocomponents or in the original restricted components.

4.4. UPPER BOUND RESTRICTIONS ON SOME OF THE COMPONENT PROPORTIONS

When one or more of the component proportions is restricted by upper bounds, the simplest modification to the simplex-lattice design consists of replacing the restricted components with mixtures consisting of combinations of the restricted components and predetermined proportions of the unrestricted components. These mixtures are then used to obtain observations from which estimates of the parameters in the standard mixture polynomials can be calculated.

Let us assume for simplicity that only one component is restricted, say $x_1 \le h$, where the system consists of the four components denoted by x_1, x_2, x_3, and x_4. We assume also that the following second-degree polynomial is to be fitted over the restricted region

$$y = \sum_{i=1}^{4} \beta_i x_i + \sum_{i<j}^{3} \sum^{4} \beta_{ij} x_i x_j + \epsilon \tag{4.11}$$

The feasible experimental region consists of the lower frusta *EFGDCB* of the tetrahedron *ABCD* shown in Figure 4.3.

Since the design points in the simplex-lattice arrangement where $x_1 > h$ cannot be used (that is, the design space is now the frustrum of the simplex satisfying $x_1 \le h$), we may consider replacing the usual lattice

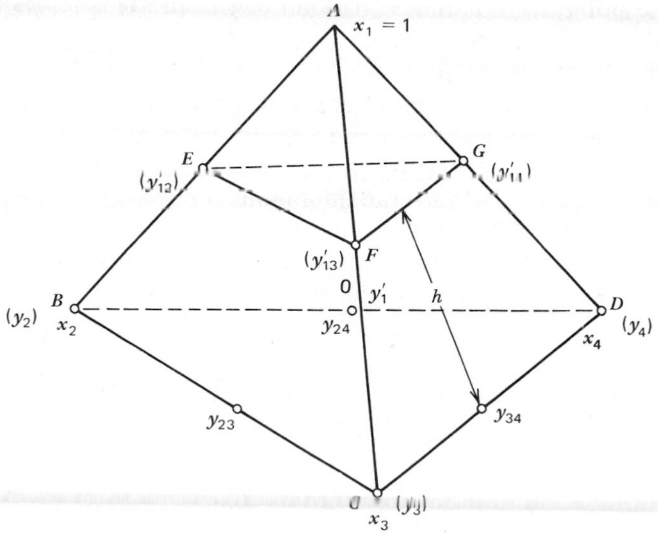

FIGURE 4.3. Lower frusta of the tetrahedron defined as $x_1 \le h$ with values of the response observed at the design points.

points, where $x_1 > h$, with component combinations consisting of the proportion h of x_1 and the proportion x_j $(2 \le j \le 4)$ of the remaining three components such that

$$\sum_{j=2}^{4} x_j = 1 - h \tag{4.12}$$

For each blend, if two or more components other than x_1 are present, then they are equally represented, so that, with the standard notation of binary, ternary and quartenary blends, some of the combinations which might be used are

$$\begin{array}{lll} x_1x_i & x_1 = h, & x_i = 1 - h \\ x_1x_ix_j & x_1 = h, & x_i = x_j = \dfrac{1-h}{2} \\ x_1x_ix_jx_k & x_1 = \dfrac{h}{2}, & x_i = x_j = x_k = \dfrac{(1-h/2)}{3} \end{array} \tag{4.13}$$

where the subscripts i, j, and $k = 2$, 3, and 4, $i \ne j \ne k$. The blends x_1x_i are represented by the points E, F, and G in Figure 4.3, whereas the blends $x_1x_ix_j$ are located at the midpoints of the edges joining the points E, F, and G. The point $x_1x_2x_3x_4$ that is defined in Eq. (4.13) is located on the x_1-axis midway between the triangle EFG and the triangle BCD.

In addition to the design points above, we may also consider taking the standard $\{q-1, 2\}$ lattice points where $x_1 = 0$ and the other components are set at the six combinations $x_i = 1$, $x_j = x_k = 0$ and $x_i = 0$, $x_j = x_k = \frac{1}{2}$, $i, j, k = 2, 3, 4$, $i \neq j \neq k$. These are the three vertices associated with pure components 2, 3, and 4 and the midpoints of the three edges (i.e., $x_i = x_j = \frac{1}{2}$, $i < j$, $i, j, k \neq 1$) connecting the vertices associated with the components 2, 3, and 4. If these latter six points plus the seven points in Eq. (4.13) make up the design, the estimates of the parameters β_i, β_{ij} $(j > i > 1)$ in Eq. (4.11) can be obtained as follows: let the observed response at $x_i = 1$, $x_1 = x_j = x_k = 0$ $(i \neq j \neq k)$ be denoted by y_i, and let the observed response at $x_i = x_j = \frac{1}{2}$, $x_1 = x_k = 0$, be denoted by y_{ij}. Upon substituting these values of x_1, x_i, x_j, x_k, y_i, and y_{ij} into Eq. (4.11), the estimates are

$$
\begin{aligned}
b_i &= y_i, \qquad i = 2, 3, 4 \\
b_{ij} &= 4y_{ij} - 2y_i - 2y_j, \qquad j > i > 1
\end{aligned} \tag{4.14}
$$

These formulas are identical to the formulas presented in Section 2.4 as we know they must be since we are fitting the parameters associated with the components 2, 3, and 4 to the $\{3, 2\}$ simplex-lattice that defines the base of the frustum in Figure 4.3.

Still to be determined are the estimates of β_1 and β_{1j} $(j > 1)$ in Eq. (4.11). If the observed response taken at the combination $x_1 = h$, $x_j = 1 - h$ $(j = 2, 3, 4)$ is denoted by y'_{1j}, (see Figure 4.3) and the observed response taken at the combination $x_1 = h/2$, $x_2 = x_3 = x_4 = (1 - h/2)/3$ is denoted by y'_1, then the estimates b_1 and b_{1j} can be found by solving the following equations:

$$
hb_1 + h(1-h)b_{1j} = y'_{1j} - (1-h)b_j, \qquad j = 2, 3, 4 \tag{4.15}
$$

$$
\frac{h}{2}b_1 + \frac{h}{2}\frac{(1-h/2)}{3}\sum_{j=2}^{4} b_{1j} = y'_1 - \frac{(1-h/2)}{3}\sum_{j=2}^{4} b_j - \frac{(1-h/2)^2}{9}\sum_{i<j}^{3}\sum^{4} b_{ij} \tag{4.16}
$$

We now illustrate the use of the estimating formulas in Eqs. (4.14)–(4.16) for modeling the shape characteristics associated with the flavor surface of a tropical beverage.

4.5. AN EXAMPLE OF THE PLACING OF AN UPPER BOUND ON A SINGLE COMPONENT: THE FORMULATION OF A TROPICAL BEVERAGE

A tropical beverage was formulated by combining juices of watermelon (x_1), orange (x_2), pineapple (x_3), and grapefruit (x_4). It was decided to restrict the percentage of watermelon in all blends to at most 80%. However, the feasibility of a punch consisting of 80% watermelon was of interest because watermelon is so much more inexpensive than the other fruits. Thus several combinations of $x_1 = 0.80$ with $x_2 + x_3 + x_4 = 1.0 - 0.80 = 0.20$ were studied. The response of interest for this example is the average flavor score (based on a hedonic scale of 1–9) where the flavor scores in Table 4.2 represent average values taken over 40 samples of each blend.

In Table 4.2, the first four blends designated as points 1–4, respectively, are from Eq. (4.13),

$$x_1 x_i \qquad \text{where } x_1 = 0.80,\ x_i = 0.20, \qquad i = 2, 3, \text{ and } 4$$

$$x_1 x_2 x_3 x_4 \qquad \text{where } x_1 = 0.40,\ x_2 = x_3 = x_4 = 0.60/3 = 0.20$$

TABLE 4.2. Tropical beverage data

Design Point	Watermelon (x_1)	Orange (x_2)	Pineapple (x_3)	Grapefruit (x_4)	Average Flavor Score (y)
1	0.80	0.20	0	0	6.50
2	0.80	0	0.20	0	6.96
3	0.80	0	0	0.20	6.00
4	0.40	0.20	0.20	0.20	6.82
5	0	1.00	0	0	$5.80 = y_2$
6	0	0	1.00	0	$5.65 = y_3$
7	0	0.50	0.50	0	$5.93 = y_{23}$
8	0	0	0	1.00	$5.05 = y_4$
9	0	0.50	0	0.50	$5.36 = y_{24}$
10	0	0	0.50	0.50	$5.72 = y_{34}$
11	0.80	0.10	0.10	0	7.25
12	0.80	0	0.10	0.10	6.20
13	0.80	0.10	0	0.10	6.47
14	0.40	0.30	0.30	0	7.21
15	0.40	0.30	0	0.30	6.53
16	0.40	0	0.30	0.30	6.88

The six additional blends designated as points 5–10 in Table 4.2 are

$x_i x_j$ where $x_i = x_j = 0.50$, $i, j = 2, 3$, and 4, $i \neq j$ and

$x_i = 1$ where $i = 2, 3$, and 4

The average flavor scores for the 10 blends, as well as for six additional blends to be mentioned shortly, are presented in Figure 4.4.

Fitted to the 10 observations is the second-degree model

$$y = \sum_{i=1}^{4} \beta_i x_i + \sum_{i<j}^{3} \sum^{4} \beta_{ij} x_i x_j + \epsilon \qquad (4.17)$$

The estimates of the parameters are computed using Eqs. (4.14)–(4.16). From Eq. (4.14),

$$b_2 = y_2 = 5.80, \qquad b_3 = y_3 = 5.65, \qquad b_4 = y_4 = 5.05$$

$$b_{23} = 4(5.93) - 2(5.80 + 5.65) = 0.82, \qquad b_{24} = 4(5.36) - 2(5.80 + 5.05) = -0.26$$

$$b_{34} = 4(5.72) - 2(5.65 + 5.05) = 1.48$$

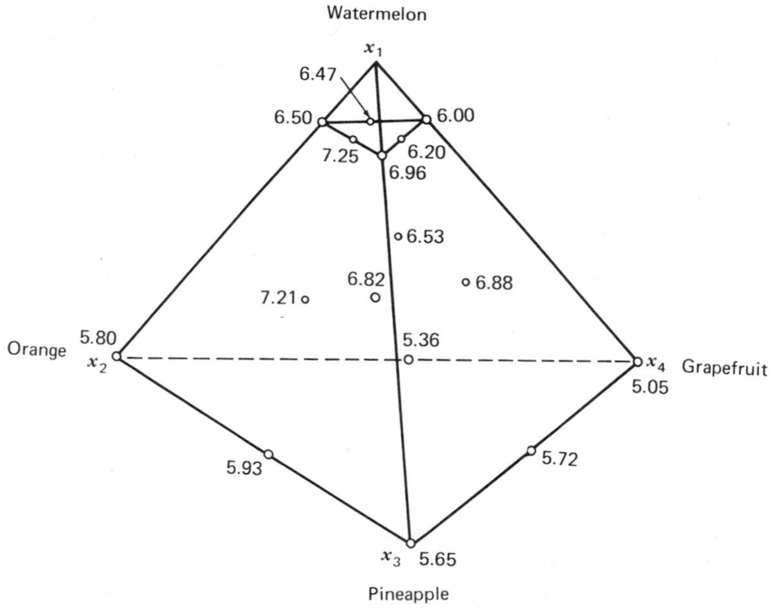

FIGURE 4.4. Average flavor scores at the 16 juice blends.

From Eqs. (4.15) and (4.16), the four equations to solve are

$$0.80b_1 + 0.80(1.0-0.80)b_{12} = 6.50 - (1.0-0.80)b_2$$

$$0.80b_1 + 0.80(1.0-0.80)b_{13} = 6.96 - (1.0-0.80)b_3$$

$$0.80b_1 + 0.80(1.0-0.80)b_{14} = 6.00 - (1.0-0.80)b_4$$

$$0.40b_1 + 0.40\frac{(1.0-0.40)}{3}[b_{12}+b_{13}+b_{14}] = 6.82 - \frac{(1.0-0.40)}{3}[b_2+b_3+b_4] - \frac{(1.0-0.40)^2}{9}[b_{23}+b_{24}+b_{34}]$$

or in the case of this last equation

$$0.40b_1 + 0.40\frac{(0.60)}{3}[b_{12}+b_{13}+b_{14}] = 6.82 - (0.20)[16.50] - 0.40[2.04] = 3.4384$$

and therefore

$$b_1 = 5.77$$

$$b_{12} = 4.37, \qquad b_{13} = 7.43, \qquad b_{14} = 2.18$$

The 10-term prediction equation for flavor score is

$$\begin{aligned}\hat{y}(\mathbf{x}) = \underset{(0.57)}{5.77}x_1 + \underset{(0.33)}{5.80}x_2 + \underset{(0.33)}{5.65}x_3 + \underset{(0.33)}{5.05}x_4 + \underset{(2.85)}{4.37}x_1x_2 + \underset{(2.85)}{7.43}x_1x_3 + \underset{(2.85)}{2.18}x_1x_4 \\ + \underset{(1.64)}{0.82}x_2x_3 - \underset{(1.64)}{0.26}x_2x_4 + \underset{(1.64)}{1.48}x_3x_4 \end{aligned} \tag{4.18}$$

where the quantities in parentheses below the coefficient estimates are the standard errors of the coefficient estimates. The estimate of the variance of each average response value is $s_{\bar{y}}^2 = 0.112$ which was calculated using the pooled within blend variance $s_{\bar{y}}^2 = 5.0/40$ where the number 40 represents the sample size used for calculating each y_u value in Table 4.2.

Prior to using Eq. (4.18) to predict the flavor score of blends other than the ten blends used in the experiment, the fit of the model in Eq. (4.18) should be verified at other blends. To do so would require that data be collected on additional blends in the factor space and the model refitted. This is because Eq. (4.18) contains ten terms and was fitted to data collected at exactly ten blends. If data is collected at other points in the composition space and the fitted model performance is measured at these points, this would provide us with a measure of "lack of fit" or "inadequacy of the fitted model" if the model does not perform well.

Six additional blends were chosen because of their importance and the second-degree equation (4.17) was fitted to the data at all 16 points resulting in

$$\begin{aligned}\hat{y}(\mathbf{x}) = {} & \underset{(0.37)}{5.87x_1} + \underset{(0.36)}{5.79x_2} + \underset{(0.36)}{5.65x_3} + \underset{(0.36)}{5.05x_4} + \underset{(1.95)}{5.29x_1x_2} + \underset{(1.95)}{7.13x_1x_3} \\ & + \underset{(1.95)}{1.90x_1x_4} + \underset{(1.73)}{0.68x_2x_3} - \underset{(1.73)}{0.16x_2x_4} + \underset{(1.73)}{1.53x_3x_4} \end{aligned} \tag{4.19}$$

The estimated standard errors in parentheses were calculated using the new pooled error estimate $s_{\bar{y}}^2 = [5.0(390) + 5.23(234)]/624(40) = 0.13$ where the second within estimate $s_{\bar{y}}^2 = 5.23$ was obtained from the 40 responses to each of the six new blends.

The analysis of variance table constructed from the 16 data values in Table 4.2 is presented as Table 4.3. The pure error sum of squares in Table 4.3 was divided by 40 to give 79.35 and thus guaranteeing the mean squares for pure error and residual are on a per-average basis. An F-test comparing the mean square for residual to mean square for pure error was valued at $F = 0.058/0.13 < 1$. Since there was no reason to suspect that the residual mean square contained any source of variation other than error variation, the model was thought to be adequate.

TABLE 4.3. Analysis of variance for the tropical beverage data

Source of Variation	Degrees of Freedom	Sums of Squares	Mean Square
Regression	9	6.22	0.69
Residual	6	0.35	0.058
Pure Error from Replicates	624	79.35	0.13

4.6. THE PLACING OF BOTH UPPER AND LOWER BOUNDS ON THE COMPONENT PROPORTIONS

In Section 4.2 pseudocomponents were introduced for the primary purpose of simplifying the design problem, as well as for modeling the surface over the interior region, which was defined from the placing of lower bounds on the component proportions. In Section 4.4 we discussed the estimation of the model coefficients (as well as the selection of certain component blends) for the case where upper bounds are placed on some of the component proportions. Quite often, in practice, one is faced with both of these situations at the same time because some of the component proportions are constrained by upper bounds while other component proportions are restricted by lower bounds and still others are restricted by both upper and lower bounds. Such situations arise for example when to form a valid blend we require or need at least a_i, but no more than b_i, of component i, and similar bounds may be specified for the other components as well.

As an example, in the formulations of the fruit punch of the previous section, let us assume that we want the punch to be at least 40% watermelon but not more than 80%; we want at least 10% orange, and we insist that pineapple and grapefruit contribute at least 5% each but not more than 30% each. We can write these restrictions as

$$\begin{array}{ll} 40\% \leq \text{Watermelon} \leq 80\% & 0.40 \leq x_1 \leq 0.80 \\ 10\% \leq \text{Orange} \leq 100\% & 0.10 \leq x_2 \leq 1.0 \\ 5\% \leq \text{Pineapple} \leq 30\% & 0.05 \leq x_3 \leq 0.30 \\ 5\% \leq \text{Grapefruit} \leq 30\% & 0.05 \leq x_4 \leq 0.30 \end{array} \tag{4.20}$$

With q components, the multiple constraints are written as

$$0 \leq a_i \leq x_i \leq b_i \leq 1, \qquad i = 1, 2, \ldots, q \tag{4.21}$$

When only one or two of the component proportions are restricted, the resulting factor space is not so difficult to define. However, if nearly all of the component proportions are constrained above and below then the resulting factor space will take the form of a hyperpolyhedron (convex polyhedron) which will often be considerably more complicated in form than the simplex.

Depending on the degree of the equation that is to be used to model the surface over the region defined in Eq. (4.21), most often we shall require for our design points, at least some of the vertices of the region as well as some of the centroids of the faces, sides and edges of the region. For example, if we assume that a second-degree model of the form is to be fitted

$$\eta = \sum_{i=1}^{q} \beta_i x_i + \sum_{i<j}^{q} \sum \beta_{ij} x_i x_j \tag{4.22}$$

then a minimum of $q + q(q-1)/2$ points will be needed at which to collect observations since this is the number of parameters in Eq. (4.22) to be estimated.

Once the constraints on the component proportions are specified, most of the points of the base design are decided on. These points are represented by the vertices of the polyhedron and some of the edges and faces formed by taking feasible combinations of the component proportions. The vertices of the polyhedron, particularly those that are listed as specific combinations of the constraints, are obtained by combining certain boundaries which are formed by the upper or lower constraints of $q-1$ factors. The vertices of the polyhedron can be obtained by the following two-step procedure that was first introduced in McLean and Anderson (1966):

1. List all possible combinations (as in a two-level factorial arrangement) of the proportions of $q-1$ factors at a time using the a_i and b_i levels for all of the factors except one and this one is left blank. This procedure produces 2^{q-1} points. With three factors for example, where a_1, a_2, and a_3 are the lower bounds and b_1, b_2, and b_3 are the upper bounds, the two-level combinations are a_1a_2__, a_1b_2__, b_1a_2__, b_1b_2__ where the proportion due to component 3 is left blank. The process is repeated q times allowing each x_i to be the component whose level is left blank and is therefore to be computed. This list will consist of $q \times (2^{q-1})$ possible combinations.
2. Go through all possible combinations generated in 1. and fill in those blanks that are admissible, that is, fill in the level (necessarily falling within the constraints of the missing factor) which will make the total of the levels for that treatment combination sum to unity.

After the above procedure is used to establish which vertices of the convex polyhedron are feasible, next we define a variety of centroids or points centrally positioned on the faces and the edges and so on, which are referred to as the centroids. There is one centroid point located in each bounding two-dimensional face, three-dimensional face, . . . , k-dimensional face where $k \leq q-2$, as well as the "overall" centroid of the polyhedron. This latter point is the component combination obtained by averaging all of the factor levels which define the existing feasible vertices and it may or may not coincide with the true centroid of the polyhedron and that is why quotation marks are used. The centroids of the two-dimensional faces are found by isolating all vertices which have $q-3$ factor levels identical and by averaging the factor levels of each of the three remaining factors. All remaining centroids are found in similar fashion using all vertices which have $q-r-1$ factor levels identical for an r-dimensional face where $r \leq k \leq q-2$. It should be noted that k may be less than $q-2$ since the constraints listed and applied to the proportions may reduce the dimensionality of the hyper-polyhedron to less than $q-1$.

To illustrate the procedure for locating the extremities of a region which is defined from the placing of constraints on the x_i, let us refer to the tropical beverage example of the previous section where in addition to insisting that between 40% and 80% watermelon be present we require at least 10%, 5%, and 5% of orange, pineapple, and grapefruit, respectively, be present, but at most 30% of each of the latter two fruit. The constraints on the proportions were listed previously in Eqs. (4.20) as

$$0.40 \leq x_1 \leq 0.80$$

$$0.10 \leq x_2 \leq 1.00$$

$$0.05 \leq x_3 \leq 0.30$$

$$0.05 \leq x_4 \leq 0.30$$

To generate the $q \times (2^{q-1}) = 4 \times (2^{4-1}) = 32$ possible component combinations which may or may not be admissible, we begin our listing as

	x_1	x_2	x_3	x_4		x_1	x_2	x_3	x_4
	0.40	0.10	0.05	___	(3)	0.40	0.50	0.05	0.05
(1)	0.40	0.10	0.30	0.20	(4)	0.40	0.25	0.05	0.30
	0.40	1.00	0.05	___	(5)	0.40	0.25	0.30	0.05
	0.40	1.00	0.30	___		0.40	___	0.30	0.30
(2)	0.80	0.10	0.05	0.05	(2)	0.80	0.10	0.05	0.05
	0.80	0.10	0.30	___		0.80	___	0.05	0.30
	0.80	1.00	0.05	___		0.80	___	0.30	0.05
	0.80	1.00	0.30	___		0.80	___	0.30	0.30
	0.40	0.10	___	0.05	(2)	0.80	0.10	0.05	0.05
(6)	0.40	0.10	0.20	0.30	(7)	0.55	0.10	0.05	0.30
	0.40	1.00	___	0.05	(8)	0.55	0.10	0.30	0.05
	0.40	1.00	___	0.30		___	0.10	0.30	0.30
	0.80	0.10	0.05	0.05		___	1.00	0.05	0.05
	0.80	0.10	___	0.30		___	1.00	0.05	0.30
	0.80	1.00	___	0.05		___	1.00	0.30	0.05
	0.80	1.00	___	0.30		___	1.00	0.30	0.30

Eight admissible vertices are present and are numbered as (1), (2), . . . , (8). The two-dimensional faces are found by grouping the vertices of the polyhedron into groups of three or more vertices where each vertex has the same value x_i for one of the components. For example, the vertices (2), (3), (4), and (7) each have $x_3 = 0.05$ and thus these four vertices define a face. There are six two-dimensional faces to the polyhedron and the coordinates of the centroids of the six faces are

TABLE 4.4. Coordinates of the 15 design points for the constrained beverage example

Design Point Designation	Type of Boundary	Mixture Component Proportions (Coordinates) x_1	x_2	x_3	x_4	Combination of Vertices
1	Vertex	0.40	0.10	0.30	0.20	
2	Vertex	0.80	0.10	0.05	0.05	
3	Vertex	0.40	0.50	0.05	0.05	
4	Vertex	0.40	0.25	0.05	0.30	
5	Vertex	0.40	0.25	0.30	0.05	
6	Vertex	0.40	0.10	0.20	0.30	
7	Vertex	0.55	0.10	0.05	0.30	
8	Vertex	0.55	0.10	0.30	0.05	
9	Face centroid	0.58	0.24	0.05	0.13	2, 3, 4, 7
10	Face centroid	0.54	0.24	0.17	0.05	2, 3, 5, 8
11	Face centroid	0.40	0.24	0.18	0.18	1, 3, 4, 5, 6
12	Face centroid	0.54	0.10	0.18	0.18	1, 2, 6, 7, 8
13	Face centroid	0.45	0.15	0.10	0.30	4, 6, 7
14	Face centroid	0.45	0.15	0.30	0.10	1, 5, 8
15	Overall centroid	0.49	0.19	0.16	0.16	1, 2, . . . , 8

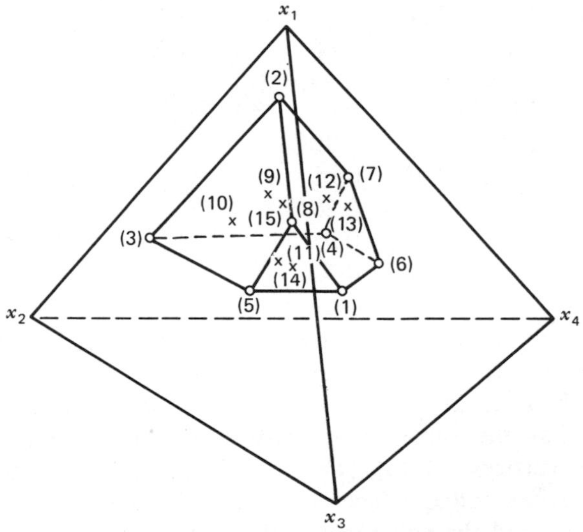

FIGURE 4.5. The constrained factor space inside the tetrahedron.

listed as the design points (9)–(14) in Table 4.4. The overall centroid, which is defined as the average of the eight vertices, is listed as point (15). The resulting factor space and the design point designations are presented in Figure 4.5.

Fitted to the values of the response observed at the design points might be the second-degree model of the form in Eq. (4.22). The centroids of the faces, points (9)–(14), were obtained, because centroid points are necessary in determining if the surface is nonlinear.

The model in Eq. (4.22) contains 10 terms and yet the number of vertices and centroids obtained is 15, five more than the number of coefficients to be estimated. We might ask ourselves whether or not some of the centroids and some of the vertices could be removed from consideration prior to fitting the model without sacrificing the precision of the parameter estimates in terms of their variances, as well as the precision of the predictor $\hat{y}(\mathbf{x})$. On the other hand, we might use all of the 15 points so that there is information on the residual variation that we can use for an estimate of the error variance or as a measure of model inadequacy. Although we would probably use all of the 15 points in small examples like the previous one, the choice of which design points to use and not use may not always be so straightforward. A method for choosing which points of an extreme vertices design to use when fitting a first-degree model over the constrained region inside a simplex is presented in Snee and Marquardt (1974). In a later paper, Snee (1975) discusses designs for fitting quadratic models in constrained mixture spaces. We now discuss the XVERT algorithm introduced by Snee and Marquardt in their 1974 paper.

4.7. THE XVERT ALGORITHM FOR LOCATING EXTREME VERTICES AND FOR SELECTING SUBSETS OF EXTREME VERTICES

As an alternative procedure to the algorithm presented in the previous section that was suggested by McLean and Anderson (1966) for locating the vertices of a constrained region inside the simplex, Snee and Marquardt proposed the XVERT algorithm. This algorithm generates the coordinates of all of the extreme vertices and selects subsets of the extreme vertices to serve as design points for fitting a first-degree canonical polynomial model of the Scheffé type.

The XVERT algorithm operates mainly on the principle of choosing a design that has points as spread out as possible over the region. When fitting the first-degree model for example,

$$\mathbf{y} = \mathbf{X}\boldsymbol{\beta} + \boldsymbol{\epsilon} \tag{4.23}$$

where $\mathbf{y}$ is the $N \times 1$ vector of observations, $\mathbf{X}$ is the $N \times q$ matrix of component proportions corresponding to the N blends and where the uth row is $\mathbf{x}'_u = (x_{u1}, x_{u2}, \ldots, x_{uq})$, $\boldsymbol{\beta}$ is the $q \times 1$ vector of coefficients to be estimated and $\boldsymbol{\epsilon}$ is the $N \times 1$ vector of random errors with zero mean and variance $\sigma^2\mathbf{I}$, the XVERT algorithm chooses the design from a class of possible designs, whose trace of the matrix $(\mathbf{X}'\mathbf{X})^{-1}$ is the lowest in value. The trace of $(\mathbf{X}'\mathbf{X})^{-1}$ is the sum of the diagonal elements of $(\mathbf{X}'\mathbf{X})^{-1}$ and since the ith element on the diagonal, say c_{ii}, is used in $\text{var}(b_i) = c_{ii}\sigma^2$, then the trace is the sum of the variances of the elements of $\mathbf{b}$, where $\mathbf{b}$ is the estimator of $\boldsymbol{\beta}$ in Eq. (4.23). In other words,

$$\mathbf{b} = (\mathbf{X}'\mathbf{X})^{-1}\mathbf{X}'\mathbf{y}, \qquad \text{var}(\mathbf{b}) = (\mathbf{X}'\mathbf{X})^{-1}\sigma^2$$

$$\sigma^2 \, \text{trace}(\mathbf{X}'\mathbf{X})^{-1} = \sum_{i=1}^{q} \text{var}(b_i) \tag{4.24}$$

and by concentrating on minimizing the trace $(\mathbf{X}'\mathbf{X})^{-1}$, one is intent on minimizing the sum of the variances of the individual coefficient estimates b_i, $i = 1, 2, \ldots, q$.

In order to help us describe the steps performed by the XVERT algorithm, we shall refer to and make use of the four-component fruit punch example of the previous section. We shall attempt only to locate the eight extreme vertices of the constrained region which was shown in Figure 4.5 so as to be able to compare the simplicity of this algorithm with the previous procedure. We shall not concern ourselves with choosing the best subset of the eight vertices for fitting the linear model

$$y = \beta_1 x_1 + \beta_2 x_2 + \beta_3 x_3 + \beta_4 x_4 + \epsilon \tag{4.25}$$

We recall that the restrictions on the component proportions were

$$0.40 \le x_1 \le 0.80, \quad 0.10 \le x_2 \le 1.0, \quad 0.05 \le x_3 \le 0.30, \quad 0.05 \le x_4 \le 0.30$$

To generate the extreme vertices of the feasible region, the steps of the XVERT algorithm are as follows:

***Step* 1.** Rank the components in order of increasing range sizes $b_i - a_i$:

$$b_3 - a_3 = 0.25, \qquad b_4 - a_4 = 0.25$$

$$b_1 - a_1 = 0.40, \qquad b_2 = 1.0 - 0.10 = 0.90$$

List the ordered components as $X_1, X_2, \ldots, X_q$ where X_1 is the component with the smallest range:

$$X_1 = x_3, \quad X_2 = x_4, \quad X_3 = x_1, \quad X_4 = x_2$$

***Step* 2.** Set up a two-level design arrangement using the lower and upper bounds of the $q-1$ components X_i having the smallest ranges. There are 2^{q-1} combinations.

1. $a_3a_4a_1\_\_$	5. $b_3a_4a_1\_\_$
2. $a_3a_4b_1\_\_$	6. $b_3a_4b_1\_\_$
3. $a_3b_4a_1\_\_$	7. $b_3b_4a_1\_\_$
4. $a_3b_4b_1\_\_$	8. $b_3b_4b_1\_\_$

***Step* 3.** Determine the level of the omitted component X_q (x_2 in our case), with each of the 2^{q-1} combinations in step 2 using $X_q = 1.0 - \Sigma_{i=1}^{q-1} X_i$.

1. 0.05, 0.05, 0.40, 0.50	5. 0.30, 0.05, 0.40, 0.25
2. 0.05, 0.05, 0.80, 0.10	6. 0.30; 0.05, 0.80, −0.15
3. 0.05, 0.30, 0.40, 0.25	7. 0.30, 0.30, 0.40, 0
4. 0.05, 0.30, 0.80, −0.15	8. 0.30, 0.30, 0.80, −0.40

***Step* 4.** If the value of X_q in step 3 falls within its acceptable limits (in our example, if X_4 lies between 0.10 and 1.00), then the combination is an extreme vertex of the inscribed region. If X_q does not fall inside its acceptable limits, then set X_q equal to its upper or lower limit, whichever is closer to the computed value.

Points 1, 2, 3, and 5 are extreme vertices
Points 4, 6, 7, and 8 need to be adjusted

***Step* 5.** With each of the points in which X_q is outside of its limits initially, we are to generate additional points by adjusting the levels of the other components, one at a time, by an amount equal to the difference between the computed value of X_q in step 3 and the substituted upper or lower limit. For each point in step 4 which is initially outside the region, we can generate at most $q-1$ additional points. These points then are to be adjusted.

		feasible blend
Adjusting 4:	0.05 − 0.25 = −0.20, 0.30, 0.80, 0.10	no
	0.05, 0.30 − 0.25 = 0.05, 0.80, 0.10	yes, point 2 above
	0.05, 0.30, 0.80 − 0.25 = 0.55, 0.10	yes, ⑤
Adjusting 6:	0.30 − 0.25 = 0.05, 0.05, 0.80, 0.10	yes, point 2 above
	0.05, 0.05 − 0.25 = −0.20, 0.80, 0.10	no
	0.30, 0.05, 0.80 − 0.25 = 0.55, 0.10	yes, ⑥
Adjusting 7:	0.30 − 0.10 = 0.20, 0.30, 0.40, 0.10	yes, ⑦
	0.30, 0.30 − 0.10 = 0.20, 0.40, 0.10	yes, ⑧
	0.30, 0.30, 0.40 − 0.10 = 0.30, 0.10	no
Adjusting 8:	0.30 − 0.50 = −0.20, 0.30, 0.80, 0.10	no
	0.30, 0.30 − 0.50 = −0.20, 0.80, 0.10	no
	0.30, 0.30, 0.80 − 0.50 = 0.30, 0.10	no

Thus from step 4 we acquire points 1, 2, 3, and 5 which correspond to the design points (3), (2), (4), and (5), respectively, in Figure 4.5, and from step 5, we acquire the adjusted points or feasible blends, ⑤, ⑥, ⑦, and ⑧, which correspond to the design points (7), (8), (6), and (1), respectively, in Figure 4.5. Points 1, 2, 3, and 5, which are obtained without adjusting, form the *core* of the design (as termed by Snee and Marquardt), while remaining points ⑤, ⑥, and ⑦ or ⑧ are from each of the three candidate subgroups. The three subgroups result from the adjusting of point 4, the adjusting of point 6 and the adjusting of point 7.

The number of vertices of the restricted factor space is eight. The number of parameters to be estimated in the first-degree equation (4.25) is four. If data is taken at all eight points, then the additional or extra four observations can be used to obtain a measure of the model lack of fit (assuming an estimate of the error variance has been obtained previously) or the four observations may be used to obtain an estimate of the error variance. On the other hand, if we wanted to use less than the eight vertices for our design, that is, if we wanted to use only four points or if we wanted to use at most six points, then the XVERT algorithm can be instructed to search through the core points and the candidate subgroups and calculate the quantity trace $(\mathbf{X}'\mathbf{X})^{-1}$ for each design. In the end, the design which chooses the largest ranges for the x_i or which uses points spread out at the extremes will be favored since these designs produce a smaller value for the trace of $(\mathbf{X}'\mathbf{X})^{-1}$ than do designs whose points are closer to one another.

The following five-component example is discussed at length in Snee and Marquardt (1974), and was presented later by Snee (1976). This example very nicely exemplifies the efficient use of the XVERT algorithm in a rather complicated industrial setting.

4.8. AN XVERT DESIGN FOR FITTING A FIVE-COMPONENT GASOLINE BLENDING LINEAR MODEL

The objective of the following gasoline blending study was to develop a five-component linear blending model subject to satisfying the following restrictions on the component proportions.

Component	Restriction on Proportion	
Butane	$0.00 \leq x_1 \leq 0.10$	
Alkylate	$0.00 \leq x_2 \leq 0.10$	
Light straight run	$0.05 \leq x_3 \leq 0.15$	(4.26)
Reformate	$0.20 \leq x_4 \leq 0.40$	
Cat cracked	$0.40 \leq x_5 \leq 0.60$	

The response of interest is the gasoline research octane number at 2.0 grams of lead per gallon. In addition to obtaining a fitted first-degree model, an estimate of residual error variance is wanted so that a measure of the model residual standard deviation could be obtained. The residual standard deviation represents the sum of lack of fit variation plus observation error variation. If the model residual standard deviation is

TABLE 4.5. Gasoline blends selected by the XVERT algorithm

Extreme Vertex	Butane x_1	Alkylate x_2	Light Straight Run x_3	Reformate x_4	Catalytically Cracked x_5	Octane y
1	0.10	0.10	0.05	0.20	0.55	95.1
2	0.10	0.00	0.15	0.20	0.55	93.4
3	0.00	0.10	0.15	0.20	0.55	93.3
4	0.10	0.10	0.15	0.20	0.45	94.1
5	0.00	0.00	0.05	0.40	0.55	91.8
6	0.10	0.00	0.05	0.40	0.45	91.8
7	0.00	0.10	0.05	0.40	0.45	92.5
8	0.00	0.00	0.15	0.40	0.45	90.5
9	0.00	0.00	0.05	0.35	0.60	92.7
10	0.10	0.10	0.15	0.25	0.40	93.5
11*	0.10	0.00	0.05	0.25	0.60	94.8
12	0.10	0.00	0.10	0.20	0.60	
13	0.10	0.05	0.05	0.20	0.60	
14*	0.00	0.10	0.05	0.25	0.60	93.7
15	0.00	0.10	0.10	0.20	0.60	
16	0.05	0.10	0.05	0.20	0.60	
17*	0.00	0.00	0.15	0.25	0.60	92.5
18	0.00	0.05	0.15	0.20	0.60	
19	0.05	0.00	0.15	0.20	0.60	
20*	0.10	0.10	0.05	0.35	0.40	93.1
21	0.10	0.05	0.05	0.40	0.40	
22	0.05	0.10	0.05	0.40	0.40	
23*	0.10	0.00	0.15	0.35	0.40	91.8
24	0.10	0.00	0.10	0.40	0.40	
25	0.05	0.00	0.15	0.40	0.40	
26*	0.00	0.10	0.15	0.35	0.40	91.6
27	0.00	0.10	0.10	0.40	0.40	
28	0.00	0.05	0.15	0.40	0.40	

*Denotes blend from candidate subgroup selected by the XVERT algorithm.
Source: Snee and Marquardt (1974), Table 3, p. 406; reprinted with permission from R. D. Snee.

close to the research octane rating standard deviation, which according to the authors is known ahead of time to be approximately of the magnitude 0.30, then the fitted linear model would be satisfactory and could be used to determine the relative importance of the various components in the study, as well as used also to predict octane ratings for blends other than those used in the design. A 16-point design was chosen to estimate the model parameters and to obtain an estimate of experimental error variance.

The core points and the candidate subgroup points are presented in Table 4.5. The six points that were selected by the XVERT algorithm, one from each candidate subgroup are denoted by an asterisk (*). The research octane rating at each of the sixteen blends is also listed.

The five-component linear blending model fitted to the 16 blends is

$$\begin{array}{l} \hat{y}(\mathbf{x}) = 102.4x_1 + 100.7x_2 + 85.2x_3 + 84.7x_4 + 97.6x_5 \\ \qquad\qquad (1.5) \qquad\quad (1.5) \qquad (1.4) \qquad (0.7) \qquad (0.5) \end{array} \tag{4.27}$$

where the number in parentheses below each coefficient estimate is the estimated standard error of the estimate. The analysis of variance table is presented in Table 4.6. The estimate of the error variance (or observation variance) is $s^2 = 0.10$ with 11 degrees of freedom and because this estimate is close to the octane rating error variance of $(0.30)^2$, it is concluded that the linear model in Eq. (4.27) provides an adequate description of the research octane within the constrained blending region defined in Eqs. (4.26). The highly significant F-test value of 54.2 supported this view.

A final word on this example concerns the addition of points for the purpose of fitting a second-degree model. The 16 vertices listed in Table 4.5 would not allow the estimation of all the coefficients in a complete 15-term second-degree model. Additional points are required, presumably at the centroids of some of the two-dimensional faces or at the centroids of the edges joining the vertices of the polyhedron. In a later

TABLE 4.6. Analysis of variance table for octane data

Source of Variation	Degrees of Freedom	Sum of Squares	Mean Square	F
Regression	4	21.68	5.42	54.2[a]
Residual	11	1.10	0.10	
Total	$16-1=15$	22.78		

[a]Highly significant at $\alpha = 0.01$ level.

paper (Snee, 1975), Snee addresses the problem of finding designs for fitting quadratic models and suggests using other algorithms in addition to the XVERT algorithm for these second-order designs.

4.9. THE USE OF SYMMETRIC SIMPLEX DESIGNS FOR FITTING SECOND-ORDER MODELS IN RESTRICTED REGIONS

In a highly constrained region where the proportions of one or more of the components are restricted as in Eq. (4.21) where $a_i \le x_i \le b_i$, even a small change in one or more of the proportions may produce a curvilinear change in the response surface. To model such a change would require the use of a second-degree equation. A procedure for obtaining the settings of a second-degree design, which uniformly covers the highly constrained region, is presented now.

In Saxena and Nigam (1977) the following procedure is outlined. Suppose p components are restricted by $0 < a_i \le x_i \le b_i < 1$ where $p \le q$. Select q other variables z_i that are restricted by $0 \le z_i \le 1$ and $z_1 + z_2 + \cdots + z_q = 1$ but that we shall use for design purposes only. The z_i are called *design variables*. Corresponding to the restrictions $0 < a_i \le x_i \le b_i < 1$ on the mixture component proportions, the region of the simplex for the z_i's is defined to be $0 \le B \le z_i \le B' \le 1$ so that B and B' represent the minimum and maximum proportions of *any* component z_i in the design.

In this design strategy the restricted polyhedron is covered with a uniform array of points. To do so, we select a symmetric-simplex design in the z_i's. From the symmetric design configuration in the z_i's, a transformation is made from the coordinates in the z_i's to produce a symmetrical design in the x_i's as follows.

Consider the linear transformation from z_{ui} to x_{ui}

$$x_{ui} = \lambda_i + \mu_i z_{ui} \tag{4.28}$$

where z_{ui} is the uth setting of the ith design variable $u = 1, 2, \ldots, N$; $i = 1, 2, \ldots, q$, and λ_i and μ_i are scalar constants. Let the x_{ui} be one of the p mixture component proportions constrained by

$$0 \le a_i \le x_i \le b_i \le 1, \qquad i = 1, 2, \ldots, p \tag{4.29}$$

Since x_{ui} must satisfy Eq. (4.29), then $x_{ui} = a_i$ when $z_{ui} = B$ and $x_{ui} = b_i$ when $z_{ui} = B'$ so that the values of λ_i and μ_i in Eq. (4.28) are easily found to be

$$\lambda_i = \frac{a_i B' - b_i B}{B' - B}, \qquad \mu_i = \frac{b_i - a_i}{B' - B} \tag{4.30}$$

To obtain the design proportions corresponding to the coordinates in the z_{ui}, we substitute the expressions for λ_i and μ_i in Eq. (4.30) into Eq. (4.28) and the restricted proportions for component i say is given by

$$x_{ui} = \frac{(a_iB' - b_iB)}{(B' - B)} + \left(\frac{b_i - a_i}{B' - B}\right) z_{ui}, \qquad i = 1, 2, \ldots, p \tag{4.31}$$

Each of the p restricted components in Eq. (4.29) is transformed in Eq. (4.31) and the remaining $q - p$ components are adjusted to have the proportions

$$x_{uj} = \left[\frac{1 - (x_{u1} + x_{u2} + \cdots + x_{up})}{1 - (z_{u1} + z_{u2} + \cdots + z_{up})}\right] z_{uj}, \qquad j = p + 1, \ldots, q \tag{4.32}$$

To illustrate the transformations of Eqs. (4.28)–(4.32), let us choose a small three-component example where only the components 1 and 2 are restricted by

$$0.1 \leq x_1 \leq 0.6$$
$$0.2 \leq x_2 \leq 0.8$$
$$0 \leq x_3 \leq 1.0$$

Let us choose an arbitrary symmetric-simplex design in the z_i. The

TABLE 4.7. Symmetric-simplex coordinates to coordinates in the constrained region

	Symmetric-Simplex Coordinates			Mixture Component Proportions		
Point	z_1	z_2	z_3	x_1	x_2	x_3
1	$\frac{3}{4}$	$\frac{1}{8}$	$\frac{1}{8}$	0.60	0.30	0.10
2	$\frac{1}{8}$	$\frac{3}{4}$	$\frac{1}{8}$	0.18	0.80	0.02
3	$\frac{1}{8}$	$\frac{1}{8}$	$\frac{3}{4}$	0.18	0.30	0.52
4	$\frac{1}{2}$	$\frac{1}{2}$	0	0.43	0.57[a]	0.00
5	$\frac{1}{2}$	0	$\frac{1}{2}$	0.43	0.20	0.37
6	0	$\frac{1}{2}$	$\frac{1}{2}$	0.10	0.60	0.30

[a]The value of x_2 for $z_2 = \frac{1}{2}$ is $x_2 = 0.60$, but with $x_1 = 0.43$ already, the value of x_2 was lowered to 0.57.

coordinates of the z_i are listed in Table 4.7. Now $B = 0$, $B' = \frac{3}{4}$ and x_i will assume the value a_i if $z_i = B = 0$ and $x_i = a_i + \frac{4}{3}(b_i - a_i)z_i$ if $0 < z_i \le \frac{3}{4}$. At point 1 for example

$$
\begin{aligned}
z_{11} = \tfrac{3}{4}, \qquad x_{11} &= a_1 + \tfrac{4}{3}(b_1 - a_1)z_{11} \\
&= 0.1 + \tfrac{4}{3}(0.6 - 0.1)\tfrac{3}{4} = 0.6 \\
z_{12} = \tfrac{1}{8}, \qquad x_{12} &= a_2 + \tfrac{4}{3}(b_2 - a_2)z_{12} \\
&= 0.2 + \tfrac{4}{3}(0.8 - 0.2)\tfrac{1}{8} = 0.3 \\
x_{13} &= \frac{(1 - 0.6 - 0.3)}{(1 - \frac{3}{4} - \frac{1}{8})}\tfrac{1}{8} = 0.1 = 1 - x_{11} - x_{12}
\end{aligned}
$$

The transformed mixture proportions in the x_i corresponding to the coordinates of the six points in the z_i are listed in Table 4.7 and are drawn in Figure 4.6.

A couple of comments about the transformation above, which uses Eqs. (4.28)–(4.32) are worth mentioning. The formula in Eq. (4.31) for going from z_{ui} to x_{ui} is simplified when $B = 0$. Thus, when choosing the symmetric-simplex design in the z_i's, a boundary point where $z_i = 0$ making $B = 0$ may simplify the transformation calculations. However, points on the boundaries of the simplex in the z_i's can also produce extreme values for the x_i that do not fall inside the bounds specified in Eq. (4.29). In particular, the vertex points of the simplex in the z_i correspond to the extreme vertices of the polyhedron in the x_i and the points of the simplex where $z_i = B$ and $z_i = B'$ correspond to points

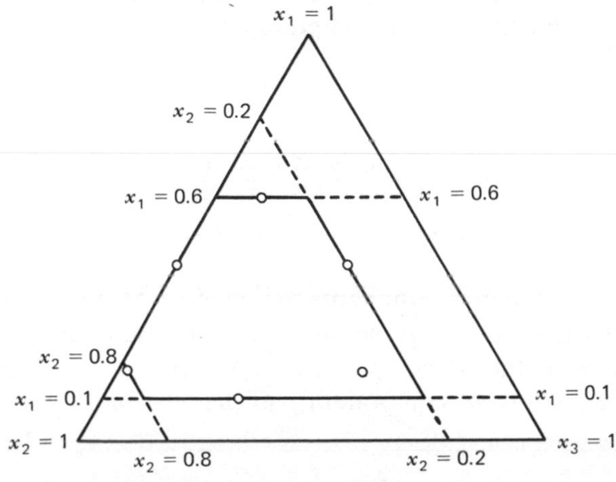

FIGURE 4.6. The symmetric-simplex design in the constrained region.

whose coordinates are convex combinations of the coordinates of the extreme vertices of the polyhedron. The freedom in selecting any arbitrary symmetric-simplex design in the z_i's is a pleasant feature of this approach to choosing a second-degree design but the efficiency of the final design choice relative to the selection of an extreme vertices design with additional boundary points still remains in doubt.

We shall now discuss mixture experiments in which the components are categorized. The categories are classified as major components and each category consist of some number of member components which are referred to as minor components. The designs used for studying the blending characteristics of all of the components are known as multiple-lattice designs.

4.10. MULTIPLE LATTICES FOR MAJOR AND MINOR COMPONENT CLASSIFICATIONS

Some mixture experiments involve two or more classes of components. For example, we might have a major class of chemical salt types (or categories) to be blended where each salt category consists of one or more subset salts which differ only slightly from one another. When there are only two types of component classifications the members of the respective classes can be referred to as "major" components and "minor" components. The "major" components will be designated as M-components and the "minor" components will be designated as m-components.

Let us confine our discussion to only two classes of components where there are p M-components present in the mixture system under study. Let us define the proportion for M-component i to be

$$c_i \geq 0 \qquad \text{for } i = 1, 2, \ldots, p$$

All of the M-components are blended together so that

$$c_1 + c_2 + \cdots + c_p = 1$$

The number of minor components assigned to M-component i can be denoted by n_i, $i = 1, 2, \ldots, p$, so that the total number of minor components over all of the M-components is $\Sigma_{i=1}^{p} n_i = q$. Furthermore, if the proportion assigned to m-component j in the M-component i is denoted by

$$x_{ij} \geq 0, \qquad 1 \leq j \leq n_i; \qquad 1 \leq i < p$$

then for the n_i m-components in M-component i

$$x_{i1} + x_{i2} + \cdots + x_{in_i} = 1, \qquad i = 1, 2, \ldots, p \tag{4.33}$$

Each M-component can be thought of as a blend of the n_i m-components as seen from Eq. (4.33) and thus the factor space for each M-component may be geometrically represented by a regular simplex of dimensionality $n_i - 1$, $i = 1, 2, \ldots, p$.

To any blend in the overall mixture system of q components, the proportion contributed by m-component j from M-component i is defined as the product of c_i and x_{ij}, that is,

$$X_{ij} = c_i x_{ij} \tag{4.34}$$

The nonnegative proportions X_{ij} must satisfy the condition, $\sum_i^p \sum_j^{n_i} X_{ij} = 1$.

To cover the factor space of feasible mixtures of all q components, a lattice design can be defined. The design is constructed by setting up combinations of the points of the separate simplex lattices associated with the individual M-components. To see this, first let us consider the simple case where members of only two M-components are to be combined to form the mixtures and let us assume the respective M-component proportions are equal so that $c_1 = \frac{1}{2}$ and $c_2 = \frac{1}{2}$. If we consider fitting a polynomial of degree m_1 over the simplex-lattice associated with the M-component 1, and a model of degree m_2 over the lattice associated with M-component 2, then the proportions x_{ij} in each M-component will take the values

$$x_{ij} = 0, \frac{1}{m_i}, \frac{2}{m_i}, \ldots, 1, \qquad i = 1, 2$$

As an example, if $m_1 = 2$ and $m_2 = 3$, then

$$x_{1j} = 0, \tfrac{1}{2}, 1, \qquad j = 1, 2, \ldots, n_1$$

$$x_{2j} = 0, \tfrac{1}{3}, \tfrac{2}{3}, 1, \qquad j = 1, 2, \ldots, n_2$$

Note that the m_i, $i = 1, 2, \ldots, p$ with the individual M-components, do not have to be equal; in fact, they probably will not be, meaning that one may consider fitting polynomials of different degrees over the individual M-component lattices.

To form the point arrangement of the multiple-lattice design the design points from each of the individual M-component lattices are combined.

The dimensionality of the new factor space will be the sum of the dimensionalities of the individual lattices. For example, let us consider the two M-components mentioned previously and for simplicity, let $n_1 = 2$ and $n_2 = 2$, that is, let each M-component consist of two "minor" components. As before, let us fit a $m_1 = 2$ or second-degree model over the lattice of M-component 1 and this we can do with a $\{2, 2\}$ lattice, and suppose we choose a $m_2 = 3$ or third-degree model to fit to the points of a $\{2, 3\}$ lattice of M-component 2. The multiple lattice will be the result of combining the points of the $\{n_1, m_1\} = \{2, 2\}$ simplex-lattice with those of the $\{n_2, m_2\} = \{2, 3\}$ simplex-lattice and the combined arrangement is called a $\{n_1, n_2; m_1, m_2\}$ multiple-lattice, in this case, a $\{2, 2; 2, 3\}$ double lattice. Since each $\{2, m_i\}$ simplex-lattice is a line of dimensionality one, the $\{2, 2; 2, 3\}$ double lattice will be of dimensionality $1 + 1 = 2$. The number of points in the multiple lattice will be

$$\prod_{i=1}^{p} \binom{n_i + m_i - 1}{m_i} = \prod_{i=1}^{p} \frac{(n_i + m_i - 1)!}{m_i!(n_i - 1)!}.$$

To illustrate the construction of a double lattice, let us define the proportions of the individual m-components from M-component 1 to be X_{11} and X_{12} and similarly, the components in M-component 2 to be X_{21} and X_{22}. The possible mixture combinations formed by combining the minor components from each of the two M-components are

$$\begin{array}{ll} X_{1j}X_{2j} & \\ X_{11}X_{12}X_{2j} & \\ X_{1j}X_{21}X_{22} & j = 1, 2 \\ X_{11}X_{12}X_{21}X_{22} & \end{array}$$

where $X_{1j}X_{2j}$ represents a mixture or blend consisting of one component from each M-component.

The individual m-component proportions of Eq. (4.34) that are present in blends are determined by recalling that $X_{11} + X_{12} = \frac{1}{2}$ and $X_{21} + X_{22} = \frac{1}{2}$, and therefore the mixture $X_{1j}X_{2j}$ which represents a single component from each M-component, requires that each m-component proportion X_{ij} $(i = 1, 2)$ be equal to $\frac{1}{2}$. Each of the four combinations $X_{11}X_{21}$, $X_{12}X_{21}$, $X_{11}X_{22}$ and $X_{12}X_{22}$ thus consists of a single m-component from each M-component and each m-component contributes $\frac{1}{2}$ towards the mixture. The following combinations comprise the remainder of the $\{2, 2; 2, 3\}$ double-lattice,

$$
\begin{array}{lll}
X_{11}X_{12}X_{2j} & \text{where } X_{11} = \tfrac{1}{4},\ X_{12} = \tfrac{1}{4},\ X_{2j} = \tfrac{1}{2} & j = 1, 2 \\
X_{1j}X_{21}X_{22} & X_{1j} = \tfrac{1}{2},\ X_{21} = \tfrac{1}{6},\ X_{22} = \tfrac{2}{6} & j = 1, 2 \\
 & X_{1j} = \tfrac{1}{2},\ X_{21} = \tfrac{2}{6},\ X_{22} = \tfrac{1}{6} & j = 1, 2 \qquad (4.35) \\
X_{11}X_{12}X_{21}X_{22} & X_{11} = X_{12} = \tfrac{1}{4},\ X_{21} = \tfrac{1}{6},\ X_{22} = \tfrac{2}{6} & \\
 & X_{11} = X_{12} = \tfrac{1}{4},\ X_{21} = \tfrac{2}{6},\ X_{22} = \tfrac{1}{6} &
\end{array}
$$

The 12 mixture compositions are displayed in the form of the $\{2, 2; 2, 3\}$ double-lattice in Figure 4.7.

A regression function that can be used to model the response over the multiple lattice is formed by combining the terms from the individual models which would be used separately to represent the response over the simplex space corresponding to each of the M-components. In other words, the regression function that would be fitted to the points of the double lattice is formed by combining the terms from each of the two polynomials that would be used over the two single lattices. For example, let us refer to our previous two M-component example where with

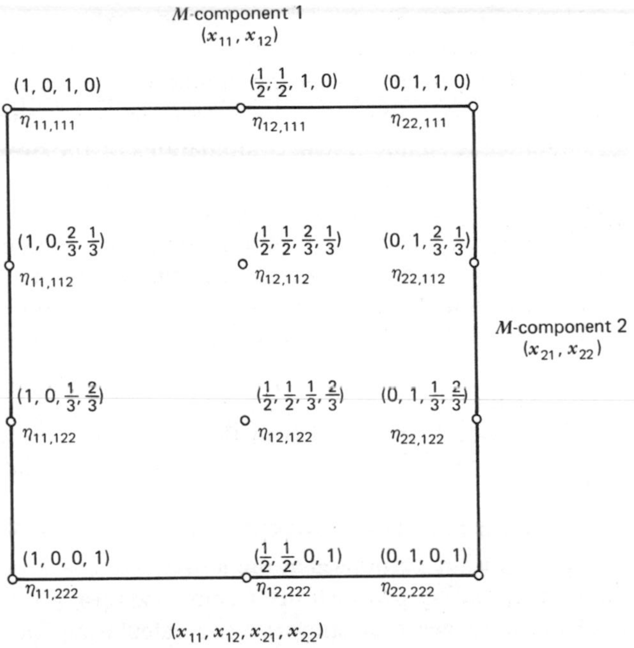

FIGURE 4.7. The minor component proportions and the expected responses at the points of the $\{2, 2; 2, 3\}$ double lattice.

M-component 1, a {2, 2} lattice was set up to accommodate the second-degree polynomial

$$\eta_{S_1} = d_1x_{11} + d_2x_{12} + d_{12}x_{11}x_{12}, \qquad S_1 = 11,\ 12,\ 22$$

whereas to the points of the {2, 3} lattice for M-component 2, we consider fitting the cubic polynomial

$$\eta_{S_2} = e_1x_{21} + e_2x_{22} + e_{12}x_{21}x_{22} + e_{112}x_{21}x_{22}(x_{21} - x_{22}), \qquad S_2 = 111,\ 112,\ 122,\ 222$$

Setting up crossproducts between each of the terms in η_{S_1} and each of the terms in η_{S_2}, that is, $\eta_{S_1,S_2} = (\eta_{S_1})(\eta_{S_2})$, the resulting 12-term regression function which is used to represent the response over the {2, 2; 2, 3} double lattice is

$$\begin{aligned}\eta_{S_1,S_2} = &\sum_{j=1}^{2}\sum_{l=1}^{2} \gamma_{jl}x_{1j}x_{2l} + \sum_{j=1}^{2} \gamma_{j12}x_{1j}x_{21}x_{22} + \sum_{j=1}^{2} \gamma_{j112}x_{1j}x_{21}x_{22}(x_{21} - x_{22}) \\ &+ \sum_{l=1}^{2} \gamma_{12l}x_{11}x_{12}x_{2l} + \gamma_{1212}x_{11}x_{12}x_{21}x_{22} + \gamma_{12112}x_{11}x_{12}x_{21}x_{22}(x_{21} - x_{22})\end{aligned} \tag{4.36}$$

where the subscripts S_1 and S_2 are defined in the next paragraph. The polynomial equation (4.36) of degree $m_1 + m_2 = 2 + 3 = 5$ contains the same number of terms as there are points on the double lattice.

The estimates of the parameters in the double lattice model in Eq. (4.36) are calculated using simple linear combinations of the observations collected at the points of the double lattice. To see this, let us refer to Figure 4.7 where each response is denoted by a suffixed letters η_{S_1,S_2}. Each of the suffixes S_1 and S_2 consists of a sequence of numbers. In S_j, $j = 1, 2$, the number l ($l = 1$ or 2) appears m_jx_{jl} times indicating the lth m-component is present with proportion x_{jl}. For example, the response $\eta_{11,111}$ is from the blend $X_{11} = \frac{1}{2}$, $X_{21} = \frac{1}{2}$ which in the m-components is $x_{11} = 1$, $x_{21} = 1$, because $x_{jl} = X_{jl}/c_j$ and $c_j = \frac{1}{2}$ and therefore we have the suffices $S_1 = 2x_{11} = 11$ and $S_2 = 111 = 3x_{21}$; the response $\eta_{12,111}$ belongs to the blend $X_{11} = \frac{1}{4}$, $X_{12} = \frac{1}{4}$, $X_{21} = \frac{1}{2}$, which in the m-components is $x_{11} = \frac{1}{2}$, $x_{12} = \frac{1}{2}$, $x_{21} = 1$; the response $\eta_{12,112}$ corresponds to the blend $X_{11} = \frac{1}{4}$, $X_{12} = \frac{1}{4}$, $X_{21} = \frac{2}{6}$, $X_{22} = \frac{1}{6}$, which in the m-components is $x_{11} = \frac{1}{2}$, $x_{12} = \frac{1}{2}$, $x_{21} = \frac{2}{3}$, $x_{22} = \frac{1}{3}$, and so on.

If the averages $\bar{y}_{S_1,S_2}$ of the responses at the lattice points are substituted along with the corresponding proportions x_{1j} into Eqs. (4.36), the linear combinations of the averages that comprise the calculating formulas for the parameter estimates result. The coefficients of the averages that are used for estimating the parameters in Eq. (4.36) are listed in Table 4.8.

To set up an equation to predict the response over the double lattice,

TABLE 4.8. Coefficients of the average response values at the lattice points used to estimate the parameters in the double-lattice model

Parameter Estimate												Mean Response at	$(x_{11}, x_{12}:x_{21}, x_{22})$
g_{11}	g_{12}	g_{21}	g_{22}	g_{112}	g_{212}	g_{121}	g_{122}	g_{1112}	g_{2112}	g_{1212}	g_{12112}		
1				$-\frac{9}{4}$		-2		$-\frac{9}{4}$		$\frac{9}{2}$	$\frac{9}{2}$	$y_{11,111}$	$(1, 0:1, 0)$
				$\frac{9}{4}$				$\frac{27}{4}$		$-\frac{9}{2}$	$\frac{27}{2}$	$\bar{y}_{11,112}$	$(1, 0:\frac{2}{3}, \frac{1}{3})$
				$\frac{9}{4}$				$-\frac{27}{4}$		$-\frac{9}{2}$	$-\frac{9}{2}$	$\bar{y}_{11,122}$	$(1, 0:\frac{1}{3}, \frac{2}{3})$
	1			$-\frac{9}{4}$			-2	$\frac{9}{4}$		$\frac{9}{2}$	$-\frac{27}{2}$	$\bar{y}_{11,222}$	$(1, 0:0, 1)$
						4				-9	-9	$\bar{y}_{12,111}$	$(\frac{1}{2}, \frac{1}{2}:1, 0)$
										9	-27	$\bar{y}_{12,112}$	$(\frac{1}{2}, \frac{1}{2}:\frac{2}{3}, \frac{1}{3})$
										9	9	$\bar{y}_{12,122}$	$(\frac{1}{2}, \frac{1}{2}:\frac{1}{3}, \frac{2}{3})$
							4			-9	27	$\bar{y}_{12,222}$	$(\frac{1}{2}, \frac{1}{2}:0, 1)$
		1			$-\frac{9}{4}$	-2			$-\frac{9}{4}$	$\frac{9}{2}$	$\frac{9}{2}$	$\bar{y}_{22,111}$	$(0, 1:1, 0)$
					$\frac{9}{4}$				$\frac{27}{4}$	$-\frac{9}{2}$	$\frac{27}{2}$	$\bar{y}_{22,112}$	$(0, 1:\frac{2}{3}, \frac{1}{3})$
					$\frac{9}{4}$				$-\frac{27}{4}$	$-\frac{9}{2}$	$-\frac{9}{2}$	$\bar{y}_{22,122}$	$(0, 1:\frac{1}{3}, \frac{2}{3})$
			1		$-\frac{9}{4}$		-2		$\frac{9}{4}$	$\frac{9}{4}$	$-\frac{27}{2}$	$\bar{y}_{22,222}$	$(0, 1:0, 1)$

the estimates $g_{ij}, \ldots, g_{12112}$ are substituted into Eq. (4.36) or one may use the average responses directly. In this latter case

$$\hat{y}(\mathbf{x}) = \sum_{j=1}^{2}\sum_{l=1}^{2} C_{jj,lll}\bar{y}_{jj,lll} + \sum_{j=1}^{2} C_{jj,112}(\bar{y}_{jj,112} + \bar{y}_{jj,122}) + \sum_{l=1}^{2} C_{12,lll}\bar{y}_{12,lll}$$
$$+ C_{12,112}\bar{y}_{12,112} + C_{12,122}\bar{y}_{12,122} \tag{4.37}$$

where the coefficients of the $\bar{y}_{s_1,s_2}$ are

$$\begin{aligned}
C_{11,111} &= x_{11}x_{21}(1-2x_{12})[1-\tfrac{9}{4}x_{22}(1+x_{12}-x_{22})] \\
&= \{x_{11}(2x_{11}-1)\}\{\tfrac{1}{2}x_{21}(3x_{21}-2)(3x_{21}-1)\} \\
&= a_{11}b_{111} \\
C_{jj,lll} &= \{x_{ij}(2x_{1j}-1)\}\{\tfrac{1}{2}x_{2l}-2)(3x_{2l})\} = a_{jj}b_{lll}, & j, l = 1, 2 \\
C_{jj,112} &= \tfrac{9}{2}x_{1j}x_{21}x_{22}(2x_{1j}-1)(3x_{21}-1) = a_{jj}b_{112}, & j = 1, 2 \\
C_{jj,122} &= \tfrac{9}{2}x_{1j}x_{21}x_{22}(2x_{1j}-1)(3x_{22}-1) = a_{jj}b_{122}, & j = 1, 2 \\
C_{12,lll} &= 2x_{11}x_{12}x_{2l}(3x_{2l}-2)(3x_{2l}-1) = a_{12}b_{lll}, & l = 1, 2 \\
C_{12,112} &= 4x_{11}x_{12}\{\tfrac{9}{2}x_{21}x_{22}(3x_{21}-1)\} = a_{12}b_{112} \\
C_{12,122} &= 4x_{11}x_{12}\{\tfrac{9}{2}x_{21}x_{22}(3x_{22}-1)\} = a_{12}b_{122}
\end{aligned} \tag{4.38}$$

The constants C_{S_1,S_2} are products of the orthogonal polynomials a_{S_1} and b_{S_2} as shown in (4.38). These polynomials are functions of the coordinates of the points on the {2, 2} and {2, 3} simplex-lattices, respectively, which produce the points or mixtures on the {2, 2; 2, 3} double lattice. For example the coefficient $C_{11,111}$ is the product of $a_{11} = x_{11}(2x_{11} - 1)$ and $b_{111} = \frac{1}{2}[x_{21}(3x_{21} - 2)(3x_{21} - 1)]$. The other orthogonal polynomial expressions are

$$a_{jj} = x_{ij}(2x_{1j} - 1), \qquad b_{lll} = \tfrac{1}{2}[x_{2l}(3x_{2l} - 2)(3x_{2l} - 1)]$$

$$a_{12} = 4x_{11}x_{12}, \qquad b_{112} = \tfrac{9}{2}[x_{21}x_{22}(3x_{21} - 1)]$$

$$b_{122} = \tfrac{9}{2}[x_{21}x_{22}(3x_{22} - 1)]$$

The allocation of observations collected at the points of the multiple lattice for the purpose of minimizing the expected variance of $\hat{y}(\mathbf{x})$ over the factor space is discussed in Lambrakis (1968).

We continue with the mixture blends consisting of categories of components but now the region of experimentation is an ellipsoidal region located completely inside the polyhedron defined by the additional constraints on the component proportions. The methodology in this section is similar to that of Section 3.3 where the ellipsoidal region inside the simplex was introduced and the designs and polynomial models were set up in the system of the independent variables.

4.11. CATEGORIZING THE MIXTURE COMPONENTS: AN ELLIPSOIDAL REGION OF INTEREST

Let us again consider the blending of categories of components such as a category of acid components, a category of bases, and so on, but where the categories are not considered as mixtures in themselves (as was the case in the previous section). Each category of components contributes a fixed proportion to each mixture and every category is required to be present in every mixture.

To form the mixtures, let us assume there are p distinct categories ($p \geq 2$) present and that each category must be represented by one or more of its member components. Let us denote by n_l the number of components in the lth category, $l = 1, 2, \ldots, p$, where $n_l \geq 2$ and $q \geq 2p$. If the sum of the components up to the jth category is written as

$$S_j = \sum_{l=1}^{j} n_l \qquad \text{so that } S_p = q \tag{4.39}$$

then in addition to the usual constraints $x_i \geq 0$ and $x_1 + \cdots + x_q = 1$ placed on the component proportions, the following constraints are also considered now,

$$0 \leq x_i \leq 1/p, \qquad i = 1, 2, \ldots, q \tag{4.40}$$

and

$$\sum_{i=S_{j-1}+1}^{S_j} x_i = \frac{1}{p}, \qquad j = 1, 2, \ldots, p \tag{4.41}$$

The constraint (4.41) implies that every category contributes an equal proportion (p categories each contributing $1/p$) to every mixture. This assumption is unessential and in Figure 4.8, the result of not making this assumption is shown. However, we shall assume that each category contributes an equal proportion to each blend.

As a result of the additional constraints in Eqs. (4.40) and (4.41) being placed on the component proportions, the factor space takes the form of a convex polyhedron which is more complicated than the simplex. The dimensionality of the polyhedron is the same as that of the multiple-lattice polyhedron of the previous section. To show this, let us consider initially that only two categories (1 and 2) of components are present in forming the mixtures and further, that each category contains only two components. Let the component proportions from category 1 be denoted by x_1 and x_2 while those from category 2 be denoted by x_3 and x_4. We also denote the components themselves by the symbols x_1, x_2, x_3, and x_4. Then a valid mixture can be represented by any of the following

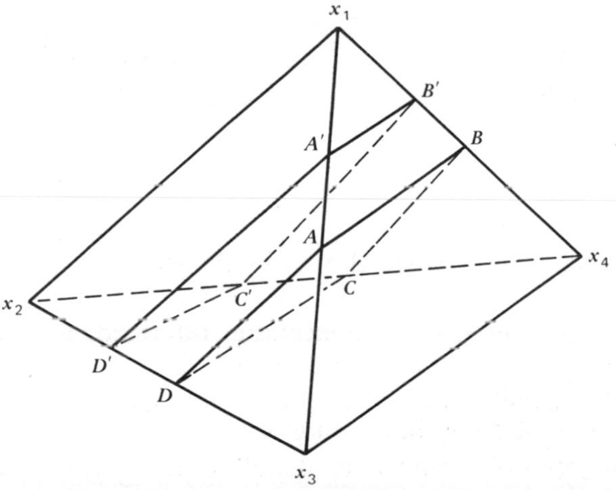

FIGURE 4.8. Factor space of the mixture system in two dimensions. Reproduced from Cornell and Good (1970), with permission of the American Statistical Association.

combinations:

$$\begin{array}{lllll} x_i x_j & i=1,2; & j=3,4 & & \\ x_i x_{i'} x_j & i=1,2; & i'=1,2; & i'\neq i; & j=3,4 \\ x_i x_j x_{j'} & i=1,2; & j=3,4; & j'=3,4; & j\neq j' \\ x_i x_{i'} x_j x_{j'} & i=1,2; & i'=1,2; & i\neq i'; & j=3,4; \\ & j'=3,4; & j\neq j' & & \end{array} \tag{4.42}$$

where

$$x_1+x_2=\tfrac{1}{2}, \qquad x_3+x_4=\tfrac{1}{2} \tag{4.43}$$

The factor space (composition space) of the components is represented by the boundary and interior of the square *ABCD* in Figure 4.8. If instead of Eq. (4.43) we have

$$x_1+x_2=\tfrac{3}{4}, \qquad x_3+x_4=\tfrac{1}{4}$$

then the factor space is represented by the rectangle $A'B'C'D'$ in Figure 4.8.

Turning our attention very briefly to the dimensionality and structure of the factor space in the general case, because of the constraint equation (4.41), we see that the number of independent components in the jth category is $S_j - S_{j-1} - 1 = n_j - 1$. Hence the total number of independent components in all p categories is $\Sigma_{j=1}^{p}(n_j - 1) = q - p$, and this number also represents the dimensionality of the factor space. The geometrical structure of the factor space can be partly specified by giving the number N_k of its boundaries that have dimensionality k ($k = 0, 1, 2, \ldots, q-p$). In Table 4.9 a few cases where $p = 2$ and $p = 3$ are presented.

In Section 3.3, an ellipsoidal region of interest inside the simplex for general q was expressed analytically in Eq. (3.20) as

$$\sum_{i=1}^{q}\left(\frac{x_i - x_{0i}}{h_i}\right)^2 \leq 1$$

where x_{0i} and h_i ($1 \leq i \leq q$) are chosen by the experimenter so as to give appropriate location and spread to the interval of interest for the ith component in the particular application. Intermediate variables $v_i = (x_i - x_{0i})/h_i$, $1 \leq i \leq q$, were defined and the ellipsoidal region was transformed to the boundary and interior of a unit sphere of dimensionality $q-1$ in the space of the v_i with center at $v_i = 0$.

With p categories of components, the ellipsoidal region is defined as in Eq. (3.20) but because of the additional restriction in Eq. (4.41), namely that the proportions of the components in each category must sum to $1/p$

TABLE 4.9. The number N_k of boundaries of dimensionality k for some factor spaces with $p = 2$ and $p = 3$

p	n_1	n_2	k	N_k	p	n_1	n_2	n_3	k	N_k
2	2	2	2	1	3	2	2	2	3	1
			1	4					2	6
			0	4					1	12
									0	8
2	3	2	3	1						
			2	5	3	3	2	2	4	1
			1	9					3	7
			0	6					2	19
									1	24
2	4	2	4	1					0	12
			3	6						
			2	14	3	3	3	2	5	1
			1	16					4	8
			0	8					3	27
									2	48
2	3	3	4	1					1	45
			3	6					0	18
			2	15						
			1	18						
			0	9						

Source: Cornell and Good (1970), with permission of the American Statistical Association.

and this introduces $p-1$ additional restrictions on the q-components, the dimensionality of the unit spherical region in the intermediate variables is $q-p$. Since the factor space has $q-p$ dimensions, it is desirable that there should be only $q-p$ coordinates instead of q, particularly since $p \geq 2$. Therefore we shall reparametrize the model $\mathbf{y} = \mathbf{V}\boldsymbol{\gamma} + \boldsymbol{\epsilon}$ in the intermediate variables [see Eq. (3.25)] to one of full rank. As in Section 3.3, we shall transform from the v's to new coordinates, the w's, where now the w's occupy a space of only $q-p$ dimensions.

In matrix notation, the first-degree model in the categorized mixture components is

$$\mathbf{y} = \mathbf{X}_c\boldsymbol{\beta} + \boldsymbol{\epsilon} \tag{4.44}$$

where $\mathbf{y}$ is the $N \times 1$ vector of observations, $\mathbf{X}_c$ is an $N \times q$ matrix of rank $q-p$ and with elements $x_{ui} - x_{0i}$ $(1 \le u \le N, 1 \le i \le q)$, $\boldsymbol{\beta}$ is a $q \times 1$ vector of unknown parameters and $\boldsymbol{\epsilon}$ is an $N \times 1$ vector of random errors. In the intermediate variables, Eq. (4.44) is

$$\mathbf{y} = \mathbf{V}\boldsymbol{\gamma} + \boldsymbol{\epsilon} \tag{4.45}$$

where $\mathbf{V} = \mathbf{X}_c\mathbf{H}^{-1}$, $\boldsymbol{\gamma} = \mathbf{H}\boldsymbol{\beta}$ and $\mathbf{H}$ is the diagonal matrix $\mathbf{H} = \text{diag}(h_1, h_2, \ldots, h_q)$. To get to the system in the w_i's, we choose a $q \times q$ orthogonal matrix $\mathbf{T}$ (as in Eq. (3.26)) so that

$$\mathbf{VT} = [\mathbf{W} \quad \mathbf{0}] \tag{4.46}$$

where $\mathbf{W}$ is an $N \times (q-p)$ matrix of rank $q-p$ and $\mathbf{0}$ is an $N \times p$ matrix of zeros. The form of the transformation matrix $\mathbf{T}$ in Eq. (4.46) is similar to the $\mathbf{T}$ described previously in Eqs. (3.26)–(3.27) except that in partitioning $\mathbf{T} = [\mathbf{T}_1, \mathbf{T}_2]$ now, the matrix $\mathbf{T}_1$ is $q \times (q-p)$, $\mathbf{T}_2$ is $q \times p$ and the sufficient conditions on the matrix $\mathbf{T}$ are that $\mathbf{T}$ is orthogonal and $\mathbf{VT}_2 = \mathbf{0}$. A derivation of the form for the matrix $\mathbf{T}$ is provided in Appendix B of Cornell and Good (1970).

The model in Eq. (4.45) can be expressed as

$$\begin{aligned} \mathbf{y} &= \mathbf{VTT}'\boldsymbol{\gamma} + \boldsymbol{\epsilon} \\ &= \mathbf{VT}_1\mathbf{T}_1'\boldsymbol{\gamma} + \boldsymbol{\epsilon} \\ &= \mathbf{W}\boldsymbol{\alpha} + \boldsymbol{\epsilon} \end{aligned} \tag{4.47}$$

since $\mathbf{VT}_2 = \mathbf{0}$ and if we let $\mathbf{VT}_1 = \mathbf{W}$ and $\mathbf{T}_1'\boldsymbol{\gamma} = \boldsymbol{\alpha}$. The matrix $\mathbf{W}$ is an $N \times (q-p)$ matrix which contains the levels of the $q-p$ independent variables over all N experiments and $\boldsymbol{\alpha}$ is a $(q-p) \times 1$ vector of unknown parameters. Once the estimates $\mathbf{a}$ of $\boldsymbol{\alpha}$ are obtained, the estimates $\mathbf{g}$ of the coefficients in the model (4.45) in the intermediate variables are obtained by placing p constraints on the $\mathbf{g}$. This is done by insisting that $\mathbf{T}_2'\mathbf{g} = \mathbf{0}$, so that

$$\begin{aligned} \mathbf{T}'\mathbf{g} &= \mathbf{T}_1'\mathbf{g} + \mathbf{T}_2'\mathbf{g} = [\mathbf{T}_1'\mathbf{g}, \mathbf{0}] = [\mathbf{a}, \mathbf{0}] \\ \mathbf{g} &= \mathbf{TT}'\mathbf{g} = \mathbf{T}[\mathbf{a}, \mathbf{0}] = \mathbf{T}_1\mathbf{a} \end{aligned} \tag{4.48}$$

Finally, the estimates $\mathbf{b}$ of the coefficients in the model (4.44) in the categorized mixture components are obtained from the equations $\mathbf{b} = \mathbf{H}^{-1}\mathbf{g}$.

As in Section 3.3, the w_i's are used in the construction of designs as well as the fitting of the model to the observed values of the response over the region of interest. The settings of the mixture components corresponding to the settings of the levels of the w_i's are determined, as in Eqs. (3.29) and (3.30), from Eq. (4.46), with $\mathbf{D}_w$ used for $\mathbf{W}$ in

$$\mathbf{X}_c = [\mathbf{W}, \mathbf{0}]\mathbf{T}'\mathbf{H} = \mathbf{W}\mathbf{T}_1'\mathbf{H} \tag{4.49}$$

4.12. A NUMERICAL EXAMPLE OF A CATEGORIZED COMPONENT EXPERIMENT

In the production of Polyethylene Terephthalate for the manufacturing of polyester fibers, the response to be studied is the result of blending one or more acid salts with one or more glycols. As was mentioned in the previous section, the chemist might wish to learn how the fiber's properties are affected by modifying the current manufacturing conditions and so we assume the current conditions are altered by the addition of other acids and/or other glycols.

For purposes of simplicity, let us assume that the blends are the result of mixing one or two acids x_1 and x_2 with one or two glycols x_3 and x_4. A description of the shape of the response surface using a prediction equation is sought so that future predictions of the response can be made with these four components. Also, a contour plot of the response surface will be drawn, since this enables one to visually picture the surface over the region of interest.

A quadratic model is assumed to be adequate for modeling the response surface in the ellipsoidal region

$$\sum_{i=1}^{4} \left(\frac{x_i - \frac{1}{4}}{\frac{1}{4}} \right)^2 \leq 1$$

centered at $\mathbf{x}_0 = (\frac{1}{4}, \frac{1}{4}, \frac{1}{4}, \frac{1}{4})'$ which is also the centroid of the simplex. The components are of equal importance and are thus given equal spread so that $\mathbf{H} = \text{diag}(\frac{1}{4}, \frac{1}{4}, \frac{1}{4}, \frac{1}{4})$. Since a surface contour plot over the region of interest is desired, the largest possible region will be used for this example. The design points are placed on the boundary of the largest spherical region centered at $\mathbf{x}_0$, or at $\mathbf{w} = \mathbf{0}$, that will fit inside the factor space.

The formula for the radius of the largest spherical region centered at $\mathbf{w} = \mathbf{0}$ that will fit inside the polytope is identical to the formula (3.33), and therefore the radius ρ^* of the largest spherical region for our example is

$$\rho^* = \left\{ 1 + \frac{(\frac{1}{4})^2}{(\frac{1}{4})^2} \right\}^{1/2} = \sqrt{2}$$

Setting the points on the perimeter of the largest sphere is equivalent to choosing the value of the radius multiplier c in $\mathbf{D}_1$ of $\mathbf{D}_w$ in (3.39) to be $c = \sqrt{2}/\sqrt{2} = 1.0$. If we select a design matrix $\mathbf{D}_w$ of the form shown in (3.39), where the value of g is $g = \sqrt{2}$ from Eq. (3.40), then the matrix $\mathbf{W}_A$ for a central composite rotatable design with four center point replicates is of the form

$$\mathbf{W}_A = \begin{bmatrix} 1 & \overbrace{-1 & -1}^{\mathbf{D}_w} & 1 & 1 & 1 \\ 1 & 1 & 1 & 1 & 1 & 1 \\ 1 & -1 & 1 & 1 & 1 & -1 \\ 1 & 1 & -1 & 1 & 1 & -1 \\ 1 & -1.414 & 0 & 2 & 0 & 0 \\ 1 & 1.414 & 0 & 2 & 0 & 0 \\ 1 & 0 & -1.414 & 0 & 2 & 0 \\ 1 & 0 & 1.414 & 0 & 2 & 0 \\ 1 & 0 & 0 & 0 & 0 & 0 \\ 1 & 0 & 0 & 0 & 0 & 0 \\ 1 & 0 & 0 & 0 & 0 & 0 \\ 1 & 0 & 0 & 0 & 0 & 0 \end{bmatrix}$$

where the entries in each row of $\mathbf{W}_A$ are the values of $\mathbf{w}_u = (1, w_{u1}, w_{u2}, w_{u1}^2, w_{u2}^2, w_{u1}w_{u2})'$, $u = 1, 2, \ldots, 12$.

To obtain the settings of the mixture components corresponding to the design settings in w_1 and w_2, we require the form of the transformation matrix $\mathbf{T}$. If the elements of $\mathbf{T}$ are selected arbitrarily (or as shown in Appendix 4A), then one possible form of $\mathbf{T}$ is

$$\mathbf{T} = \begin{bmatrix} 0.707 & -0.707 & 0 & 0 \\ 0.707 & 0.707 & 0 & 0 \\ 0 & 0 & -0.707 & 0.707 \\ 0 & 0 & 0.707 & 0.707 \end{bmatrix}$$

and the settings of the mixture components are determined from Eq. (4.49) where $\mathbf{X}_c = \mathbf{D}_w\mathbf{T}_1'\mathbf{H}$ and where the ujth element of $\mathbf{X}_c$ is $x_{uj} - \frac{1}{4}$. In Table 4.10 are presented the values of the mixture component proportions at the design points and the corresponding observed value of the response at the design point designations, respectively.

From a least squares analysis, the prediction equation in the w's is

$$\begin{aligned}\hat{y}(\mathbf{w}) = 7.80 &+ 0.84w_1 - 1.49w_2 - 0.11w_1^2 + 0.87w_2^2 + 0.25w_1w_2 \\ (0.10)\ &\ (0.07)\qquad (0.07)\qquad (0.08)\qquad (0.08)\qquad (0.10)\end{aligned} \tag{4.50}$$

where the quantities in parentheses are the estimated standard errors of

TABLE 4.10. Design coordinates, component proportions, and the observed response values at the design points

Design Point	w_1	w_2	x_1	x_2	x_3	x_4	y_u
1	−1	−1	0.427	0.073	0.427	0.073	9.3
2	1	1	0.073	0.427	0.073	0.427	8.2
3	−1	1	0.427	0.073	0.073	0.427	6.1
4	1	−1	0.073	0.427	0.427	0.073	10.4
5	−1.414	0	0.500	0.000	0.250	0.250	6.4
6	1.414	0	0.000	0.500	0.250	0.250	8.9
7	0	−1.414	0.250	0.250	0.500	0.000	11.9
8	0	1.414	0.250	0.250	0.000	0.500	7.3
9	0	0	0.250	0.250	0.250	0.250	7.7
10	0	0	0.250	0.250	0.250	0.250	7.9
11	0	0	0.250	0.250	0.250	0.250	7.8
12	0	0	0.250	0.250	0.250	0.250	7.8

the coefficients based on the value of the standard deviation per observation being $s = 0.2$. The analysis of variance table is presented as Table 4.11 and because the value of the F-ratio is 145.00 which greatly exceeds the tabled $F_{(5,6,\alpha=0.01)} = 8.75$, we say the model (4.50) adequately fits the data. The prediction equation can also be expressed in the mixture components as

$$\begin{aligned}\hat{y}(\mathbf{x}) = \underset{(0.10)}{7.80} + \underset{(0.19)}{2.38}(x_2 - x_1) - \underset{(0.19)}{4.21}(x_4 - x_3) - \underset{(0.61)}{0.86}(x_2 - x_1)^2 + \underset{(0.61)}{6.95}(x_4 - x_3)^2 \\ + \underset{(0.77)}{1.99}(x_2 - x_1)(x_4 - x_3) \qquad (4.51)\end{aligned}$$

TABLE 4.11. Analysis of variance table for categorized component example

Source of Variation	Degrees of Freedom	Sum of Squares	Mean Square	F
Regression	5	28.99	5.80	145.00
Residual	6	0.22	0.04	
Total	11	29.21		

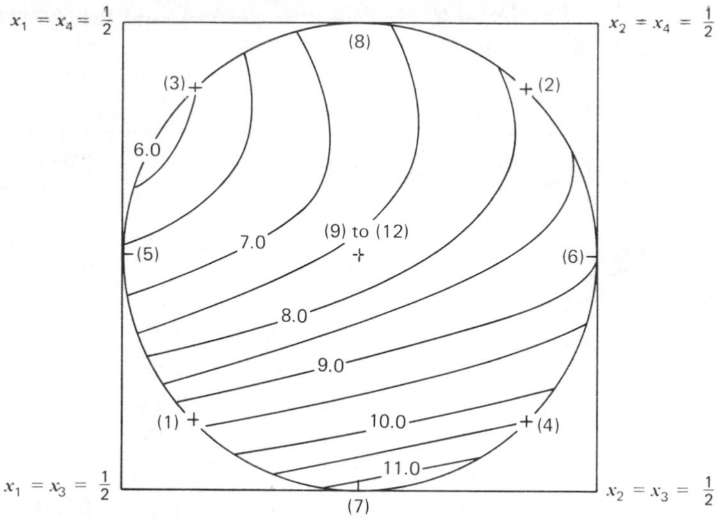

FIGURE 4.9. Response contours over the region of interest. Reproduced from Cornell and Good (1970), with permission of the American Statistical Association.

where the quantities in parentheses below the coefficient estimates are the estimated standard errors of the coefficient estimates. Equation (4.51) can be obtained directly from Eq. (4.50) by substituting the expressions $w_1 = 2\sqrt{2}(x_2 - x_1)$ and $w_2 = 2\sqrt{2}(x_4 - x_3)$ for w_1 and w_2 directly into Eq. (4.50). One can also use the y_u data from Table 4.10 fitted to the values of $x_2 - x_1$ and $x_4 - x_3$ with the model (4.51). The response contours with either Eq. (4.50) or Eq. (4.51) are plotted in Figure 4.9.

Exercise 4.1 Can you recommend values of x_1, x_2, x_3, and x_4 from Eq. (4.51) that would produce estimated response values in the range $9.0 \leq \hat{y}_u \leq 10.0$. Is the model in Eq. (4.50) easier to use than Eq. (4.51) for prediction purposes?

4.13. SUMMARY OF CHAPTER 4

In this chapter, additional constraints on the component proportions in the form of upper bounds, lower bounds, or both were considered. Pseudocomponents were introduced for those cases where lower bounds were placed on the proportions of some or all of the components. It was shown how the use of pseudocomponents can simplify the design selection procedure, even when lower bounds of different sizes are placed on the original component proportions, by illustrating the ease with which the $\{q, m\}$ simplex-lattice and simplex-centroid design can be set up in the pseudocomponent system.

When both upper and lower bound constraints are placed on some or all of the component proportions, the factor space takes the form of a convex polyhedron which in most cases is more complicated than the simplex. To locate the vertices of the polyhedra where some of the vertices can be used as design points for fitting first- and second-degree models, several algorithms have been suggested. The operational steps of two algorithms are presented and one, the XVERT algorithm, is illustrated with a gasoline-blending experimental setting.

When constraints are placed on the component proportions and in particular when the ranges $b_i - a_i$ of some of the x_i are small, the coefficient estimates in the Scheffé models may possess large variances. Rewriting the models in pseudocomponents (Gorman, 1970) may relieve the large variance property of the estimates in some instances while still another approach is to write the model in slack-variable form. The use of a "slack" variable is discussed by Marquardt and Snee (1974).

Combining categories of components is discussed in the last three sections. Multiple-lattice designs and the associated models are available for exploring all of the factor space. When only a subregion of the factor space is of interest, the subregion might take the form of an ellipsoidal region and the use of mathematically independent variables is an alternative course of action. The transformation to independent variables is illustrated with an example taken from a chemical industrial setting in the manufacture of polyester fibers.

4.14. REFERENCES AND RECOMMENDED READING

Cornell, J. A. and I. J. Good (1970). The mixture problem for categorized components. *J. Am. Stat. Assoc.*, **65**, 339–355.

Gorman, J. W. (1970). Fitting equations to mixture data with restraints on compositions. *J. Qual. Technol.*, **2**, 186–194.

Kurotori, I. S. (1966). Experiments with mixtures of components having lower bounds. *Ind. Qual. Control*, **22**, 592–596.

Lambrakis, D. P. (1968). Experiments with mixtures: A generalization of the simplex-lattice design. *J. R. Stat. Soc., B*, **30**, 123–136.

Lambrakis, D. P. (1969). Experiments with mixtures: Estimated regression function of the multiple-lattice design. *J. R. Stat. Soc., B*, **31**, 276–284.

Lund, R. E. (1975). Tables for an approximate test for outliers in linear models. *Technometrics*, **17**, 473–476.

Marquardt, D. W. and R. D. Snee (1974). Test statistics for mixture models. *Technometrics*, **16**, 533–537.

McLean, R. A. and V. L. Anderson (1966). Extreme vertices design of mixture experiments. *Technometrics*, **8**, 447–454.

Saxena, S. K. and A. K. Nigam (1977). Restricted exploration of mixtures by symmetric-simplex designs. *Technometrics*, **19**, 47–52.

Scheffé, H. (1958). Experiments with mixtures. *J. R. Stat. Soc., B*, **20**, 344–360.

Snee, R. D. (1975). Experimental designs for quadratic models in constrained mixture spaces. *Technometrics*, **17**, 149–159.

Snee, R. D. (1976). Developing models for mixture systems when the experimental region is restricted. Presented at the Gordon Research Conference on Statistics in Chemistry and Chemical Engineering, New Hampton, NH.

Snee, R. D. (1979). Experimental designs for mixture systems with multicomponent constraints. *Commun. Stat.*, **A8**, No. 4, 303–326.

Snee, R. D. and D. W. Marquardt (1974). Extreme vertices designs for linear mixture models. *Technometrics*, **16**, 399–408.

QUESTIONS FOR CHAPTER 4

4.1. In a three-component system, lower bounds for components 1, 2, and 3 are $a_1 = 0.10$, $a_2 = 0.15$, and $a_3 = 0.0$, respectively. Set up pseudocomponents x'_i, $i = 1$, 2, and 3 and construct a simplex-centroid design in the pseudocomponents. List the blending proportions in the original components corresponding to the simplex-centroid combinations in the pseudocomponents.

4.2. In the tropical beverage data of Table 4.2, the six blends numbered 1, 2, 3, 11, 12, and 13 each contain 80% watermelon. If we are asked to make final recommendations of a tropical beverage that contains exactly 80% watermelon, what suggestions do you have for the remaining 20% in terms of orange; pineapple; and grapefruit in order to produce a high flavor score?

4.3. Define upper bounds for the components 1, 2, and 3 in Exercise 4.1 to be $x_1 \leq 0.70$, $x_2 \leq 0.75$, and $x_3 \leq 1.0$. Use the symmetric-simplex coordinates for z_1, z_2, and z_3 in Table 4.7, and construct a set of mixture blends for fitting the model $y = \beta_1 x_1 + \beta_2 x_2 + \beta_3 x_3 + \epsilon$. Estimate the parameters β_i in the model if the observed response values at the points 1, 2, 3, 4, 5, and 6 are $y_1 = 22.4$, $y_2 = 9.2$, $y_3 = 20.4$, $y_4 = 16.5$, $y_5 = 18.6$, and $y_6 = 10.7$.

4.4. The minor components from three M-components are to be combined. In M-component 1 are the two m-components denoted by X_{11} and X_{12} where $c_1 = \frac{1}{2}$. In M-component 2 are the three m-components X_{21}, X_{22}, and X_{23} and $c_2 = \frac{1}{4}$, and in M-component 3 are the two m-components X_{31} and X_{32} where $c_3 = \frac{1}{4}$. List all of the blends that comprise the $\{2, 3, 2; m_1, m_2, m_3\}$ triple lattice where $m_1 = 1$, $m_2 = 2$, and $m_3 = 2$.

4.5. Three categories (alkalines, crystalines and additives) of components are to be blended together. The alkalines (x_1 and x_2) are to comprise 50% of the mixtures, the crystalines (x_3 and x_4) 35% and the additives (x_5 and x_6) 15% of the mixture. If it is felt the response of interest is affected in a quadratic manner by changing the proportions of the components in each of the categories.

a. List the settings of the component proportions that make up the 27 blends corresponding to the triple lattice $\{2, 2, 2; 2, 2, 2\}$.

b. Let us define an ellipsoidal region of interest $R = (\mathbf{x} - \mathbf{x}_0)'\mathbf{H}^{-1}(\mathbf{x} - \mathbf{x}_0)$ whose center is $\mathbf{x}_0 = (x_{01}, x_{02}, x_{03}, x_{04}, x_{05}, x_{06})' = (0.25, 0.25, 0.20, 0.15, 0.07, 0.08)'$. The spread of the intervals of interest are defined in the matrix $\mathbf{H} = \text{diag}(0.25, 0.25, 0.15, 0.15, 0.07, 0.07)$. A transformation like Eq. (4.46) to independent variables is made and a second-order rotatable design in the independent variable system is to be used. List the mixture component proportions to be run if the design point settings are located on the boundary of the largest sphere centered at $\mathbf{x}_0$.

APPENDIX 4A. AN ORTHOGONAL MATRIX FOR THE CATEGORIZED COMPONENTS PROBLEM

For the transformation in Eq. (4.46) where the intermediate variables in the $N \times q$ matrix $\mathbf{V}$ are transformed to the independent variables in the $N \times (q-p)$ matrix $\mathbf{W}$, the $q \times q$ orthogonal matrix $\mathbf{T}$ was partitioned as $\mathbf{T} = [\mathbf{T}_1 \quad \mathbf{T}_2]$ so that $\mathbf{V}\mathbf{T}_2 = \mathbf{0}$ where $\mathbf{T}_2$ is $q \times p$. Given the form of the matrix $\mathbf{T}_2$ for general p, let us construct a $q \times (q-p)$ matrix $\mathbf{T}_1$ so that the $q \times q$ matrix $\mathbf{T}$ is orthogonal.

Let us write the matrix $\mathbf{T}$ in the form

$$\mathbf{T} = \begin{bmatrix} [\;] & 0 & \cdots & 0 & \mathbf{h}^{(1)} & 0 & \cdots & 0 \\ 0 & [\;] & & 0 & 0 & \mathbf{h}^{(2)} & & 0 \\ \vdots & & \ddots & \vdots & \vdots & & \ddots & \vdots \\ 0 & 0 & \cdots & [\;] & 0 & 0 & \cdots & \mathbf{h}^{(p)} \end{bmatrix} = [\mathbf{T}_1 \quad \mathbf{T}_2] \tag{4A.1}$$

where the lth matrix in the diagonal of $\mathbf{T}_1$ has n_l rows and $n_l - 1$ columns. The last p columns of the matrix $\mathbf{T}$ comprise the matrix $\mathbf{T}_2$. Since the number of components in the lth category is n_l where from Eq. (4.41) the range is defined as $S_{l-1} + 1 \le x_j \le S_l$, then the elements of the $n_l \times 1$ vector $\mathbf{h}^{(l)}$ in $\mathbf{T}_2$ are

$$\tilde{h}_j = \frac{h_j}{h^{l*}} \qquad \begin{array}{c} S_{l-1} + 1 \le j \le S_l \\ 1 \le l \le p \end{array}$$

where $h^{l*} = \{\sum_{m=S_{l-1}+1}^{S_l} h_m^2\}^{1/2}$, $1 \le l \le p$. For example, the elements of the vector $\mathbf{h}^{(3)}$ are $\mathbf{h}^{(3)} = (\tilde{h}_{S_2+1}, \tilde{h}_{S_2+2}, \ldots, \tilde{h}_{S_3})'$.

Denote by t_{ij} the element in the ith row and jth column of the matrix $\mathbf{T}_1$. In the lth matrix in the diagonal of the matrix $\mathbf{T}_1$, the elements are

$$
\begin{aligned}
t_{ij} &= h_{i+1}, \quad i = S_{l-1}+1; \quad j = S_{l-1}+2-l \\
t_{ij} &= -h_{i-1}, \quad i = S_{l-1}+2; \quad j = S_{l-1}+2-l \\
t_{ij} &= h_i h_{j+l-1}, \quad i = S_{l-1}+1, \ldots, j+l-1; \quad j = S_{l-1}+3-l, \ldots, S_l - l \\
t_{ij} &= 0, \quad i = j+l+1, \ldots, S_l; \quad j = S_{l-1}+2-l, \ldots, S_l - l - 1 \\
t_{ij} &= -\left(\sum_{f=S_{l-1}+1}^{j+l-1} h_f^2 \right), \quad i = j+l; \quad j = S_{l-1}+3-l, \ldots, S_l - l
\end{aligned}
\tag{4A.2}
$$

Normalizing the columns in Eq. (4A.2) gives the matrix $\mathbf{T}_1$. For example, the third matrix ($l=3$) down the diagonal of the matrix $\mathbf{T}_1$, before column-normalizing is of the form

$$
\begin{bmatrix}
h_{S_2+2} & h_{S_2+1}h_{S_2+3} & h_{S_2+1}h_{S_2+4} & \cdots & h_{S_2+1}h_{S_3} \\
-h_{S_2+1} & h_{S_2+2}h_{S_2+3} & h_{S_2+2}h_{S_2+4} & \cdots & h_{S_2+2}h_{S_3} \\
0 & -\sum_{f=1}^{2} h_{S_2+f}^2 & h_{S_2+3}h_{S_2+4} & \cdots & h_{S_2+3}h_{S_3} \\
0 & 0 & -\sum_{f=1}^{3} h_{S_2+f}^2 & & h_{S_2+4}h_{S_3} \\
0 & 0 & 0 & & \vdots \\
 & & & \ddots & \\
\vdots & \vdots & \vdots & & \\
0 & 0 & 0 & \cdots & -\sum_{f=1}^{S_3-S_2-1} h_{S_2+f}^2
\end{bmatrix}
$$

CHAPTER 5

The Analysis of Mixture Data

In Section 2.6, the results of a blending experiment were presented in terms of how the polymers polyethylene, polystyrene, and polypropylene behaved singly and in combination as measured by the thread elongation of spun yarn. Briefly, a second-degree model was fitted to observed elongation values that were collected from blends of the polymers specified by the points of a $\{3, 2\}$ simplex-lattice design. Synergistic blending was assumed to be present between polyethylene and each of the other constituents because high elongation values were observed on blends containing polyethylene and a high elongation value was considered to be desirable. Polystyrene and polypropylene were suspected of being antagonistic.

In Section 2.12, the results of an experiment were discussed in which four chemical pesticides had been sprayed on strawberry plants in an attempt to control the number of mites on the plants. The data were collected at the points of a simplex-centroid design. The 15-term model seemed to adequately fit the response values, although neither an analysis of variance table nor an F-test for the fitted model was provided. The objective of the fitted model exercise was to see if any of the multiple component blends were as effective as the pure chemicals. Several two-component blends appeared to be as effective as the pure chemicals.

In Section 4.5, the experimental region was constrained from the placing of an upper bound on one of the component proportions and formulas for estimating the model parameters were presented for the constrained region problem. In the constrained space, the fitting of the model in terms of estimating the model parameters was illustrated using data from an experiment involving the formulation of a tropical beverage. Juices from watermelon (x_1), orange (x_2), pineapple (x_3), and grapefruit (x_4) were combined to make the tropical beverage. The 10-term prediction equation (4.19) for sensory flavor score was felt to be satisfactory in fitting the flavor surface because, when the residual mean square was compared to the pure error mean square from the analysis of variance table, Table 4.3, the two mean squares (variance estimates) were not significantly different. In the remainder of Chapter 4, methods were presented for analyzing data when additional restrictions were placed on the component proportions.

In this chapter, several additional techniques used in the analysis of mixture data will be discussed. The model employed will be the Scheffé-type canonical polynomials. In Chapter 6, alternative models are presented.

5.1. TECHNIQUES USED IN THE ANALYSIS OF MIXTURE DATA

Let us begin by imagining that we are faced with the task of analyzing a set of data from a mixture experiment. Initially we ask ourselves whether the objective is the fitting of some proposed model for the purpose of describing the shape of the response surface over the simplex factor space or whether we are more interested in determining the roles played by (that is, measuring the effects of) the individual components. Most of the time both objectives can be attained from the same analysis. We illustrate the partial attaining of both objectives many times in this chapter in our discussions of the results of the analysis of several data sets.

To aid us in presenting the methodology of this section, we make use of the three sets of experimental data listed in Table 5.1. These data sets are subsets of a larger, more complete set of data that was generated during a large scale experiment. The larger experiment is described in greater detail in Section 5.12.

The data values in Table 5.1 represent texture measurements which were taken on fish patties that had been formulated by blending three fish species. The species were mullet (x_1), sheepshead (x_2), and croaker (x_3). The unit of measure of the texture data is scaled (scaled value equals actual value times 10^{-3}), to facilitate the handling of the data numbers. The texture values in sets I, II, and III are listed as pairs of numbers that were collected from replicate patties. The component blends (component proportions) correspond to a simplex-centroid design. Each set (I, II, and III) of patties was prepared according to a specified cooking temperature and cooked for a specified length of time. The cooking temperatures and times differed for the three sets of patties.

The proposed model to which the observations in each of data sets I, II, and III will be fitted to initially is the Scheffé special cubic model

$$\eta = \sum_{i=1}^{3} \beta_i x_i + \sum_{i<j}^{3} \sum \beta_{ij} x_i x_j + \beta_{123} x_1 x_2 x_3 \tag{5.1}$$

TABLE 5.1. Three sets of texture measurements taken on duplicate fish patties where each set of patties was processed (prepared and cooked) differently from the patties in the other sets

Fish Component Proportions			Texture Readings (grams $\times 10^{-3}$ of force required to puncture the patty surface)					
x_1	x_2	x_3	Set I		Set II		Set III	
1	0	0	1.98	1.70	3.12	2.89	2.34	2.30
0	1	0	0.68	0.67	1.18	1.24	0.97	0.97
0	0	1	1.53	1.48	2.36	2.27	2.11	2.13
$\frac{1}{2}$	$\frac{1}{2}$	0	1.18	1.40	1.96	1.90	1.48	1.43
$\frac{1}{2}$	0	$\frac{1}{2}$	1.45	1.39	2.66	2.48	1.80	2.06
0	$\frac{1}{2}$	$\frac{1}{2}$	1.19	1.12	1.80	1.86	1.21	1.34
$\frac{1}{3}$	$\frac{1}{3}$	$\frac{1}{3}$	1.65	1.54	2.09	1.79	1.53	1.56

The special cubic model in Eq. (5.1) is the simplex-centroid model of Eq. (2.35) when $q = 3$. The purpose behind fitting the special cubic equation to each data set is to illustrate a good fit to the data of set I as well as an overfit of the data in sets II and III. By an overfit is meant that some of the terms in Eq. (5.1) are not necessarily needed when describing the texture surface and therefore the terms may be deleted. The overfit will be discovered through the testing of hypotheses which specify zero values for some of the parameters in the model. In other words, the dropping of terms from the complete model is analogous to accepting an hypothesis that states that certain terms in Eq. (5.1) are equal to zero and thus are unimportant.

Before we discuss the testing of hypotheses concerning specific parameters in the model, we shall review very briefly the strategy employed in the asking of questions about which of the component effects are likely to be present in the data. The anticipated answers to the questions prompted us to try the model form presented by Eq. (5.1).

In trying to decide on the particular form of the model to be fitted to data collected at the points of a $q = 3$ simplex-centroid design, we first recognize that the special cubic model or simplex-centroid model of Eq. (5.1) is chosen over the lower-degree models because the terms in the special cubic model not only provide a measure of each pure blend, but provide measures of the binary blends and a measure of the three-component blend as well. Nevertheless we mention the following important and relevant questions which probably were in the minds of the experimenters prior to performing the experimental runs and collecting the data at the composition points.

Q: Is the response (fish patty texture) surface likely to be planar over the triangle or are combinations of the fish expected to cause the surface to be nonplanar, that is, cause departures from linearity in the surface shape? If the blending of multiple-component mixtures is not additive which pairs of components are likely to have synergistic effects, antagonistic effects? Are complete blends (three-fish blends) likely to be firmer (have higher texture values) than the binary or pure blends? Also, if we assume the texture surface is planar, should the texture values be collected at the pure blends (vertices of the triangle) only or should we use complete mixtures consisting of all three fish types but which are very nearly located at the vertices of the triangle, such as $x_i = 0.95$ and $x_j = x_k = 0.05/2$, $i, j, k = 1, 2$, and 3, $i \neq j \neq k$? Furthermore, even if we assume the surface is planar, shouldn't additional observations be collected at other locations inside of (or on the boundary of) the triangle to enable us to check our assumptions of a planar surface (or to check the adequacy of the model)?

Ans: If the objective is to model the response surface above the triangle, it is essential that further observations be taken to check whether the regression equation adopted adequately fits the response at points other than the blends that were used to obtain the fitted model. If the planar first-degree model is fitted to vertex points for example, one should collect additional observations at several interior points or at the midpoints of the boundaries or edges (enabling the continuation to the fitting the second-degree model if desired). This latter strategy of collecting midedge observations is advantageous in the sense that if a second-degree model is required, the estimates of the coefficients β_{ij}, $i < j$, in the second-degree model will have smaller variances than if interior points were used to estimate the β_{ij}. However, some experimenters might opt for interior points since interior points represent complete blends consisting of all of the components simultaneously whereas the edge or boundary points are binary or two-component blends.

Q: If the $\{3, 2\}$ simplex-lattice arrangement is chosen for the distinct possibility that the surface is not planar, should additional observations be collected at interior points of the triangle for the purpose of checking the fit inside the triangle? If so, is the centroid of the triangle the location at which to sample the interior?

Ans: If at all possible, additional interior points should be performed with the simplex-lattice designs. Now, if we can perform only one interior point in addition to the {3, 2} simplex-lattice, this ternary blend should be the centroid of the triangle. This blend has a higher power than any other single blend for testing the significance of the extra term $\beta_{123}x_1x_2x_3$ in the special cubic model.

Quite naturally, then, once a model of the form in Eq. (5.1) has been chosen even though it was felt initially that the assumption of a planar surface was very realistic, the next step is to collect a set of data, similar in form to each of sets I, II, or III, obtain the fitted model and proceed to scrutinize the fitted data as described in the next section. Note that when collecting observations an attempt should be made to collect replicate observations at some points of the design. The replicate observations enable one to obtain an estimate of the observation variance σ^2, from which estimates of the variances (and standard errors) of the estimated model parameters can be obtained. Tests on the sizes of the individual parameter estimates can then be performed.

We now discuss the construction of test statistics for the purpose of testing hypotheses concerning the usefulness of terms in the Scheffé models. Rather than sequentially build the model by starting with the first-degree polynomial and work towards the special cubic model, we shall begin with the complete special cubic model fitted to the data in each of sets I, II, and III and work backwards by testing the usefulness of the cubic term, the crossproduct or binary terms and finally we shall test the similarity of the linear terms in the model. This approach is taken because the data sets have already been collected and assembled in Table 5.1 and also because it enables us to illustrate the test procedures of the next section. The test statistics presented are discussed in greater detail in Marquardt and Snee (1974).

5.2. TEST STATISTICS FOR TESTING THE USEFULNESS OF THE TERMS IN THE SCHEFFÉ POLYNOMIALS

When the Scheffé polynomials are used to model the response surface as well as to provide measures of the blending characteristics of the components, most often the model will include *all* of the terms up to a given degree. Usually the final form is a polynomial which contains the q terms $\beta_i x_i$, $i = 1, 2, \ldots, q$, representing the very fundamental linear blending surface or additive blending among the components, and any additional terms of higher degree such as

$$\eta = \sum_{i=1}^{q} \beta_i x_i + \sum_{i<j}^{q}\sum \beta_{ij} x_i x_j + \cdots$$

With second- and higher-degree models, all pairs of components and/or all triplets of components up to degree d are considered. (One exception to this rule is the omission of terms like $\beta_{ijk}x_ix_jx_k$ in the quartic model presented in Appendix 2B.)

In choosing the degree of the final polynomial model so that predictions of the response surface can be made, tests of hypotheses are performed on groups of parameters in the polynomial model. A group may consist of a single parameter only but more likely, the group will involve two or more parameters. For example, let us refer to the data of set I in Table 5.1 and consider the fitting of the special cubic model to the data. The model is

$$\eta = \beta_1x_1 + \beta_2x_2 + \beta_3x_3 + \beta_{12}x_1x_2 + \beta_{13}x_1x_3 + \beta_{23}x_2x_3 + \beta_{123}x_1x_2x_3 \tag{5.2}$$

Initially we choose to test the null hypothesis (H_0), which states

H_0: The response *does not* depend on the mixture components (5.3)

against the alternative hypothesis (H_A), which states

H_A: The response *does* depend on the mixture components

When the null hypothesis is true, all three linear coefficients β_1, β_2, and β_3 are equal to some constant value (say β_0), and the remaining terms in the model of Eq. (5.2) are equal to zero. For Eq. (5.2), the null hypothesis (5.3) implies

$$H_0: \quad \beta_1 = \beta_2 = \beta_3 = \beta_0 \quad \text{and} \quad \beta_{12} = \beta_{13} = \beta_{23} = \beta_{123} = 0 \tag{5.4}$$

so that Eq. (5.2) is more appropriately written as

$$\eta = \beta_0x_1 + \beta_0x_2 + \beta_0x_3 = \beta_0 \tag{5.5}$$

According to Eq. (5.5), the surface above the triangle is of constant height, which is denoted by β_0, see Figure 5.1. The least squares estimate of β_0 in Eq. (5.5) is $b_0 = \Sigma_{u=1}^{N} y_u/N = \bar{y}$ where $\bar{y}$ is the average of all N observations collected over the simplex, that is, the estimated height of the surface for all blends is $\bar{y}$.

To test the null hypothesis stated in (5.3), the model in Eq. (5.2) is fitted to the data of set I and an F-ratio is set up

$$F = \frac{\text{SSR}/(p-1)}{\text{SSE}/(N-p)} \tag{5.6}$$

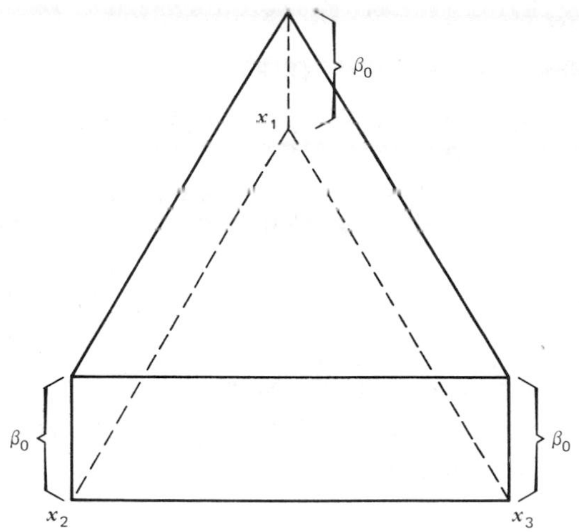

FIGURE 5.1. Response surface specified by the hypothesis H_0: $\beta_1 = \beta_2 = \beta_3 = \beta_0$, $\beta_{12} = \beta_{13} = \beta_{23} = \beta_{123} = 0$.

where $N = 14$, $p = 7$,

$$\begin{aligned} \text{SSR} &= \sum_{u=1}^{N=14} (\hat{y}_u - \bar{y})^2 \\ \text{SSE} &= \sum_{u=1}^{N=14} (y_u - \hat{y}_u)^2 \end{aligned} \tag{5.7}$$

In the sums of squares formulas, y_u is the value of the uth observation, $\hat{y}_u$ is the predicted value of the response corresponding to the uth observation where $\hat{y}_u$ uses the model in Eq. (5.2) with the parameter estimates substituted, and $\bar{y}$ is the average of the $N = 14$ observations. The degrees of freedom for the sources Regression and Error are $p - 1 = 7 - 1 = 6$ and $N - p = 14 - 7 = 7$, respectively, where p is the number of parameters that are estimated in the fitted model and N is the total number of observations. The value of the F-ratio in (5.6) is compared to the tabled value of $F_{(p-1, N-p, \alpha)}$ and the null hypothesis in (5.3) is rejected at the α-level of significance if the value of the F-ratio in (5.6) exceeds the tabled value.

Let us illustrate numerically the testing of the hypothesis in (5.3) by using the data in set I of Table 5.1. The least squares solutions to the normal equations, the fitted model with the estimated coefficient standard errors, the predicted values of the response at the design points, and the sums of squares quantities in Eq. (5.7), respectively, are given by

Fitted Model: Set I

$$\mathbf{b} = (\mathbf{X}'\mathbf{X})^{-1}\mathbf{X}'\mathbf{y}$$

$$\begin{bmatrix} 1.84 \\ 0.68 \\ 1.51 \\ 0.13 \\ -1.01 \\ 0.26 \\ 8.75 \end{bmatrix} = \begin{bmatrix} 0.5 & 0 & 0 & -1.0 & -1.0 & 0 & 1.5 \\ & 0.5 & 0 & -1.0 & 0 & -1.0 & 1.5 \\ & & 0.5 & 0 & -1.0 & -1.0 & 1.5 \\ & & & 12.0 & 2.0 & 2.0 & -30.0 \\ & \text{(same)} & & & 12.0 & 2.0 & -30.0 \\ & & & & & 12.0 & -30.0 \\ & & & & & & 594.0 \end{bmatrix} \begin{bmatrix} 7.45 \\ 4.86 \\ 6.65 \\ 1.00 \\ 1.06 \\ 0.93 \\ 0.12 \end{bmatrix}$$

$$\begin{aligned} \hat{y}(\mathbf{x}) &= (x_1,\ x_2,\ x_3,\ x_1x_2,\ x_1x_3,\ x_2x_3,\ x_1x_2x_3)'\mathbf{b} \\ &= \underset{(0.07)}{1.84x_1} + \underset{(0.07)}{0.68x_2} + \underset{(0.07)}{1.51x_3} + \underset{(0.36)}{0.13x_1x_2} - \underset{(0.36)}{1.01x_1x_3} + \underset{(0.36)}{0.26x_2x_3} + \underset{(2.52)}{8.75x_1x_2x_3} \end{aligned} \tag{5.8}$$

Estimated Response Values

$$\begin{aligned} \hat{y}(1,0,0) &= 1.84(1) + 0.68(0) + 1.51(0) + 0.13(1)(0) - 1.01(1)(0) \\ &\quad + 0.26(0)(0) + 8.75(1)(0)(0) \\ &= 1.84 \\ \hat{y}(0,1,0) &= 0.68 \\ \hat{y}(0,0,1) &= 1.51 \\ \hat{y}(\tfrac{1}{2},\tfrac{1}{2},0) &= 1.84(\tfrac{1}{2}) + 0.68(\tfrac{1}{2}) + 0.13(\tfrac{1}{2})(\tfrac{1}{2}) \\ &= 1.29 \\ \hat{y}(\tfrac{1}{2},0,\tfrac{1}{2}) &= 1.42 \\ \hat{y}(0,\tfrac{1}{2},\tfrac{1}{2}) &= 1.16 \\ \hat{y}(\tfrac{1}{3},\tfrac{1}{3},\tfrac{1}{3}) &= 1.60 \end{aligned}$$

$$\begin{aligned} \text{SSR} &= \sum_{u=1}^{14} (\hat{y}_u - \bar{y})^2 = (1.84 - 1.35)^2 + (1.84 - 1.35)^2 + (0.68 - 1.35)^2 \\ &\quad + \cdots + (1.60 - 1.35)^2 \\ &= 1.65 \end{aligned}$$

$$\begin{aligned} \text{SSE} &= \sum_{u=1}^{14} (y_u - \hat{y}_u)^2 = (1.98 - 1.84)^2 + (1.70 - 1.84)^2 + (0.68 - 0.68)^2 \\ &\quad + \cdots + (1.54 - 1.60)^2 \\ &= 0.077 \end{aligned}$$

For purposes of illustration, we are assuming the multiple observations per blend *in the sums of squares calculations* are replicates and not duplicates. This is because duplicate patties in all likelihood would not reflect a true measure of the error variance and we need an estimate of the error variance for our test. The value of the F-ratio (5.6) is

$$F = \frac{1.65/(7-1)}{0.077/7} = 25.0$$

and since $F = 25.0$ is greater than the tabled value $F_{(6,7,\alpha=0.01)} = 7.19$, we reject H_0 in Eq. (5.4) and conclude that the response *does* depend on the mixture components (that is, the magnitude of the texture or firmness measurements varies with the different fish combinations). The adjusted coefficient of determination, introduced in Section 7.6, can aid us in deciding if the model explains enough of the variation in the response values. Recalling the formulas

$$R_A^2 = 1 - \frac{\text{SSE}/(N-p)}{\text{SST}/(N-1)}$$

we find for the data in set I,

$$R_A^2 = 1 - \frac{0.077/7}{1.73/13} = 0.917$$

Since this value exceeds 0.90 which means that the error variance estimate is less than 10% of the total variance estimate, we admit to feeling confident in using the model for purposes of predicting the response.

A similar exercise in model fitting and in testing the hypothesis (5.4) with the data of sets II and III produced the following results.

Fitted Model: Set II

$$\hat{y}(\mathbf{x}) = \underset{(0.08)}{3.01x_1} + \underset{(0.08)}{1.21x_2} + \underset{(0.08)}{2.32x_3} - \underset{(0.41)}{0.71x_1x_2} - \underset{(0.41)}{0.36x_1x_3} + \underset{(0.41)}{0.27x_2x_3} - \underset{(2.87)}{3.99x_1x_2x_3} \qquad (5.9)$$

$$F = \frac{4.01/6}{0.097/7} = 48.2 \qquad \text{Reject } H_0 \text{ in Eq. (5.4)}$$

$$R_A^2 = 1 - \frac{0.097/7}{4.11/13} = 0.956$$

Fitted Model: Set III

$$\hat{y}(\mathbf{x}) = \underset{(0.06)}{2.32x_1} + \underset{(0.06)}{0.97x_2} + \underset{(0.06)}{2.12x_3} - \underset{(0.28)}{0.76x_1x_2} - \underset{(0.28)}{1.16x_1x_3} - \underset{(0.28)}{1.08x_2x_3} + \underset{(1.95)}{2.03x_1x_2x_3} \tag{5.10}$$

$$F = \frac{2.80/6}{0.045/7} = 72.6 \qquad \text{Reject } H_0 \text{ in Eq. (5.4)}$$

$$R_A^2 = 1 - \frac{0.045/7}{2.84/13} = 0.971$$

With each of the three sets of data, the null hypothesis which states that the texture value of the patties does not depend on the fish (i.e., is unaffected by the different fish combinations) is rejected. Following the rejection of H_0, we next try to see if all three of the fish types influence the patty texture, and if so, how? To this end, we look for the form of the polynomial model that fits the data values best. Before we proceed however, let us rework the data in set II and test the null hypothesis of (5.3) for the following cases. In case 1 the fitted model is quadratic and for case 2, the fitted model is linear or of the first degree.

Case 1. H_0: $\beta_1 = \beta_2 = \beta_3 = \beta_0$ and $\beta_{12} = \beta_{13} = \beta_{23} = 0$

$$\hat{y}(\mathbf{x}) = \underset{(0.09)}{3.02x_1} + \underset{(0.09)}{1.22x_2} + \underset{(0.09)}{2.33x_3} - \underset{(0.40)}{0.91x_1x_2} - \underset{(0.40)}{0.56x_1x_3} + \underset{(0.40)}{0.07x_2x_3} \tag{5.11}$$

$$F = \frac{3.98/5}{0.124/8} = 51.4$$

$$R_A^2 = 1 - \frac{0.124/8}{4.11/13} = 0.951$$

In case 1, since $F = 51.4$ exceeds $F_{(5,8,0.01)} = 6.63$ we reject H_0 and conclude that not only are the textures of the pure fish patties not the same but the textures of the two-fish patties appear to be different from the simple average of the respective textures of the single fish patties. The presence of both pure and binary blending effects is investigated in the next section on model reduction.

Case 2. H_0: $\beta_1 = \beta_2 = \beta_3 = \beta_0$

$$\hat{y}(\mathbf{x}) = \underset{(0.08)}{2.89x_1} + \underset{(0.08)}{1.16x_2} + \underset{(0.08)}{2.30x_3} \tag{5.12}$$

$$F = \frac{3.87/2}{0.23/11} = 92.5$$

$$R_A^2 = 1 - \frac{0.23/11}{4.11/13} = 0.934$$

As in case 1, for case 2 we reject H_0, since $F = 92.5 > F_{(2,11,0.01)} = 7.21$ and conclude that the texture of the patties made from each of the three individual fish types is not constant (i.e., the heights b_i of the plane at the vertices $x_i = 1$ are not the same). We can only infer at this point in the analysis that it appears as though the pure mullet (x_1) patties are firmer than the pure sheepshead (x_2) patties. The texture of the pure croaker (x_3) patties is somewhere between the textures of the others.

The question we ask ourselves now concerning the fitting of the special cubic models and the subsequent F-tests of hypothesis of the parameters in the models for data sets I, II, and III, respectively, is

"Granted the fit of the special cubic model to each of the three sets of data is significant in the sense that in each case we rejected the hypothesis that the mixture components do not influence the response. But the question we ask is: Is the special cubic model the only tool that can be used to describe the response or is it possible that a lower-degree model such as a quadratic model or even a linear first-degree model fitted to the data will describe the shape of the response surface over the factor space of the components as well as the special cubic model? If a lower-degree model does as well as the special cubic model in fitting the data, we probably would prefer to use the lower-degree model not only because it would be easier to handle when predicting the response but with a less complicated model many times it is easier to understand just how the components blend together." Also, the variance of the predicted values is less with a lower-degree model.

We now consider some techniques which can be used for reducing the form of the fitted model and again refer to the data in sets I, II, and III of Table 5.1 for illustrative purposes.

5.3. MODEL REDUCTION

In the previous section on testing hypotheses, the data in set II of Table 5.1 was fitted using a special cubic model, then a quadratic model was fitted and finally a first-degree linear model was fitted. In each case the hypothesis which stated that the texture of the patties did not depend on the fish components comprising the patties was rejected.

Now we should like to determine which of the three models does the best job of describing the texture response surface for set II. In other words, which model provides the clearest description of the shape of the texture surface while at the same time requiring the fewest number of terms in order to do so? Our feeling is that the simpler the form of the fitted model, the easier it is to use the model to predict the texture of blends, particularly of blends other than the blends that were used for fitting the model.

In choosing the simplest form of the fitted model, we must remember, however, that the simpler the form of the model the more limited is the range of usefulness of the model. If it is feared that the surface will be rather complicated in shape, normally we will not want to sacrifice the higher precision of a complete model just for simplicity of usage of the reduced model. However, for now let us assume that the surface is generated by a well-behaved system and as such the surface can be characterized by a second-degree polynomial at most. In other words, we are going to assume that a reduced model of low degree will do as good a job of describing the surface as a more complete higher-degree model.

Model reduction can be accomplished in several ways. The form of the model can be reduced by simply summing terms (i.e., summing the proportions x_i and x_j) or by setting up contrasts among the terms and using the contrasts as the new terms in the model. One might also delete terms from the model rather than combine the terms. If we use any of these approaches to come up with a reduced model form presumably then we have simplified the task of trying to interpret the mixture system.

The most obvious approach to model reduction is to remove the non-significant higher-degree terms. This can be illustrated using the data of set II in Table 5.1 by comparing the fit of the special cubic model in Eq. (5.9) to the fit of the second-degree model in Eq. (5.11). Briefly, the value of R_A^2 with the special cubic model is $R_A^2 = 0.956$, whereas with the quadratic model, the value is $R_A^2 = 0.951$. Since these values are close the usefulness of the extra cubic term $b_{123}x_1x_2x_3$ in Eq. (5.9) is questionable. In fact, if we check this result by comparing the value of the estimate $b_{123} = -3.99$ to its standard error by setting up the t-statistic, the value of the statistic $t = -3.99/2.87 = -1.39$, is not significantly different from zero and this supports our decision to remove or drop the cubic term from the fitted model.

A comparison can be made now between the fitted quadratic model in Eq. (5.11) and the fitted first-degree planar model in Eq. (5.12). With this latter model, the value of R_A^2 is $R_A^2 = 0.934$. The question that arises therefore is, "Is the increase of 0.017 in the value of R_A^2 (from 0.934 to 0.951) enough to justify the inclusion of the three second-degree terms in the fitted model?"

In order to help us answer this question, let us consider two other results. First, the increase in the estimate of the error variance when using the first-degree model rather than the quadratic model is $s_{\text{linear}}^2 - s_{\text{quadratic}}^2 = 0.021 - 0.016 = 0.005$. Although this difference represents nearly a 30% increase over the variance estimate s^2 with the quadratic model, the difference is less than half of the unit of measurement (grams $\times 10^{-3}$), and therefore is probably not significantly large. Secondly, the value of the t-statistic for testing H_0: $\beta_{12} = 0$ vs H_A: $\beta_{12} \neq 0$, is $t = -0.91/0.40 = -2.5$. This value is slightly larger than the tabled value $t_{8,0.025} = -2.306$, and as a result one might be inclined to feel that this is proof of the nonlinearity of the surface by the joint effects of components 1 and 2. However, this test result is slightly misleading because the parameter estimates are not independent (i.e., $b_{12} = 4\bar{y}_{12} - 2b_1 - 2b_2$) meaning that we are not sure that the less than zero value of b_{12} is due to the nonlinearity of the surface or because of the large size of b_1. Also, the estimate b_{12} is not independent of the estimates b_{13} and b_{23} as shown in Eq. (2.17). Since the estimates b_{12} and b_{13} are positively correlated and both contain negative signs, it is difficult to say one is significant and not the other.

In view of these arguments the planar model of Eq. (5.12) with only three terms is probably the model that would be chosen by most data analysts when fitting a polynomial to the data of set II. Model simplicity and lack of evidence of nonlinearity of the surface are probably the prevailing moods with this data set. Also, the residual plots of Figure 5.2 reflect very little difference in the three models. The same conclusion is reached even when the residuals are standardized by dividing them by their standard deviation. Nevertheless, before we proceed to discuss additional ways of reducing the number of terms in the model, we recall that the t-statistic was used previously for testing the significance of the parameter estimate b_{123} in the special cubic model. Such a test suffered from the same lack of independence as the test for the size of b_{12} in that in Eq. (5.9) the estimate b_{123} is correlated with b_i and b_{ij} in the amounts

$$\text{corr}(b_{123}, b_i) = \frac{3\sigma^2}{\sqrt{1188\sigma^2(\sigma^2)}} = \frac{1}{11.5}$$

$$\text{corr}(b_{123}, b_{ij}) = \frac{-60\sigma^2}{\sqrt{1188\sigma^2(24\sigma^2)}} = -\frac{1}{2.8}$$

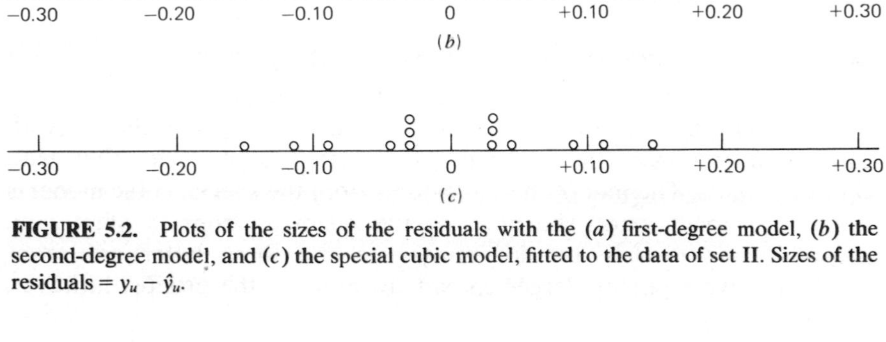

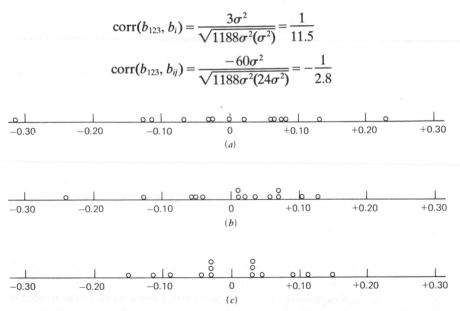

FIGURE 5.2. Plots of the sizes of the residuals with the (a) first-degree model, (b) the second-degree model, and (c) the special cubic model, fitted to the data of set II. Sizes of the residuals $= y_u - \hat{y}_u$.

Thus if the magnitudes of b_i are large and positive, while the b_{ij} are significantly less than zero, each or both of these results can contribute to make b_{123} appear to be different from zero or either or both can contribute to make b_{123} not different from zero.

A better procedure for measuring the significance of the parameter β_{123} is to measure the contribution of the term $\beta_{123}x_1x_2x_3$ to the model. (See Section 7.7.) This is done by first fitting the complete special cubic model and then calculating the regression sum of squares due to fitting the complete model as well as calculating an estimate of the error variance. Then the $\beta_{123}x_1x_2x_3$ term is removed from the proposed model, the reduced model is fitted and the regression sum of squares due to fitting the reduced model is calculated. The difference between the regression sum of squares due to the fitted models is compared to the estimate of the error variance obtained with the complete model.

Computationally, with the data of set II and the special cubic equation (5.9), we have for the sums of squares for regression and for error, respectively,

$$\text{SSR} = 4.01 \qquad \text{with 6 d.f.}$$

$$\text{SSE} = 0.097 \qquad \text{with 7 d.f.}$$

and with the quadratic model (5.11), we have

$$\text{SSR} = 3.98 \qquad \text{with 5 d.f.}$$

Taking the difference between the regression sums of squares with the two models and dividing the difference by the estimate s^2 obtained with the complete model, the F-ratio is

$$F = \frac{4.01 - 3.98}{0.097/7} = \frac{0.03}{0.014} = 2.0$$

Since $2.0 < F_{(1,7,\alpha=0.05)} = 5.59$, we conclude that the difference in the regression sums of squares with the two models (or owing to including the extra term $\beta_{123}x_1x_2x_3$) is not significant.

In Figure 5.2, the plots of the individual residuals with each of the models in Eqs. (5.9), (5.11), and (5.12) are presented. Since the special cubic model of Eq. (5.9), is forced through the average response at each design point the residuals at each point are highly correlated (with duplicated responses the residuals are reflections of each other with opposite signs). The plot of these residuals from the special cubic model is symmetric about zero. The plots of the residuals from the first- and second-degree models are not symmetric and these latter groups of residuals have a slightly larger spread. In none of the groups, however, are the sizes of the residuals correlated with the values of y_u or $\hat{y}_u$ nor are the sizes of the residuals related to the values of the x_i. Similar results appear with standardized residuals.

Another way of reducing the size of the model is to combine some of the terms in the model. The combining of the terms is achieved by summing the proportions of those components whose coefficients are approximately equal in magnitude. In deciding on which terms to combine, normally one looks for coefficient estimates of the first-degree terms in the fitted model which are approximately equal in magnitude. For example, if with the fitted model

$$\hat{y}(\mathbf{x}) = b_1x_1 + b_2x_2 + b_3x_3$$

the estimates b_1 and b_2 are approximately equal and the ranges of values for x_1 and x_2 are approximately equal, then the first two terms might be combined to produce the reduced model form

$$\eta = \beta_1(x_1 + x_2) + \beta_3x_3$$

which is now a two-component-only model where the components are $x_1 + x_2$ and x_3.

5.4. AN EXAMPLE OF REDUCING THE SYSTEM FROM THREE TO TWO COMPONENTS

To illustrate the reduction of the number of terms in a three-component quadratic model, let us introduce the following data set where the values represent average quality ratings (rating scale 1–8 but these averages are listed in units of 0.01 for illustrative purposes because we want to use continuous data) assigned to replicate yarn samples produced from the blends specified by a $\{3, 2\}$ simplex-lattice design. The data are

$(x_1, x_2, x_3) =$	$(1, 0, 0)$	$(\frac{1}{2}, \frac{1}{2}, 0)$	$(0, 1, 0)$	$(\frac{1}{2}, 0, \frac{1}{2})$	$(0, 0, 1)$	$(0, \frac{1}{2}, \frac{1}{2})$
Replicate quality =	5.77	5.49	5.97	6.17	4.36	5.86
averages	5.61	5.90	5.67	5.84	4.56	6.14

The fitted quadratic model is

$$\begin{aligned} \hat{y}(\mathbf{x}) &= b_1x_1 + b_2x_2 + b_3x_3 + b_{12}x_1x_2 + b_{13}x_1x_3 + b_{23}x_2x_3 \\ &= \underset{(0.15)}{5.69}x_1 + \underset{(0.15)}{5.82}x_2 + \underset{(0.15)}{4.46}x_3 - \underset{(0.71)}{0.24}x_1x_2 + \underset{(0.71)}{3.72}x_1x_3 + \underset{(0.71)}{3.44}x_2x_3 \end{aligned} \quad (5.13)$$

where below each coefficient estimate in parentheses is the estimated standard error of the coefficient estimate. An estimate of the observation variance from the replicate average scores is $s^2 = 0.04$.

The magnitudes of the coefficient estimates b_1 and b_2 are approximately the same. Also, because of the near zero value of b_{12} we are inclined to think that $\beta_{12} = 0$ which means that the blending of components 1 and 2 with one another is additive (i.e., the blending is neither synergistic nor antagonistic) and thus the $\beta_{12}x_1x_2$ term is dropped from the model of Eq. (5.13). Also, since the sizes of estimates $b_{13} = 3.72$ and $b_{23} = 3.44$ are approximately the same, meaning the binary blending of components 1 and 3 is approximately identical to the binary blending of components 2 and 3, this leads us to think components 1 and 2 are similar in their blending characteristics and so the x_1 and x_2 values are summed in the model. The resulting simplified model is

$$\eta = \beta_1(x_1 + x_2) + \beta_3x_3 + \beta_{13}(x_1 + x_2)x_3 \tag{5.14}$$

When Eq. (5.14) is fitted to the average quality data, the resulting equation is

$$\begin{aligned}\hat{y}(\mathbf{x}) = \underset{(0.07)}{5.74}(x_1 + x_2) + \underset{(0.12)}{4.46}x_3 + \underset{(0.45)}{3.62}(x_1 + x_2)x_3\end{aligned} \tag{5.15}$$

The estimate of the error variance with the fitted Eq. (5.15) is $s^2 = 0.03$ and the R_A^2 value is $R_A^2 = 0.91$.

Equation (5.15) represents an estimated surface for a 2-component system where the components are the sum of ingredients 1 and 2, and component 3. If future experimentation with these same three ingredients is performed, one must be careful not to assume that components 1 and 2 can be combined automatically when blended with component 3. Only the data for this example supports this action and data from future experiments will need to be retested before combining the proportions.

Exercise 5.1 Refit the quadratic model with the average ratings at $(0, \frac{1}{2}, \frac{1}{2})$ now being 4.52 and 4.06 instead of 5.86 and 6.14. Suggest a reduced model form containing at most four terms? *Ans*: $\hat{y}(\mathbf{x}) = 5.68x_1 + 5.81x_2 + 4.49x_3 + 3.56(x_1 - x_2)x_3$. $s^2 = 0.04$.

While reaching a decision on the final model form, our acceptance of the model was based on the value of R_A^2 as well as our willingness to accept the magnitude of the estimate of σ^2. With the fitted equation (5.15) the value of $s^2 = 0.03$ seemed to be acceptable to us because the units of measurement on the data were to within 0.01 and this estimate of variance was only three times the size of the actual unit of measurement which did not seem too large to us. The choice of using the magnitudes of R_A^2 and of s^2 that we could live with is strictly arbitrary. For this reason we shall discuss another criterion which can be used for deciding on the final form of the fitted model. The criterion is discussed in greater detail in Park (1978).

5.5. A CRITERION FOR SELECTING SUBSETS OF THE TERMS IN THE SCHEFFÉ MODELS

Selecting subsets of the components in the first-degree Scheffé model is useful not only for model reduction but when screening the components to find out which are the important ones. To help us in selecting subsets of the components, one strategy that we could employ and which we shall discuss now is to choose the subset that provides us with an estimator of the response which has smaller mean square error over the experimental region than does the standard least squares estimator of the complete model. Choosing an estimator involves assigning a set of linear constraints on the model parameter space and then estimating the value of the response subject to the constraints imposed.

Subset selection, model reduction, and hypothesis testing are all closely related. The hypotheses that were tested in Section 5.1 are called linear hypotheses because a value is hypothesized about some linear function of the model parameters. The hypothesis H_0: $\beta_1=\beta_2=\beta_3=0$, $\beta_{12}=\beta_{13}=\beta_{23}=\beta_{123}=0$ for example can be written using matrix notation as $\mathbf{C\beta}=\mathbf{0}$ where $\boldsymbol{\beta}$ is the 7×1 vector of parameters $\boldsymbol{\beta}=(\beta_1, \beta_2, \beta_3, \beta_{12}, \beta_{13}, \beta_{23}, \beta_{123})'$, $\mathbf{C}$ is the 7×7 identity matrix, and $\mathbf{0}$ is a 7×1 vector of zeros. This is a particular case of the general linear hypothesis $\mathbf{C\beta}=\mathbf{m}$ that is discussed in Section 7.7.

Very often the matrix $\mathbf{C}$ in $\mathbf{C\beta}=\mathbf{m}$ is a contrast matrix. In the types of contrast matrices that we shall be considering the elements in each row are scalar coefficients for some contrast involving the elements of $\boldsymbol{\beta}$. In each contrast the scalar coefficients sum to zero. The test of the hypothesis H_0: $\mathbf{C\beta}=\mathbf{0}$ versus H_A: $\mathbf{C\beta}\neq\mathbf{0}$ is performed by using an F-test. The form of the F-test is presented in Eq. (5.17) and also in Eq. (7.24) of Chapter 7.

The theory of testing linear hypotheses is most easily presented using matrix notation. To show this let us write the Scheffé general linear model as

$$\mathbf{y}=\mathbf{X\beta}+\boldsymbol{\epsilon} \tag{5.16}$$

where $\mathbf{y}$ is an $N\times 1$ vector of observations, $\mathbf{X}$ is an $N\times p$ matrix of component proportions or functions of the component proportions, $\boldsymbol{\beta}$ is the $p \times 1$ vector of model coefficients, and $\boldsymbol{\epsilon}$ is an $N \times 1$ vector of random errors where $E(\boldsymbol{\epsilon}) = \mathbf{0}$ and $E(\boldsymbol{\epsilon\epsilon}') = \sigma^2\mathbf{I}_N$. For tests of significance, we also assume that the errors ϵ_u, $u = 1, 2, \ldots, N$ are sampled from a Normal population. If Eq. (5.16) is of the first degree, then $p = q$, whereas if the model is of the second degree, then $p = q(q+1)/2$.

Suppose a set of r linear contrasts (or linear constraints) are imposed on the parameter space of the elements of $\boldsymbol{\beta}$ of the form

$$\mathbf{C}\boldsymbol{\beta} = \mathbf{0}$$

where $\mathbf{C}$ is an $r \times p$ matrix of rank r $(r \leq p)$. Let $\mathbf{b} = (\mathbf{X}'\mathbf{X})^{-1}\mathbf{X}'y$ be the standard least squares estimator of $\boldsymbol{\beta}$ in Eq. (5.16). The test of the hypothesis H_0: $\mathbf{C}\boldsymbol{\beta} = \mathbf{0}$ vs H_A: $\mathbf{C}\boldsymbol{\beta} \neq \mathbf{0}$ is performed with the F-statistic

$$F = \frac{\mathbf{b}'\mathbf{C}'[\mathbf{C}(\mathbf{X}'\mathbf{X})^{-1}\mathbf{C}']^{-1}\mathbf{C}\mathbf{b}}{rs^2} \tag{5.17}$$

where $s^2 = \mathbf{y}'(\mathbf{I} - \mathbf{X}(\mathbf{X}'\mathbf{X})^{-1}\mathbf{X}')\mathbf{y}/(N-p)$. The value of F in Eq. (5.17) is compared to the tabled value of $F_{(r, N-p, \alpha)}$ and the hypothesis H_0 is rejected if the computed F-value in (5.17) exceeds the table F-value.

An equation for the estimate of the response where the coefficient estimates are constrained as in $\mathbf{C}\boldsymbol{\beta} = \mathbf{0}$ is obtained as follows. Let $\tilde{\mathbf{b}}$ be the least squares estimator of $\boldsymbol{\beta}$ under the restriction $\mathbf{C}\boldsymbol{\beta} = \mathbf{0}$. The elements in $\mathbf{b}$ and $\tilde{\mathbf{b}}$ are related by

$$\tilde{\mathbf{b}} = \mathbf{A}\mathbf{b} \tag{5.18}$$

where $\mathbf{A} = \mathbf{I} - (\mathbf{X}'\mathbf{X})^{-1}\mathbf{C}'[\mathbf{C}(\mathbf{X}'\mathbf{X})^{-1}\mathbf{C}']^{-1}\mathbf{C}$, and, $\mathbf{C}\tilde{\mathbf{b}} = \mathbf{0}$ since $\mathbf{CA} = \mathbf{0}$. With the estimator $\tilde{\mathbf{b}}$ from Eq. (5.18), we may write the estimated response equation, $\tilde{y}(\mathbf{x}) = \mathbf{x}_p'\tilde{\mathbf{b}}$ where the elements in $\mathbf{x}_p'$ correspond to the elements in a row of $\mathbf{X}$ in Eq. (5.16).

The matrix $\mathbf{A}$ in $\tilde{\mathbf{b}} = \mathbf{A}\mathbf{b}$ makes use of the contrast matrix $\mathbf{C}$. Thus to find the best fitted model $\tilde{y}(\mathbf{x}) = \mathbf{x}_p'\tilde{\mathbf{b}} = \mathbf{x}_p'\mathbf{A}\mathbf{b}$ that can be used over the simplex region we need to find a set of contrasts where the set consists of at most r contrasts that will make up $\mathbf{C}$ to give us $\tilde{y}(\mathbf{x})$. To say that $\tilde{y}(\mathbf{x})$ is best means the estimate $\tilde{y}(\mathbf{x})$ has the smallest mean square error of any estimate when taken over the entire simplex region. This mean square error (MSE) of $\tilde{y}(\mathbf{x})$ at any arbitrary point $\mathbf{x}$ in the simplex region is

$$\begin{aligned} \text{MSE}[\tilde{y}(\mathbf{x})] &= E(\mathbf{x}_p'\tilde{\mathbf{b}} - \mathbf{x}_p'\boldsymbol{\beta})^2 \\ &= \sigma^2\mathbf{x}_p'\mathbf{A}(\mathbf{X}'\mathbf{X})^{-1}\mathbf{A}'\mathbf{x}_p + \mathbf{x}_p'\mathbf{L}\mathbf{C}\boldsymbol{\beta}\boldsymbol{\beta}'\mathbf{C}'\mathbf{L}'\mathbf{x}_p \end{aligned} \tag{5.19}$$

where the first quantity on the right-hand side of the equality sign is a measure of the variance of the estimate $\tilde{y}(\mathbf{x})$ and the right-most quantity in Eq. (5.19) is the square of the bias associated with the estimate $\tilde{y}(\mathbf{x})$. The matrix $\mathbf{L}$ in Eq. (5.19) is $\mathbf{L} = (\mathbf{X}'\mathbf{X})^{-1}\mathbf{C}'[\mathbf{C}(\mathbf{X}'\mathbf{X})^{-1}\mathbf{C}']^{-1}$.

In what follows, we shall observe that there exists a matrix $\mathbf{C}$ ($\mathbf{C}$ is not unique) which, when used in $\tilde{y}(\mathbf{x})$ to estimate η, produces an estimator $\tilde{y}(\mathbf{x})$ with a uniformly smaller variance across the simplex than if we had chosen to use the standard least squares estimator $\hat{y}(\mathbf{x}) = \mathbf{x}_p'\mathbf{b}$. However we shall have to accept some bias in the estimated value $\tilde{y}(\mathbf{x})$ in trade for the reduction in variance. Nevertheless, if the magnitude of the reduction in the variance is greater than the magnitude of the square of the bias at the point $\mathbf{x}$, then $\text{MSE}[\tilde{y}(\mathbf{x})] \leq \text{MSE}[\hat{y}(\mathbf{x})]$.

In order to be able to compare the precision of $\tilde{y}(\mathbf{x})$ with that of the standard least squares estimator $\hat{y}(\mathbf{x}) = \mathbf{x}_p'\mathbf{b}$, we assume the true model in Eq. (5.16) is of the first degree so that the estimate $\mathbf{b} = (\mathbf{X}'\mathbf{X})^{-1}\mathbf{X}'\mathbf{y}$ is unbiased. The mean square error of $\hat{y}(\mathbf{x}) = \mathbf{x}_p'\mathbf{b}$ at the point $\mathbf{x}$ is equal to the variance of $\hat{y}(\mathbf{x})$

$$\text{MSE}[\hat{y}(\mathbf{x})] = \sigma^2\mathbf{x}_p'(\mathbf{X}'\mathbf{X})^{-1}\mathbf{x}_p \tag{5.20}$$

At the point $\mathbf{x}$, the difference between the mean square errors of $\tilde{y}(\mathbf{x})$ and $\hat{y}(\mathbf{x})$ is

$$\begin{aligned} D(\mathbf{x}) &= \text{MSE}[\hat{y}(\mathbf{x})] - \text{MSE}[\tilde{y}(\mathbf{x})] \\ &= \{\text{var}[\hat{y}(\mathbf{x})] - \text{var}[\tilde{y}(\mathbf{x})]\} - \{\text{squared bias of } \tilde{y}(\mathbf{x})\} \\ &= \sigma^2\mathbf{x}_p'\{\mathbf{L}\mathbf{C}(\mathbf{X}'\mathbf{X})^{-1}\mathbf{C}'\mathbf{L}'\}\mathbf{x}_p - \mathbf{x}_p'\mathbf{L}\mathbf{C}\boldsymbol{\beta}\boldsymbol{\beta}'\mathbf{C}'\mathbf{L}'\mathbf{x}_p \end{aligned} \tag{5.21}$$

For $\tilde{y}(\mathbf{x})$ to be the better estimate of η, $D(\mathbf{x})$ must be nonnegative at all points $\mathbf{x}$. This is possible since $D(\mathbf{x})$ can be written as

$$D(\mathbf{x}) = \mathbf{x}_p'\mathbf{L}[\sigma^2\mathbf{C}(\mathbf{X}'\mathbf{X})^{-1}\mathbf{C}' - \mathbf{C}\boldsymbol{\beta}\boldsymbol{\beta}'\mathbf{C}']\mathbf{L}'\mathbf{x}_p$$

and $D(\mathbf{x}) \geq 0$ if the matrix $\sigma^2\mathbf{C}(\mathbf{X}'\mathbf{X})^{-1}\mathbf{C}' - \mathbf{C}\boldsymbol{\beta}\boldsymbol{\beta}'C'$ is positive semidefinite. A necessary and sufficient condition for the matrix $\sigma^2\mathbf{C}(\mathbf{X}'\mathbf{X})^{-1}\mathbf{C}' - \mathbf{C}\boldsymbol{\beta}\boldsymbol{\beta}'\mathbf{C}'$ to be positive semidefinite is

$$\boldsymbol{\beta}'\mathbf{C}'[\mathbf{C}(\mathbf{X}'\mathbf{X})^{-1}\mathbf{C}']^{-1}\mathbf{C}\boldsymbol{\beta} \leq \sigma^2$$

A comparison of the precisions of the estimators $\tilde{y}(\mathbf{x})$ and $\hat{y}(\mathbf{x})$ over the simplex region can be made by averaging (or integrating) the difference $D(\mathbf{x})$ over the simplex region. The integrated difference, denoted by J_D, is

$$J_D = \frac{K}{\sigma^2} \int_R D(\mathbf{x})\, dW(\mathbf{x}) \tag{5.22}$$

where $W(\mathbf{x})$ is a weighting function which could assign different weights to $D(\mathbf{x})$ at different points in the simplex region R, and $K^{-1} = \int_R dW(\mathbf{x})$ is a constant used in the averaging of $D(\mathbf{x})$ over R. If the formula (5.21) for $D(\mathbf{x})$ is substituted into J_D in (5.22), the integrated difference is

$$\begin{aligned} J_D &= \frac{K}{\sigma^2} \int_R \{\text{MSE}[\hat{y}(\mathbf{x})] - \text{MSE}[\tilde{y}(\mathbf{x})]\}\, dW(\mathbf{x}) \\ &= K \int_R \mathbf{x}_p'\{\mathbf{LC}(\mathbf{X}'\mathbf{X})^{-1}\mathbf{C}'\mathbf{L}'\}\mathbf{x}_p\, dW(\mathbf{x}) - \frac{K}{\sigma^2} \int_R \mathbf{x}_p'\mathbf{LC}\boldsymbol{\beta}\boldsymbol{\beta}'\mathbf{C}'\mathbf{L}'\mathbf{x}_p\, dW(\mathbf{x}) \\ &= \text{trace}[\mathbf{LC}(\mathbf{X}'\mathbf{X})^{-1}\mathbf{C}'\mathbf{L}'\mathbf{M}] - \frac{1}{\sigma^2} \boldsymbol{\beta}'\mathbf{C}'\mathbf{L}'\mathbf{MLC}\boldsymbol{\beta} \end{aligned} \tag{5.23}$$

where $\mathbf{M} = K \int \mathbf{x}_p \mathbf{x}_p'\, dW(\mathbf{x})$ is the matrix of region moments. (The elements of $\mathbf{M}$ are derived in Appendix 5A.)

A positive value of J_D in Eq. (5.23) implies that the drop in variance from using $\tilde{y}(\mathbf{x})$ rather than $\hat{y}(\mathbf{x})$ is greater than the amount of bias in $\tilde{y}(\mathbf{x})$. Thus there is less error associated with using the restricted estimator whose parameters are subject to the linear contrasts $\mathbf{C}\boldsymbol{\beta} = \mathbf{0}$.

To find a set of linear contrasts $\mathbf{C}\boldsymbol{\beta} = \mathbf{0}$ from which an estimator can be found and which produces a positive value of J_D in Eq. (5.23), Park (1978) suggests substituting $\mathbf{b}$ and s^2 into Eq. (5.23) for $\boldsymbol{\beta}$ and σ^2, respectively, producing a formula for $\hat{J}_D$. Next, several different forms of $\mathbf{C}$ are selected by trial and error and the value of $\hat{J}_D$, corresponding to each $\mathbf{C}$, is computed. The final choice that is made for the matrix $\mathbf{C}$ is the $\mathbf{C}$ that produces the largest positive value for $\hat{J}_D$.

5.6. A NUMERICAL EXAMPLE ILLUSTRATING THE INTEGRATED MEAN SQUARE ERROR CRITERION

To illustrate the calculations required in comparing the mean square errors of $\tilde{y}(\mathbf{x})$ and $\hat{y}(\mathbf{x})$, let us recall the data of the three-component yarn example mentioned previously in Section 5.4 where quality ratings were assigned to the replicate samples at the six points of a {3, 2} simplex lattice. A quadratic model was fitted to the ratings producing the prediction equation (5.13),

$$\begin{aligned}\hat{y}(\mathbf{x}) &= \mathbf{x}_p'\mathbf{b} \\ &= b_1x_1 + b_2x_2 + b_3x_3 + b_{12}x_1x_2 + b_{13}x_1x_3 + b_{23}x_2x_3 \\ &= 5.69x_1 + 5.82x_2 + 4.46x_3 - 0.24x_1x_2 + 3.72x_1x_3 + 3.44x_2x_3 \qquad (5.24)\end{aligned}$$

An estimate of the observation variance was $s^2 = 0.04$ with 6 degrees of freedom but with this example we use $s^2 = 0.0426$ for greater precision.

One of the first hypotheses we might consider testing is H_0: $\beta_{13} = \beta_{23}$ or H_0: $\mathbf{C\beta} = 0$ where we define the single contrast $\beta_{13} - \beta_{23} = 0$. The 1×6 matrix $\mathbf{C}$ is of the form $\mathbf{C} = (0, 0, 0, 0, 1, -1)$ and the summary calculations are

$$\mathbf{C(X'X)^{-1}C'} = 20, \qquad \mathbf{L}' = [\mathbf{C(X'X)^{-1}C'}]^{-1}\mathbf{C(X'X)^{-1}} = (-\tfrac{1}{20}, \tfrac{1}{20}, 0, 0, \tfrac{10}{20}, -\tfrac{10}{20})$$

$$\mathbf{L'ML} = \tfrac{11}{7200} = 0.0015278$$

$$\mathbf{Cb} = 0.28$$

where the matrix $\mathbf{M}$ of region moments and the $\mathbf{(X'X)^{-1}}$ matrix are of the form

$$\mathbf{M} = K \int_R \mathbf{x}_p\mathbf{x}_p'\,d\mathbf{x}$$

$$= \frac{1}{180}\begin{bmatrix} 30 & 15 & 15 & 6 & 6 & 3 \\ 15 & 30 & 15 & 6 & 3 & 6 \\ 15 & 15 & 30 & 3 & 6 & 6 \\ 6 & 6 & 3 & 2 & 1 & 1 \\ 6 & 3 & 6 & 1 & 2 & 1 \\ 3 & 6 & 6 & 1 & 1 & 2 \end{bmatrix}$$

$$\mathbf{(X'X)^{-1}} = \frac{1}{2}\begin{bmatrix} 1 & 0 & 0 & -2 & -2 & 0 \\ 0 & 1 & 0 & -2 & 0 & -2 \\ 0 & 0 & 1 & 0 & -2 & -2 \\ -2 & -2 & 0 & 24 & 4 & 4 \\ -2 & 0 & -2 & 4 & 24 & 4 \\ 0 & -2 & -2 & 4 & 4 & 24 \end{bmatrix}$$

The weighting function, $W(\mathbf{x})$, used in calculating the elements of $\mathbf{M}$ was taken to be uniform over the entire simplex region. (See Appendix 5A.)

The value of the F-statistic in Eq. (5.17), for testing H_0: $\beta_{13} - \beta_{23} = 0$, is

$$F = \frac{0.28[20]^{-1}0.28}{0.0426} = 0.092$$

The test of the hypothesis is clearly not significant and so we accept $\mathbf{C\beta} = 0$ and continue formulating our restricted estimator $\tilde{y}(\mathbf{x}) = \mathbf{x}_p'\tilde{\mathbf{b}} = \mathbf{x}_p'\mathbf{A}\mathbf{b}$ according to Eq. (5.18). The matrix $\mathbf{A}$ and the vector of restricted estimates $\tilde{\mathbf{b}}$ are

$$\mathbf{A} = \mathbf{I} - (\mathbf{X}'\mathbf{X})^{-1}\mathbf{C}'[\mathbf{C}(\mathbf{X}'\mathbf{X})^{-1}\mathbf{C}']^{-1}\mathbf{C} = \begin{bmatrix} 1 & 0 & 0 & 0 & \frac{1}{20} & -\frac{1}{20} \\ 0 & 1 & 0 & 0 & -\frac{1}{20} & \frac{1}{20} \\ 0 & 0 & 1 & 0 & 0 & 0 \\ 0 & 0 & 0 & 1 & 0 & 0 \\ 0 & 0 & 0 & 0 & \frac{1}{2} & \frac{1}{2} \\ 0 & 0 & 0 & 0 & \frac{1}{2} & \frac{1}{2} \end{bmatrix}$$

$$\tilde{\mathbf{b}} = \mathbf{A}\mathbf{b} = \begin{bmatrix} b_1 + (b_{13} - b_{23})/20 \\ b_2 + (b_{23} - b_{13})/20 \\ b_3 \\ b_{12} \\ (b_{13} + b_{23})/2 \\ (b_{13} + b_{23})/2 \end{bmatrix}$$

so that the restricted estimator $\tilde{y}(\mathbf{x}) = \mathbf{x}_p'\tilde{\mathbf{b}}$ is

$$\begin{aligned} \tilde{y}(\mathbf{x}) &= \tilde{b}_1x_1 + \tilde{b}_2x_2 + \tilde{b}_3x_3 + \tilde{b}_{12}x_1x_2 + \tilde{b}_{13}x_1x_3 + \tilde{b}_{23}x_2x_3 \\ &= 5.70x_1 + 5.81x_2 + 4.46x_3 - 0.24x_1x_2 + 3.58(x_1 + x_2)x_3 \end{aligned} \quad (5.25)$$

Substituting the estimates $\mathbf{b}$ and s^2 into Eq. (5.23) for $\hat{J}_D$, a measure of the gain from using $\tilde{y}(\mathbf{x})$ in Eq. (5.25) rather than $\hat{y}(\mathbf{x})$ in Eq. (5.24) over

TABLE 5.2. Some selected contrasts of the model parameters and the corresponding values of *IV*, *IB*, $\hat{J}_D$, and *F* using $s^2 = 0.0426$.

Contrast	r	*IV*	*IB*	$\hat{J}_D$	F	$F_{(r,6,0.05)}$
$\beta_1 - \beta_2 = 0$	1	0.0194	0.0077	0.0117	0.40	5.99
$\beta_1 - \beta_3 = 0$	1	0.0194	0.6901	−0.6707	35.51	5.99
$\beta_1 + \beta_2 - 2\beta_3 = 0$	1	0.0194	1.0206	−1.0012	52.51	5.99
$\beta_1 - \beta_2 = 0$ $\beta_{13} - \beta_{23} = 0$	2	0.0639	0.0074	0.0565	0.20	5.14
$\beta_1 + \beta_2 - 2\beta_3 = 0$ $\beta_{12} + \beta_{13} - 2\beta_{23} = 0$	2	0.0529	0.9913	−0.9384	26.38	5.14
$\beta_1 - \beta_3 = 0$ $\beta_{12} - \beta_{23} = 0$	2	0.0639	0.5921	−0.5282	18.85	5.14

the simplex region, is

$$\hat{J}_D = \text{trace}[\mathbf{LC(X'X)^{-1}C'L'M}] - \mathbf{b'C'L'MLCb}/s^2$$
$$= 0.03055 - 0.00281$$
$$= 0.0277$$

Since the value of $\hat{J}_D$ is greater than zero, which supports our earlier acceptance of $\mathbf{C\beta} = \mathbf{0}$, we are encouraged to use $\tilde{y}(\mathbf{x})$ as an estimator of η. Several other contrasts and the corresponding values of $IV = \text{trace}[\mathbf{LC(X'X)^{-1}C'L'M}]$, $IB = \mathbf{b'C'L'MLCb}/s^2$, $\hat{J}_D = IV - IB$ and F are presented in Table 5.2.

5.7. SCREENING COMPONENTS

In some areas of mixture experiments specifically in certain chemical industries, many times there is present a large number $q \geq 6$ of potentially important components that can be considered candidates in an experiment. If at all possible, a reduction in the number of necessary components is sought not only from the standpoint that it is easier to understand a system that contains only a small number of components but also for reasons of economics. Initially the strategy is to identify all of the components and then to try to single out the ones which are most important. If, to identify the components, it becomes necessary to actually perform experimental runs and decide on the most important components from the sizes of their effects, then the reduction of the number of components so that only the most important components are considered further is known as *screening the components.*

The construction of screening designs and the setting up of screening models quite often begins with the Scheffé first-degree model

$$\eta = \beta_1 x_1 + \beta_2 x_2 + \beta_3 x_3 + \cdots + \beta_q x_q \qquad (5.26)$$

When possible, the ranges of the values of the component proportions x_i are set equal to each other so that the relative effects of the components can be assessed by ranking the parameter estimates b_i, $i = 1, 2, \ldots, q$. In other words, if the ranges of the values of the x_i are equal, one can infer that the larger the value of b_i, relative to b_j, $i \neq j$, the more important x_i is, relative to x_j. Of course, in order to determine the significance of component i, an approximate test of the magnitude of b_i is performed which consists of comparing b_i to its estimated standard error s.e.(b_i), $i = 1, 2, \ldots, q$.

Snee and Marquardt (1976) discuss several strategies which can be used when determining the most important components in a q-component blending system. Designs, called *simplex screening designs*, are recommended in those cases where it is possible to experiment over the total simplex region (or for the special case when the experimental region is a smaller simplex inside the original simplex). Extreme vertices designs are suggested when the proportions of some or all of the components are constrained by upper and lower bounds. Two methods for locating the coordinates of the extreme vertices on a subregion of the simplex were discussed in Sections 4.6 and 4.7.

To screen out the unimportant components or to single out the important components, it is necessary to know how to measure the effects of the individual components. We present the following definition.

Definition The effect of component i on the response is the change in the value of the response resulting from a change in the proportion of component i while holding constant the relative proportions of the other components.

The size of the change that is made in the proportion x_i is not specified at this time.

The largest change that can be made in the proportion x_i is one unit when going from $x_i = 0$ to $x_i = 1$. If at the time when making this change in x_i we select to keep the proportions x_j, $j \neq i$, of the other components equal to one another, then the change must be made along the axis of component i. (The axis of component i was defined in Section 2.13 when introducing axial designs.) In Figure 5.3 are shown the three x_i-axes inside the three-component triangle. The values of the proportions of each of the other $q-1$ components along the x_i-axis are $x_j = (1 - x_i)/(q-1)$, for $j \neq i$.

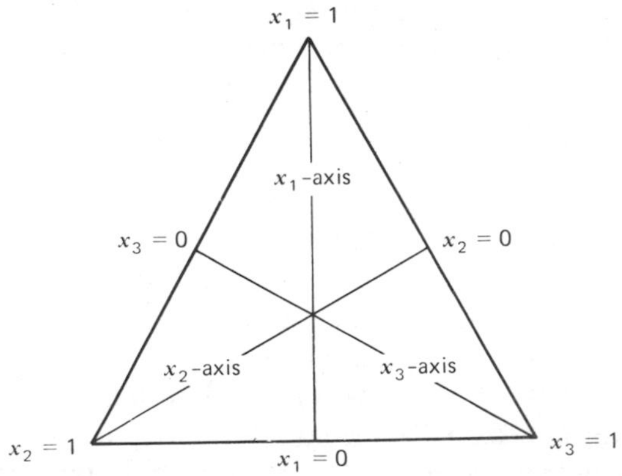

FIGURE 5.3. The x_i-axes, $i = 1, 2$, and 3.

Along the x_i-axis, the relative proportions of the other $q-1$ components are unchanged for different values of x_i. Thus, to measure the effect of component i, observations are collected at points on the axis of component i and the observations are used to calculate the effect. In fact, if we wish to measure the effects of all of the components simultaneously as with Cox's model in Section 6.8, then observations are collected at points located on all of the axes to permit the estimation of all of the parameters in Eq. (5.26).

A formula for estimating the effect of component i is derived as follows. Let us assume that an axial design of the type introduced in Section 2.13 is to be used with observations collected at the points on the axes to produce the fitted equation $\hat{y}(\mathbf{x}) = b_1x_1 + b_2x_2 + \cdots + b_qx_q$. At the base of the x_i-axis where $x_i = 0$ and $x_j = (1 - x_i)/(q-1) = 1/(q-1)$, $j \neq i$, the estimate of the response is

$$\hat{y}(\mathbf{x}) = \sum_{j \neq i}^{q} b_j(1/(q-1)) = (q-1)^{-1} \sum_{j \neq i}^{q} b_j \tag{5.27}$$

which is the average height of the response surface associated with the other $q-1$ pure components. Similarly at the oppositive end of the x_i-axis, that is, at the vertex where $x_i = 1$, $x_j = 0$, $j \neq i$, the estimate of the response is

$$\hat{y}(\mathbf{x}) = b_i \tag{5.28}$$

which is the estimated height of the surface at $x_i = 1$. Recalling that the effect of component i is defined as the change in the response resulting from a change in the proportion of component i while holding constant the relative proportions of the other components, then an intuitive comparison to make to estimate the effect of component i would be to compute the difference between the estimates $\hat{y}(\mathbf{x})$ in Eqs. (5.28) and (5.27). Such a comparison is particularly meaningful if observations are collected at $x_i = 0$, $x_j = 1/(q-1)$, $j \neq i$, and at $x_i = 1$. However, if observations are not collected at these points but the ranges of all the x_i are equal, then still we can define

$$\text{Effect of component } i = b_i - (q-1)^{-1} \sum_{j \neq i}^{q} b_j, \; i = 1, 2, \ldots, q \tag{5.29}$$

When with the help of Eq. (5.29) it is discovered the effects of two or more components are equal, which is when $b_i = b_j = b_k$, the proportions of the equivalent components are summed and the sum is then considered to be a component by itself. An example of two components having approximately equal effects is the three-component yarn-rating experiment of Section 5.4 where the proportions x_1 and x_2 corresponding to components 1 and 2 were summed to form the single component $(x_1 + x_2)$. As a result, the six-term quadratic model in the original three components was rewritten as the simplified three-term model of Eq. (5.14).

Formula (5.29) for measuring the effect of component i can be used for purposes of model reduction. For example, if the true surface is planar (i.e., of the first degree) and we find that $b_i = (q-1)^{-1} \Sigma_{j \neq i}^{q} b_j$, then component i is said to have a "zero effect" or "no effect." When this happens, the term $\beta_i x_i$ is deleted from the model and component i is removed from further consideration.

When the ranges of the component proportions x_i are not equal, which is often the case when the x_i are constrained above and/or below by upper and/or lower bounds such as $a_i \leq x_i \leq b_i$, $i = 1, 2, \ldots, q$, the formula for calculating the effect of component i contains an adjustment factor for the differences in the spreads or ranges in the values of the x_i. The adjustment consists of using the range of x_i, say $R_i = b_i - a_i$, to weight the effect of component i in Eq. (5.29). The adjusted effect is

$$\text{Adjusted effect of component } i = R_i \left[b_i - (q-1)^{-1} \sum_{j \neq i}^{q} b_j \right] \tag{5.30}$$

We see here that components with the largest ranges receive the greatest weight when defining their effects where the greatest weight is of size 1.0 for those components where x_i goes from 0 to 1.0.

Tests of significance on the magnitudes of the effects of components can be performed in much the same way as the linear hypotheses were tested using Eq. (5.17) in the previous Section 5.5. This is because the formulas for estimating the component effects can be written in matrix notation as $\mathbf{C}\boldsymbol{\beta} = 0$ where $\mathbf{C}$ is a $1 \times q$ matrix of scalar coefficients. For example, if we denote by E_1 the estimate of the effect of component 1, the contrast for calculating E_1 using Eq. (5.29) is

$$E_1 = \mathbf{Cb} = [1, -1/(q-1), -1/(q-1), \ldots, -1/(q-1)]\mathbf{b}$$

and the estimate of the adjusted effect of component 1, E_{A1} say, according to Eq. (5.30) is

$$E_{A1} = R_1\mathbf{Cb} = R_1[1, -1/(q-1), -1/(q-1), \ldots, -1/(q-1)]\mathbf{b}$$

Further, if the effects of all q components are to be estimated simultaneously then $\mathbf{C}$ is written as a $q \times q$ matrix and the *vector* of estimated component effects is

$$\mathbf{E} = \begin{bmatrix} E_1 \\ E_2 \\ \vdots \\ E_q \end{bmatrix} = \frac{1}{q-1}[q\mathbf{I}_q - \mathbf{J}_q]\mathbf{b} = \mathbf{Cb} \tag{5.31}$$

where $\mathbf{I}_q$ is the identity matrix of order q and $\mathbf{J}_q$ is the $q \times q$ matrix of ones. The vector of adjusted component effects is

$$\mathbf{E}_A = \mathbf{RE} = [\text{diag}(R_1, R_2, \ldots, R_q)]\mathbf{E} \tag{5.32}$$

where $\mathbf{R}$ is a diagonal matrix whose nonzero diagonal elements are the ranges of the x_i.

The variance–covariance matrix of the vector $\mathbf{E}$ of estimated effect contrasts is $\text{var}(\mathbf{E}) = \mathbf{C}(\mathbf{X}'\mathbf{X})^{-1}\mathbf{C}'\sigma^2$, since the variance–covariance matrix of the parameter estimates $\mathbf{b}$ is $(\mathbf{X}'\mathbf{X})^{-1}\sigma^2$. If the observations y_u, $u = 1, 2, \ldots, N$, are assumed to be sampled from a Normal Population, the hypothesis H_0: $\mathbf{E} = \mathbf{0}$ is tested using the F-test of Eq. (5.17)

$$F = \frac{\mathbf{E}'[\mathbf{C}(\mathbf{X}'\mathbf{X})^{-1}\mathbf{C}']^{-}\mathbf{E}}{(q-1)s^2} \tag{5.33}$$

The value of F in Eq. (5.33) is compared against the tabled value of $F_{(q-1, f, \alpha)}$ where f is the degrees of freedom used in estimating σ^2. Since the $q \times q$ matrix $\mathbf{C}(\mathbf{X}'\mathbf{X})^{-1}\mathbf{C}'$ is at most of rank $q-1$, $[\mathbf{C}(\mathbf{X}'\mathbf{X})^{-1}\mathbf{C}']^{-}$ for use with all of the effects is a generalized inverse matrix. To avoid the reliance on a generalized inverse matrix, one of the effects is dropped from $\mathbf{E}$ in (5.31) so that the test involves only $q-1$ of the E_i's. Then the $(q-1) \times (q-1)$ matrix $\mathbf{C}(\mathbf{X}'\mathbf{X})^{-1}\mathbf{C}'$ is of full rank and the standard inverse matrix $[\mathbf{C}(\mathbf{X}'\mathbf{X})^{-1}\mathbf{C}']^{-1}$ is used.

When a test on an individual effect E_i is performed, the test suffers from the property that the E_i are not independent and thus tests on the E_i are not independent. However, if the test of the effect $E_i = \mathbf{C}_i\boldsymbol{\beta} = 0$ where $\mathbf{C}_i$ is the ith row of the matrix $\mathbf{C}$ in Eq. (5.31) is preplanned, such a test can be made using Students' t-test

$$t = \frac{\mathbf{C}_i\mathbf{b}}{\{s^2 c_{ii}\}^{1/2}} \tag{5.34}$$

where c_{ii} is the iith element of the matrix $\mathbf{C}(\mathbf{X}'\mathbf{X})^{-1}\mathbf{C}'$. The number of degrees of freedom for the t-test in Eq. (5.34) is the number of degrees of freedom associated with the estimate s^2. The test in Eq. (5.34) is equivalent to the F-test in Eq. (5.33) for a single effect but for more than one effect, the F-test is the more general test statistic.

We now present a seven-component blending experiment for the purpose of illustrating the computing of the component effects and the testing of the component effects. The screening of the seven components involves actually combining the proportions of several components rather than deleting the components. The numerical example helps to show how correlations between individual effects can mask the sizes of the effects and leave the impression that the component effects are small or non-significant. When the correlations are reduced by the combining of terms in the model, the remaining component effects become significantly different from zero.

5.8. A SEVEN-COMPONENT OCTANE BLENDING EXPERIMENT: AN EXERCISE IN MODEL REDUCTION

The motor octane ratings from twelve different blends were recorded in an effort to determine the effects of the following gasoline-blending components within the specified ranges:

Catalytically cracked naphtha (x_4): $0 \le x_4 \le 0.62$
Polymer (x_5): $0 \le x_5 \le 0.12$
Natural gasoline (x_7): $0 \le x_7 \le 0.08$; Reformate (x_2): $0 \le x_2 \le 0.62$
Thermally cracked naphtha (x_3): $0 \le x_3 \le 0.12$
Alkylate (x_6): $0 \le x_6 \le 0.74$; Straightrun (x_1): $0 \le x_1 \le 0.21$

The response values and the component settings are presented in Table 5.3.

In matrix notation, the first-degree model $y = \beta_1 x_1 + \cdots + \beta_7 x_7 + \epsilon$, over the 12 observations, is

$$\begin{array}{ccccc} \mathbf{y} & = & [x_1 \quad x_2 \quad x_3 \quad x_4 \quad x_5 \quad x_6 \quad x_7] & \boldsymbol{\beta} & + \quad \boldsymbol{\epsilon} \end{array}$$

$$\begin{bmatrix} 98.7 \\ 97.8 \\ \vdots \\ 88.1 \end{bmatrix} = \begin{bmatrix} 0 & 0.23 & 0 & 0 & 0 & 0.74 & 0.03 \\ 0 & 0.10 & 0 & 0 & 0.12 & 0.74 & 0.04 \\ & & & \vdots & & & \\ 0 & 0 & 0 & 0.55 & 0 & 0.37 & 0.08 \end{bmatrix} \begin{bmatrix} \beta_1 \\ \beta_2 \\ \vdots \\ \beta_7 \end{bmatrix} + \begin{bmatrix} \epsilon_1 \\ \epsilon_2 \\ \vdots \\ \epsilon_{12} \end{bmatrix}$$

The estimates of the model parameters and the vectors of unadjusted and

TABLE 5.3. Gasoline motor octane ratings

Gasoline Components							Motor Octane at 1.5 ml Pb/gal
x_1	x_2	x_3	x_4	x_5	x_6	x_7	y_u
0	0.23	0	0	0	0.74	0.03	98.7
0	0.10	0	0	0.12	0.74	0.04	97.8
0	0	0	0.10	0.12	0.74	0.04	96.6
0	0.49	0	0	0.12	0.37	0.02	92.0
0	0	0	0.62	0.12	0.18	0.08	86.6
0	0.62	0	0	0	0.37	0.01	91.2
0.17	0.27	0.10	0.38	0	0	0.08	81.9
0.17	0.19	0.10	0.38	0.02	0.06	0.08	83.1
0.17	0.21	0.10	0.38	0	0.06	0.08	82.4
0.17	0.15	0.10	0.38	0.02	0.10	0.08	83.2
0.21	0.36	0.12	0.25	0	0	0.06	81.4
0	0	0	0.55	0	0.37	0.08	88.1

adjusted component effects are

$$\mathbf{b} = (\mathbf{X}'\mathbf{X})^{-1}\mathbf{X}'\mathbf{y}$$

$$\underset{\mathbf{b}}{\begin{bmatrix} 34.3 \\ 85.9 \\ 141.3 \\ 77.2 \\ 87.8 \\ 100.3 \\ 116.9 \end{bmatrix}} = \underset{(\mathbf{X}'\mathbf{X})^{-1}}{\begin{bmatrix} 62547 & 13 & -112060 & -1589 & -546 & -587 & 14521 \\ & 2 & -39 & -3 & -1 & -2 & 31 \\ & & 201471 & 3067 & 1048 & 1137 & -27998 \\ & & & 121 & 34 & 44 & -1060 \\ & & & & 48 & 9 & -321 \\ & \text{(symmetric)} & & & & 17 & -389 \\ & & & & & & 9406 \end{bmatrix}}$$

$$\times \underset{\mathbf{X}'\mathbf{y}}{\begin{bmatrix} 73.3 \\ 231.1 \\ 42.8 \\ 257.8 \\ 48.1 \\ 351.1 \\ 58.8 \end{bmatrix}}$$

$$\mathbf{E} = \tfrac{1}{6}[7\mathbf{I}_7 - \mathbf{J}_7]\mathbf{b} = (-67.3, -7.0, 57.5, -17.2, -4.9, 9.7, 29.1)'$$

$$\mathbf{E}_A = \mathbf{RE} = \text{diag}(0.21, 0.62, 0.12, 0.62, 0.12, 0.74, 0.08)\mathbf{E}$$

$$= (-14.1, -4.4, 6.9, -10.7, -0.6, 7.2, 2.3)'$$

The variance–covariance matrix of the vector of adjusted effects is

$$\text{var}(\mathbf{E}_A) = \sigma^2\mathbf{RC}(\mathbf{X'X})^{-1}\mathbf{C'R'}$$

$$= \sigma^2 \begin{bmatrix} 4431 & 1044 & -3968 & 745 & 182 & 1113 & 485 \\ & 260 & -918 & 211 & 46 & 287 & 91 \\ & & 3585 & -612 & -157 & -957 & -473 \\ & & & 231 & 41 & 261 & 12 \\ & & & & 9 & 52 & 13 \\ & \text{(symmetric)} & & & & 333 & 73 \\ & & & & & & 101 \end{bmatrix} \qquad (5.35)$$

The test of the hypothesis H_0: $\mathbf{E}_A = \mathbf{0}$ versus H_A: $\mathbf{E}_A \neq \mathbf{0}$, where $\mathbf{E}_A$ consists of the adjusted effects for the first six components, is a test that all of the effects are simultaneously zero. Although most of the time we would not perform the test except maybe to satisfy our curiosity because our interest is mainly in the individual component effects, for the completeness of the example, we calculate the value of the F-ratio. The F-ratio value is

$$F = \frac{(-14.1, -4.4, 6.9, -10.7, -0.6, 7.2)\begin{bmatrix} 4431 & 1044 & -3968 & 745 & 181 & 1113 \\ 1044 & 260 & -918 & 211 & 46 & 287 \\ -3968 & -918 & 3585 & -612 & -157 & -957 \\ 745 & 211 & -612 & 231 & 41 & 261 \\ 181 & 46 & -157 & 41 & 9 & 52 \\ 1113 & 287 & -957 & 261 & 52 & 333 \end{bmatrix}\begin{bmatrix} -14.1 \\ -4.4 \\ 6.9 \\ -10.7 \\ -0.6 \\ 7.2 \end{bmatrix}}{6(0.70)}$$

$$= 110.5$$

where $s^2 = 0.70$ is obtained from the analysis of variance Table 5.4. The value $F = 110.5$ exceeds $F_{(6,5,0.01)} = 10.67$ and H_0: $\mathbf{E}_A = \mathbf{0}$ is rejected.

Since the hypothesis that all effects are zero is rejected, we ask ourselves "Which of the effects are different from zero?" In an attempt at answering this question, we perform the approximate tests on the individual adjusted effects as is shown with the E_i in Eq. (5.34) to get

TABLE 5.4. The analysis of the motor octane rating data

Source of Variation	Degrees of Freedom	Sum of Squares	Mean Square	R_A^2
Regression	6	464.3	77.40	
Residual	5	3.5	0.70	0.984
Total	11	467.8		

$$\frac{R_1E_1}{\widehat{\text{s.e.}}(R_1E_1)}=\frac{-14.1}{\{4431(0.70)\}^{1/2}}=-0.25 \qquad \frac{R_4E_4}{\widehat{\text{s.e.}}(R_4E_4)}=\frac{-10.7}{\{231(0.70)\}^{1/2}}=-0.84$$

$$\frac{R_2E_2}{\widehat{\text{s.e.}}(R_2E_2)}=\frac{-4.4}{\{260(0.70)\}^{1/2}}=-0.32 \qquad \frac{R_5E_5}{\widehat{\text{s.e.}}(R_5E_5)}=\frac{-0.6}{\{9(0.70)\}^{1/2}}=-0.24$$

$$\frac{R_3E_3}{\widehat{\text{s.e.}}(R_3E_3)}=\frac{6.9}{\{3585(0.70)\}^{1/2}}=0.14 \qquad \frac{R_6E_6}{\widehat{\text{s.e.}}(R_6E_6)}=\frac{7.2}{\{333(0.70)\}^{1/2}}=0.46 \tag{5.36}$$

where $\widehat{\text{s.e.}}(R_iE_i)$ represents the estimated standard error of the ith adjusted effect.

None of the t-test values is significantly different from zero. We suspect these nonsignificant results are due to the several large elements of the var($\mathbf{E}_A$) matrix in (5.35). Particularly noticeable are the quantities $\widehat{\text{var}}(R_1E_1)=4431(0.70)=3102$, $\widehat{\text{var}}(R_3E_3)=3585(0.70)=2510$ and $\widehat{\text{cov}}(R_1E_1, R_3E_3)=-3968(0.70)=-2778$.

A partial solution to the problem of large values of $\widehat{\text{var}}(R_1E_1)$, $\widehat{\text{var}}(R_3E_3)$ and $\widehat{\text{cov}}(R_1E_1, R_3E_3)$ is found by removing the dependency between components 1 and 3. This is done by summing the proportions x_1 and x_3 to form the pseudocomponent, say (x_1+x_3). The new six-term model is then fitted to the 12 data values and the estimates are

$$\underset{\mathbf{b}}{\begin{bmatrix}72.6\\85.9\\76.2\\87.4\\99.9\\126.2\end{bmatrix}} = \underset{(\mathbf{X}'\mathbf{X})^{-1}}{\begin{bmatrix}90.3 & -5.4 & 76.5 & 24.6 & 29.6 & -688.0\\ & 2.2 & -2.6 & -1.0 & -1.8 & 26.3\\ & & 77.0 & 18.5 & 27.2 & -654.9\\ & & & 43.2 & 3.7 & -181.8\\ \text{(symmetric)} & & & & 11.2 & -238.5\\ & & & & & 5702.1\end{bmatrix}} \underset{\mathbf{X}'\mathbf{y}}{\begin{bmatrix}116.1\\231.1\\257.8\\48.1\\351.1\\58.8\end{bmatrix}}$$

The adjusted effects of the components and the estimated variance–

covariance matrix $\widehat{\text{var}}(\mathbf{E}_A)$ are

$$\mathbf{E}_A = \text{diag}(0.33, 0.62, \ldots, 0.08)\{\tfrac{1}{5}[6\mathbf{I}_6 - \mathbf{J}_6]\}\mathbf{b}$$
$$= (-7.4, -4.0, -11.3, -0.6, 7.6, 3.3)'$$

$$\widehat{\text{var}}(\mathbf{E}_A) = \begin{bmatrix} 51 & 44 & 91 & 11 & 75 & -45 \\ & 41 & 82 & 10 & 67 & -40 \\ & & 170 & 20 & 138 & -83 \\ & & & 3 & 16 & -10 \\ & \text{(symmetric)} & & & 114 & -68 \\ & & & & & 41 \end{bmatrix} (s^2 = 0.59)$$

Again none of the individual t-tests are significant prompting us to consider combining x_7 with $(x_1 + x_3)$, mainly because the range of x_7 values is only 0.08.

The five-term model, fitted to the 12 motor octane rating values, is

$$\hat{y}(\mathbf{x}) = \underset{(1.8)}{78.4}(x_1 + x_3 + x_7) + \underset{(1.1)}{85.7}x_2 + \underset{(1.2)}{81.6}x_4 + \underset{(4.6)}{88.9}x_5 + \underset{(0.8)}{101.9}x_6 \qquad (5.37)$$

where the numbers in parentheses are the estimated standard errors of the parameter estimates computed using $s^2 = 0.56$ with 7 degrees of freedom. The adjusted component effects, the $\widehat{\text{var}}(\mathbf{E}_A)$ matrix, and the subsequent individual t-tests are

$$\mathbf{E}_A = \text{diag}(0.41, 0.62, 0.62, 0.12, 0.74)\{\tfrac{1}{4}[5\mathbf{I}_5 - \boldsymbol{J}_5]\}\mathbf{b}$$
$$= (-4.6, -1.3, -4.4, 0.2, 13.5)'$$

$$\widehat{\text{var}}(\mathbf{E}_A) = \begin{bmatrix} 1.5 & -0.6 & -0.7 & -0.3 & 0.8 \\ & 2.0 & 1.4 & -0.6 & 0.9 \\ & & 2.8 & -0.9 & 1.6 \\ & \text{(symmetric)} & & 0.6 & -1.2 \\ & & & & 3.1 \end{bmatrix} (s^2 = 0.56)$$

$$\frac{R_{(1+3+7)}E_{(1+3+7)}}{\widehat{\text{s.e.}}\{R_{(1+3+7)}E_{(1+3+7)}\}} = \frac{-4.6}{\{1.5(0.56)\}^{1/2}} = -5.0 \quad \frac{R_4E_4}{\widehat{\text{s.e.}}(R_4E_4)} = \frac{-4.4}{1.25} = -3.5$$

$$\frac{R_6E_6}{\text{s.e.}(R_6E_6)} = \frac{13.5}{1.32} = 10.2 \quad \frac{R_2E_2}{\widehat{\text{s.e.}}(R_2E_2)} = \frac{-1.3}{1.06} = -1.2 \quad \frac{R_5E_5}{\widehat{\text{s.e.}}(R_5E_5)} = \frac{0.2}{0.58} = 0.3$$

The t-tests values for the adjusted effects of components $(1 + 3 + 7)$, 4, and 6, are significantly different from zero, and therefore our next course of action is to fit a model including these three terms only. We leave this exercise to the reader.

At the beginning of Section 5.7, simplex screening designs were mentioned as being recommended by Snee and Marquardt (1976) when the region of interest is the entire simplex. These designs consists of at least $2q+1$ points that are symmetrically positioned on the component axes at the following locations; the q *vertices* $x_i = 1$, $x_j = 0$, $j \neq i$, at the *centroid* $x_1 = x_2 = \cdots = x_q = 1/q$, and, at the q *interior* locations $x_i = (q+1)/2q$, $x_j = 1/2q$, $j \neq i$. All of the $2q+1$ observations are used to estimate the coefficients β_i, $i = 1, 2, \ldots, q$, in the first-degree model of Eq. (5.26). In addition, under the normality assumption for the y_u's, $u = 1, 2, \ldots, N$, a test for surface curvature can be made by comparing the average value of the observations at the q vertices against the average value of the observations collected inside the simplex region. The test consists of comparing the calculated t-value with a tabled t-value

$$t = \frac{\bar{y}_{\text{vertices}} - \bar{y}_{\text{interior}}}{\{s^2(1/q + 1/(q+1))\}^{1/2}}$$

where in the denominator the estimate s^2 of the observation variance is an overall pooled estimate. The variation among the vertex observations is combined or pooled with the variation among the interior observations in the form $s^2 = ((q-1)s^2_{\text{vertices}} + qs^2_{\text{interior}})/(2q-1)$. The number of degrees of freedom for s^2 is $2q-1$.

Finally, for those cases where the number of components is small, say $q \leq 6$, one might consider including in the screening design the base point of each axis where $x_i = 0$, $x_j = 1/(q-1)$, $j \neq i$, in addition to the $2q+1$ points listed thus far. However, for those cases where the number of components exceeds six, the overall centroid $x_i = 1/q$ for all i, and the base points of the axes $x_i = 0$, $x_j = 1/(q-1)$, $j \neq i$, are not very different in the proportions of the x_i and therefore using the extra base points probably would not be worthwhile.

5.9. THE SLOPE OF THE RESPONSE SURFACE ALONG THE COMPONENT AXES

In this section we discuss how to measure the rate of change of the mixture surface along the component axes with the help of the Scheffé polynomials. The reason for measuring the rate of change of the surface along the axes is to learn more about the mixture systems' surface characteristics in terms of the shape of the surface, the proximity of the location of the maximum or minimum of the surface and so on. Hereafter, the rate of change of the surface will be called the *slope* of the surface.

The general form of the second-degree Scheffé polynomial in q components is

$$\eta = \sum_{i=1}^{q} \beta_i x_i + \sum_{i<j}\sum^{q} \beta_{ij} x_i x_j \tag{5.38}$$

The parameters β_i and β_{ij}, $i<j$, were defined previously as the height of the surface above the simplex at $x_i = 1$, $x_j = 0$, $j \neq i$, whereas β_{ij} is a measure of the departure of the surface from the plane along the edge $0 \leq x_i + x_j = 1$, respectively.

In order to write an expression for the slope of the surface along the x_i-axis we shall first need to write an expression for the equation for the response along the axis of component i. On the x_i-axis at the proportion x_i, the proportions of the other $q-1$ components are equal to $x_j = (1-x_i)/(q-1)$, for all $j \neq i$. Substituting this expression for x_j into Eq. (5.38), the expected response at x_i on the x_i-axis is

$$\begin{aligned}\eta = \beta_i x_i + \sum_{j\neq i}^{q} \beta_j \frac{(1-x_i)}{q-1} + \sum_{l=1}^{i-1} \beta_{li} \frac{(1-x_i)}{q-1} x_i + \sum_{j=i+1}^{q} \beta_{ij} x_i \frac{(1-x_i)}{q-1} \\ + \sum_{\substack{j<k \\ j,k\neq i}}\sum^{q} \beta_{jk} \frac{(1-x_i)^2}{(q-1)^2}\end{aligned} \tag{5.39}$$

The slope of η with respect to component i, evaluated at x_i, is

$$\text{slope(at } x_i) = \frac{\partial \eta}{\partial x_i} = \gamma_0 + \gamma_1 x_i \tag{5.40}$$

where $$\gamma_0 = \frac{1}{q-1}\left[q\beta_i - \sum_{i=1}^{q} \beta_i + \sum_{l=1}^{i-1} \beta_{li} + \sum_{j=i+1}^{q} \beta_{ij} - \frac{2}{q-1} \sum_{\substack{j<k \\ j,k\neq i}}\sum^{q} \beta_{jk}\right] \tag{5.41}$$

$$\gamma_1 = \frac{2}{q-1}\left[\sum_{\substack{j<k \\ j,k\neq i}}\sum^{q} \frac{\beta_{jk}}{q-1} - \sum_{l=1}^{i-1} \beta_{li} - \sum_{j=i+1}^{q} \beta_{ij}\right] \tag{5.42}$$

The formulas for γ_0 and γ_1 in Eqs. (5.41) and (5.42) are valid as long as the model that is used is the second-degree Scheffé polynomial. The slope formula (5.40) is a first-degree equation in the magnitude of the proportion x_i of component i.

For the special cubic model

$$\eta = \sum_{i=1}^{q} \beta_i x_i + \sum_{i<j}\sum^{q} \beta_{ij} x_i x_j + \sum_{i<j<k}\sum\sum^{q} \beta_{ijk} x_i x_j x_k \tag{5.43}$$

along the axis of component i, the slope of the mixture surface at x_i is

$$\text{slope(at } x_i) = \gamma_0 + \gamma_1 x_i + \left\{\frac{(1-x_i)(1-3x_i)}{(q-1)^2}\right\} \sum_{i<j<k} \sum \sum^{q} \beta_{ijk} - \frac{3(1-x_i)^2}{(q-1)^3} \sum_{\substack{j<k<l \\ j,k,l \neq i}} \sum \sum^{q} \beta_{jkl} \tag{5.44}$$

The parameters γ_0 and γ_1 are defined in Eqs. (5.41) and (5.42), respectively, and the β_{ijk}, $i, j, k = 1, 2, \ldots, q$, $i \neq j \neq k$, are parameters associated with the ternary blends involving component i.

Writing the formulas for the slope of the surface at x_i, in terms of the model parameters, is straightforward. Corresponding to the second-degree surface in Eq. (5.38), the slope expressions are, for $q = 3$

$$\begin{aligned}
\text{slope(at } x_1) &= \beta_1 - \tfrac{1}{2}\{\beta_2 + \beta_3 + \beta_{23} - (\beta_{12} + \beta_{13})\} + \{\tfrac{1}{2}\beta_{23} - \beta_{12} - \beta_{13}\}x_1 \\
\text{slope(at } x_2) &= \beta_2 - \tfrac{1}{2}\{\beta_1 + \beta_3 + \beta_{13} - (\beta_{12} + \beta_{23})\} + \{\tfrac{1}{2}\beta_{13} - \beta_{12} - \beta_{23}\}x_2 \\
\text{slope(at } x_3) &= \beta_3 - \tfrac{1}{2}\{\beta_1 + \beta_2 + \beta_{12} - (\beta_{13} + \beta_{23})\} + \{\tfrac{1}{2}\beta_{12} - \beta_{13} - \beta_{23}\}x_3
\end{aligned} \tag{5.45}$$

If the surface is more correctly specified with the special cubic model, then from Eq. (5.44) and if $i = 1$, for example,

$$\begin{aligned}
\text{slope(at } x_1) &= \gamma_0 + \gamma_1 x_1 + \frac{\beta_{123}}{4} - \beta_{123}x_1 + \tfrac{3}{4}\beta_{123}x_1^2 \\
&= \beta_1 - \tfrac{1}{2}\{\beta_2 + \beta_3 + \beta_{23} - (\beta_{12} + \beta_{13}) - \tfrac{1}{2}\beta_{123}\} \\
&\quad + \{\tfrac{1}{2}\beta_{23} - (\beta_{12} + \beta_{13} + \beta_{123})\}x_1 + \tfrac{3}{4}\beta_{123}x_1^2
\end{aligned}$$

An estimate for the value of the slope(at x_i) of the second-degree surface is obtained by substituting the parameter estimates b_i and b_{ij} for β_i and β_{ij}, respectively, into Eqs. (5.41) and (5.42). The estimate is

$$\widehat{\text{slope}}(\text{at } x_i) = g_0 + g_1 x_i \tag{5.46}$$

where g_0 and g_1 are estimators of γ_0 and γ_1, respectively. The variance of the estimate of the slope(at x_i) is

$$\text{var}[\widehat{\text{slope}}(\text{at } x_i)] = \text{var}(g_0) + x_i^2 \, \text{var}(g_1) + 2x_i \, \text{cov}(g_0, g_1)$$

A more general formula for the estimate of the slope is written using vector notation

$$\widehat{\text{slope}}(\text{at } x_i) = \mathbf{s}_i'\mathbf{b} \qquad i = 1, 2, \ldots, q \tag{5.47}$$

where $\mathbf{s}_i'$ is a $1 \times p$ vector of scalar coefficients for multiplying the elements of the $p \times 1$ vector $\mathbf{b} = (\mathbf{X}'\mathbf{X})^{-1}\mathbf{X}'\mathbf{y}$. The subscript i associates $\mathbf{s}_i$ with component i. For example, if $q = 3$ and the second-degree equation $\hat{y}(\mathbf{x}) = \mathbf{x}_p'\mathbf{b}$ contains $\mathbf{b} = (b_1, b_2, b_3, b_{12}, b_{13}, b_{23})$, then for components 1, 2, and 3, respectively, from Eqs. (5.45),

$$\mathbf{s}_1' = [1, -\tfrac{1}{2}, -\tfrac{1}{2}, (1-2x_1)/2, (1-2x_1)/2, (x_1-1)/2]$$
$$\mathbf{s}_2' = [-\tfrac{1}{2}, 1, -\tfrac{1}{2}, (1-2x_2)/2, (x_2-1)/2, (1-2x_2)/2]$$
$$\mathbf{s}_3' = [-\tfrac{1}{2}, -\tfrac{1}{2}, 1, (x_3-1)/2, (1-2x_3)/2, (1-2x_3)/2]$$

The variance of the estimate $\widehat{\text{slope}}(\text{at } x_i) = \mathbf{s}_i'\mathbf{b}$ is

$$\text{var}[\widehat{\text{slope}}(\text{at } x_i)] = \mathbf{s}_i'(\mathbf{X}'\mathbf{X})^{-1}\mathbf{s}_i\sigma^2 \tag{5.48}$$

For example, if duplicate observations are collected at the points of a $\{3, 2\}$ lattice and $i = 1$, then

$$\widehat{\text{slope}}(\text{at } x_1) = \mathbf{s}_1'\mathbf{b}$$
$$= [1, -\tfrac{1}{2}, -\tfrac{1}{2}, (1-2x_1)/2, (1-2x_1)/2, (x_1-1)/2]\mathbf{b}$$

and

$$\text{var}[\widehat{\text{slope}}(\text{at } x_1)] = \mathbf{s}_1' \begin{bmatrix} \tfrac{1}{2} & 0 & 0 & -1 & -1 & 0 \\ 0 & \tfrac{1}{2} & 0 & -1 & 0 & -1 \\ 0 & 0 & \tfrac{1}{2} & 0 & -1 & -1 \\ -1 & -1 & 0 & 12 & 2 & 2 \\ -1 & 0 & -1 & 2 & 12 & 2 \\ 0 & -1 & -1 & 2 & 2 & 12 \end{bmatrix} \mathbf{s}_1\sigma^2$$

5.10. A NUMERICAL EXAMPLE ILLUSTRATING THE SLOPE CALCULATIONS FOR A THREE-COMPONENT SYSTEM: STUDYING THE FLAVOR SURFACE WHERE PEANUT MEAL IS CONSIDERED A SUBSTITUTE OF BEEF IN SANDWICH PATTIES

An experiment was performed to see if defatted peanut meal could be substituted in patties as a beef replacement when used in combination with ground beef. Two brands of peanut meal (denoted by PM_A and PM_B) each were blended separately and together with pure ground beef (GB) to form the patties. The patties were rated subjectively on flavor, using a 1–9 scale, when compared to a standard pure beef patty. The

TABLE 5.5. Average flavor scores of ground beef—peanut meal patties

Design Point	Ground Beef (%)	Peanut Meal *A* (%)	Peanut Meal *B* (%)	x_1	x_2	x_3	Average Flavor Score
1	100	0	0	1	0	0	5.20, 5.10
2	75	25	0	$\frac{1}{2}$	$\frac{1}{2}$	0	3.61, 3.80
3	50	50	0	0	1	0	4.20, 4.05
4	75	0	25	$\frac{1}{2}$	0	$\frac{1}{2}$	4.53, 4.65
5	50	0	50	0	0	1	3.85, 4.15
6	50	25	25	0	$\frac{1}{2}$	$\frac{1}{2}$	4.49, 4.35

scoring reflected a measure of the degree of preference relative to the beef patty where a 1 represented "extreme dislike of sample patty compared to reference patty" while a 9 meant "like sample patty extremely better than reference." The data in Table 5.5 represent average scores for replicate patty formulations where each average was computed from thirty patties of each blend. The six blends ranged from pure or 100% ground beef to 50% : 50% ground beef and peanut meal. The values of the component proportions x_1, x_2, and x_3 were computed as

$$x_1 = \frac{\%\text{ ground beef} - 50\%}{50\%}, \qquad x_2 = \frac{\%\ PM_A}{50\%}, \qquad x_3 = \frac{\%\ PM_B}{50\%}$$

since all patty blends consisted of at least 50% ground beef.

A second-degree model was fitted to the average flavor scores resulting in

$$\hat{y}(\mathbf{x}) = \underset{(0.09)}{5.15}x_1 + \underset{(0.09)}{4.13}x_2 + \underset{(0.09)}{4.00}x_3 - \underset{(0.44)}{3.73}x_1x_2 + \underset{(0.44)}{0.06}x_1x_3 + \underset{(0.44)}{1.43}x_2x_3 \qquad (5.49)$$

The R_A^2 value and the estimate of the error variance using the replicate formulations only were $R_A^2 = 0.934$ and $s^2 = 0.016$. Since the value of $R_A^2 = 0.934$ exceeded 0.90, the model presented in Eq. (5.49) was believed to be an adequate representation of the surface and therefore estimates of the slope of the surface, measured along the axes, were to be computed from the model.

Using the formulas of Eqs. (5.45) for the slope measures with the estimates b_i and b_{ij} replacing the β_i and β_{ij}, respectively, the formulas for the slope estimates are

$$\widehat{\text{slope}}(\text{at } x_1) = b_1 - \tfrac{1}{2}\{b_2 + b_3 + b_{23} - (b_{12} + b_{13})\} + \{\tfrac{1}{2}b_{23} - b_{12} - b_{13}\}x_1$$
$$= 5.15 - \tfrac{1}{2}\{4.13 + 4.00 + 1.43 - (-3.73 + 0.06)\}$$
$$+ \{0.72 - (-3.73) - 0.06\}x_1$$
$$= -1.47 + 4.39x_1$$

$$\widehat{\text{slope}}(\text{at } x_2) = b_2 - \tfrac{1}{2}\{b_1 + b_3 + b_{13} - (b_{12} + b_{23})\} + \{\tfrac{1}{2}b_{13} - b_{12} - b_{23}\}x_2$$
$$= 4.13 - \tfrac{1}{2}\{5.15 + 4.00 + 0.06 - (-3.73 + 1.43)\}$$
$$+ \{0.03 - (-3.73) - 1.43\}x_2$$
$$= -1.63 + 2.33x_2$$

$$\widehat{\text{slope}}(\text{at } x_3) = 1.97 - 3.36x_3$$

Plots of the estimated slope equations along the x_i-axes are drawn in Figure 5.4. The corresponding plots of the estimated flavor score along

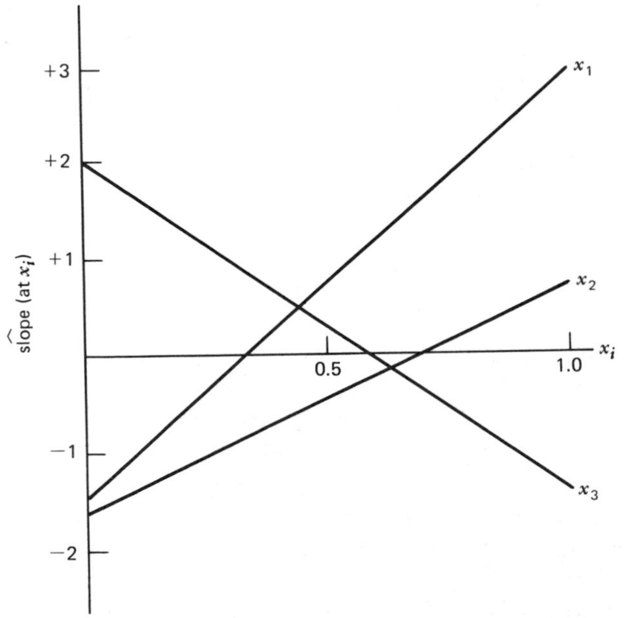

FIGURE 5.4. Plots of $\widehat{\text{slope}}$ (at x_i) versus x_i.

the x_i-axes are drawn in Figure 5.5, where

$$\begin{aligned}\hat{y}(x_1) &= 5.15x_1 + 4.13(1-x_1)/2 + 4.00(1-x_1)/2 - 3.73x_1(1-x_1)/2 \\ &\quad + 0.06x_1(1-x_1)/2 + 1.43(1-x_1)^2/4 \\ &= 4.42 - 1.47x_1 + 2.19x_1^2 \\ \hat{y}(x_2) &= 4.59 - 1.63x_2 + 1.17x_2^2 \\ \hat{y}(x_3) &= 3.71 + 1.97x_3 - 1.68x_3^2\end{aligned}$$

From the plots of the estimated slopes in Figure 5.4 (as well as the plots of $\hat{y}(x_i)$ in Figure 5.5), we see, beginning at $x_1 = 0$, that along the x_1-axis the slope is negative meaning the surface drops off until x_1 reaches 0.33 at

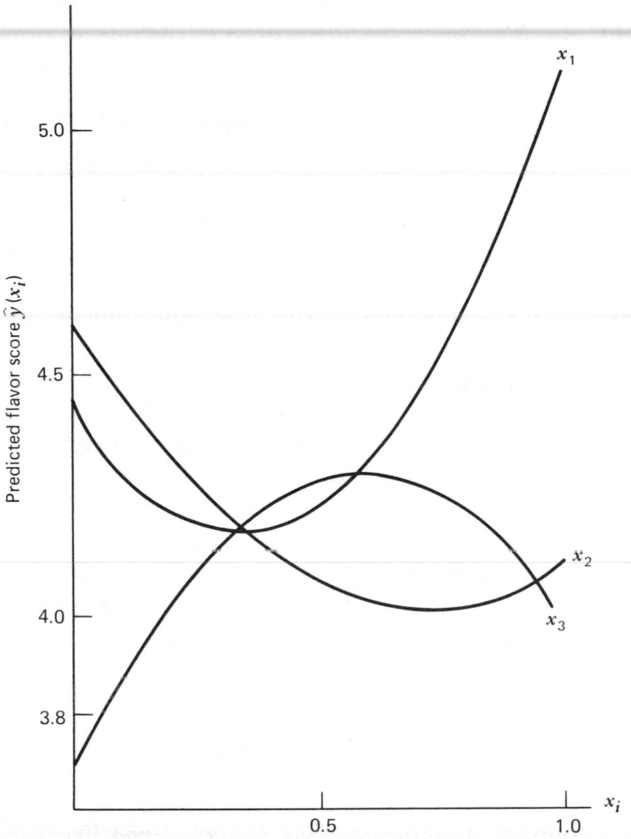

FIGURE 5.5. Plots of $\hat{y}(x_i)$ versus x_i along the x_i-axes.

which point the value of $\widehat{\text{slope}}(0.33) = 0$. For values of $x_1 > 0.33$ the slope is positive meaning the height of the surface increases. The same is true along the x_2-axis except that the surface starts rising only after $x_2 = 0.70$. Along the x_3-axis, the surfaces rises initially (for values of $x_3 > 0$) until $x_3 = 0.59$. Above this value of x_3, the height of the surface drops.

The plots of the $\widehat{\text{slope}}(\text{at } x_i)$ and of $\hat{y}(x_i)$ along the x_i-axis complement each other. The estimated surface plots in Figure 5.5 perhaps present a clearer picture of the magnitude of the surface curvature along the axes while the slope plots are more informative about the location where the surface shape changes (slope = 0) along the axes. Also, the slope plots provide quantitative measures of the rate of change of the surface at different values of x_i along the x_i-axes.

Additional discussion on the usefulness of the slope estimate as well as some suggestions for the placement of design points along the x_i-axes to increase the precision of the estimate of the slope is presented in Cornell and Ott (1975).

5.11. THE INCLUSION OF PROCESS VARIABLES: COMBINING LATTICE DESIGNS AND FACTORIAL ARRANGEMENTS

Process variables were introduced into mixture experiments in Section 3.10. Initially, an ellipsoidal region of interest was set up in the q-component mixture system, and then the mixture system was transformed to an independent variable system. The $q-1$ independent mixture related variables were combined with n independent process variables to form a $(q-1+n)$-dimensional design space. Only two-level factorial arrangements were considered for the process variables. This restriction was made on the type of design for the process variables in an attempt to keep the methodology as simple as possible.

At the beginning of Section 3.12, a polynomial model, presented as Eq. (3.53), was written to include three mixture components whose proportions were denoted by x_1, x_2, and x_3, and a single process variable defined as z_1. Assuming only linear blending to be present in the x_i system, along with the possible linear effect of z_1 on the linear blending of the x_i, the model was written as

$$y(\mathbf{x}, z_1) = \sum_{i=1}^{3} [\gamma_i^0 + \gamma_i^1 z_1]x_i + \epsilon \tag{5.50}$$

where the parameter γ_i^0 represents the average (over the two levels of z_1) expected response to pure component i (the expected response is depicted as the height of the surface above the vertex $x_i = 1$), and the parameter γ_i^1

represents the effect of z_1 on the expected response to pure component i. The effect of z_1 is taken to be one-half the difference in the heights of the surface at $x_i = 1$ measured at $z_1 = +1$ and at $z_1 = -1$.

Let us generalize Eq. (5.50) to include a second-degree model in x_1, x_2, and x_3 along with a second-degree model in two process variables z_1 and z_2. Writing these separate models as

$$y(\mathbf{x}) = \beta_1 x_1 + \beta_2 x_2 + \beta_3 x_3 + \beta_{12} x_1 x_2 + \beta_{13} x_1 x_3 + \beta_{23} x_2 x_3 + \epsilon \tag{5.51}$$

$$y(\mathbf{z}) = \alpha_0 + \alpha_1 z_1 + \alpha_2 z_2 + \alpha_{12} z_1 z_2 + \epsilon \tag{5.52}$$

the combined model is

$$\begin{aligned} y(\mathbf{x}, \mathbf{z}) = &\sum_{i=1}^{3} \left[\gamma_i^0 + \sum_{l=1}^{2} \gamma_i^l z_l + \gamma_i^{12} z_1 z_2 \right] x_i \\ &+ \sum_{i<j}^{3} \sum \left[\gamma_{ij}^0 + \sum_{l=1}^{2} \gamma_{ij}^l z_l + \gamma_{ij}^{12} z_1 z_2 \right] x_i x_j + \epsilon \end{aligned} \tag{5.53}$$

where $\gamma_i^0 = \alpha_0 \beta_i$, $\gamma_i^l = \alpha_l \beta_i$, $\gamma_i^{12} = \alpha_{12} \beta_i$, $\gamma_{ij}^0 = \alpha_0 \beta_{ij}$, $\gamma_{ij}^l = \alpha_l \beta_{ij}$, and $\gamma_{ij}^{12} = \alpha_{12} \beta_{ij}$. Equation (5.53) consists of $6 \times 4 = 24$ terms and is said to be of degree $2+2=4$ in the x's and z's.

Since we are restricting the number of levels of the process variables to two each, let us also restrict the design in the mixture components to be of the simplex-centroid class. The simplex-centroid designs contain the class of $\{q, 2\}$ simplex-lattice designs and there should not be any reason why anyone who wishes to use the $\{q, 2\}$ lattice cannot employ the same procedure that we present in this section. The use of the simplex-centroid design enables us to specify the model in the mixture components as

$$y(\mathbf{x}) = \sum_{i=1}^{q} \beta_i x_i + \sum_{i<j}^{q} \sum \beta_{ij} x_i x_j + \cdots + \beta_{12\cdots q} x_1 x_2 \cdots x_q + \epsilon$$

(see Section 2.11), whereas if we denote the process variables by z_l, $l = 1, 2, \ldots, n$, then a model in the process variables only is written as

$$y(\mathbf{z}) = \alpha_0 + \sum_{l=1}^{n} \alpha_l z_l + \sum_{l<m}^{n} \sum \alpha_{lm} z_l z_m + \cdots + \alpha_{12\cdots n} z_1 z_2 \cdots z_n + \epsilon$$

A complete simplex-centroid by 2^n factorial design is one in which at each of the $2^q - 1$ points of the simplex-centroid design, a complete 2^n factorial arrangement is set up. See Figure 5.6a for $q = 3$ and $n = 3$. Conversely, we could think of the complete design as being one where a program of mixture experiments is set up using the simplex-centroid

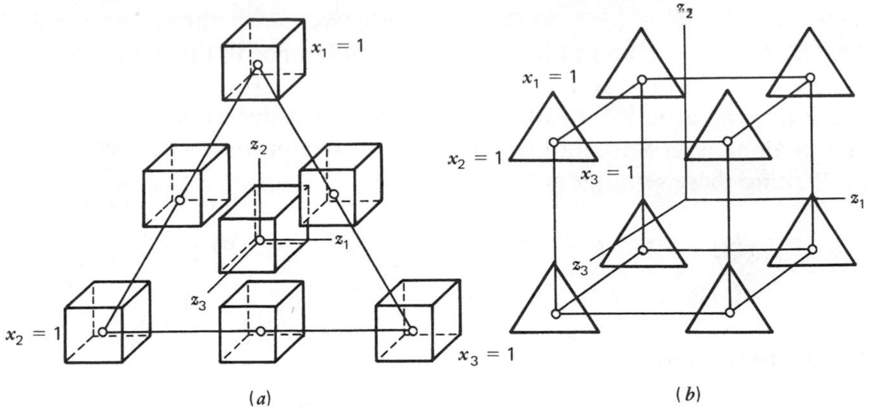

FIGURE 5.6. The complete simplex-centroid $\times 2^3$ factorial arrangement.

design at each of the 2^n combinations of the z_l, as in Figure 5.6*b*. In either case, the combined model in the q mixture components and the n process variables to be fitted to the $2^{q+n} - 2^n$ response values can be written as

$$\begin{aligned} y(\mathbf{x}, \mathbf{z}) = & \sum_{i=1}^{q} \left[\gamma_i^0 + \sum_{l=1}^{n} \gamma_i^l z_l + \cdots + \gamma_i^{12\cdots n} z_1 z_2 \cdots z_n \right] x_i \\ & + \sum_{i<j}^{q} \sum \left[\gamma_{ij}^0 + \sum_{l=1}^{n} \gamma_{ij}^l z_l + \cdots + \gamma_{ij}^{12\cdots n} z_1 z_2 \cdots z_n \right] x_i x_j + \cdots \\ & + \left[\gamma_{12\cdots q}^0 + \sum_{l=1}^{n} \gamma_{12\cdots q}^l z_l + \cdots + \gamma_{12\cdots q}^{12\cdots n} z_1 z_2 \cdots z_n \right] x_1 x_2 \cdots x_q + \epsilon \end{aligned} \tag{5.54}$$

where γ_i^0 replaces $\alpha_0\beta_i$, γ_i^l replaces $\alpha_l\beta_i, \ldots, \gamma_{Sr}^{\lambda}$ replaces $\alpha_\lambda\beta_{Sr}$, and so on; $\lambda \in \{1, 2, \ldots, n\}$ and $Sr \in \{1, 2, \ldots, q\}$. For example, when $q = 3$ and $n = 2$, Eq. (5.54) is

$$\begin{aligned} y(\mathbf{x}, \mathbf{z}) = & \sum_{i=1}^{3} [\gamma_i^0 + \gamma_i^1 z_1 + \gamma_i^2 z_2 + \gamma_i^{12} z_1 z_2] x_i \\ & + \sum_{i<j}^{3} \sum [\gamma_{ij}^0 + \gamma_{ij}^1 z_1 + \gamma_{ij}^2 z_2 + \gamma_{ij}^{12} z_1 z_2] x_i x_j \\ & + [\gamma_{123}^0 + \gamma_{123}^1 z_1 + \gamma_{123}^2 z_2 + \gamma_{123}^{12} z_1 z_2] x_1 x_2 x_3 + \epsilon \end{aligned}$$

The experimental arrangement for the complete design was defined as a 2^n factorial set up at each of the $2^q - 1$ points of the simplex-centroid, or, as a group of $2^q - 1$ mixtures set up at each combination of the factor levels of a 2^n factorial. This description of the configuration was given only to help clarify how the design configuration would look. We do not wish to suggest the performance of the experimental runs will be carried out in this manner. In an actual experiment, the $2^{q+n}-2^n$ runs would be performed in a completely random order, because if the mixtures in one simplex-centroid pattern are made at one of the factorial settings and then the simplex-centroid mixtures are made at another of the factorial settings and so on as in Figure 5.6*b*, this restricted randomization arrangement would resemble a split-plot design where the mixture blends are the subplot treatments randomized in each main plot treatment (settings of the process variables). Consequently, to perform tests of significance on the terms in the model, we would require an estimate of the subplot error variance and an estimate of the main plot error variance. To acquire estimates of these error variances, we would need to replicate the $2^{q+n}-2^n$ experimental runs more than once.

When fitting the model in the x_i's and the z_j's presented by Eq. (5.54), there does not appear to be any simple procedure to suggest for testing the individual parameters or subsets of the parameters. If computing facilities are available, what we usually do is we fit the complete model and obtain the variance–covariance matrix of all of the estimates g^{λ}_{Sr}. Tests on the individual parameters or linear functions of the parameters are then performed to see if the form of the complete model can be reduced or simplified. Procedures for testing and simplifying the model were discussed in Section 5.1.

Formulas for the variances and covariances of the parameter estimates g^{λ}_{Sr} in Eq. (5.54) are obtained as follows. We recall from Section 2.11 that the magnitudes of the variances and covariances of the parameter estimates b_{Sr}, $b_{Sr'}$ in the mixture component only model depended on the number h of common elements in Sr and Sr', where Sr and Sr' are nonempty subsets of $\{1, 2, \ldots, q\}$ of r and r' elements respectively. When $h=0$, the estimates are independent. When $h>0$ and $r \leq r'$, then $\text{cov}(b_{Sr}, b_{Sr'}) = rr'\{f(r+r', h)\}\sigma^2$ where

$$f(s, h) = (-1)^s \sum_{t=1}^{h} \binom{h}{t} t^{s-2}$$

If $r = r' = h$, then b_{Sr} and $b_{Sr'}$ are the same and from Eq. (2.39)

$$\text{var}(b_{Sr}) = \tilde{g}(r)\sigma^2 = r^2 f(2r, r)\sigma^2 = r^2 \sum_{t=1}^{r} \binom{r}{t} t^{2r-2}\sigma^2$$

As an example, if $r=2$ and $Sr = 12$, then $\text{var}(b_{12}) = 4[2(1^2)\sigma^2 + 1(2^2)\sigma^2] = 24\sigma^2$. Values of $\tilde{g}(r)$ up to $r=7$ are listed in Table 2.8.

When the model consists of x_i's and z_l's, the fitting of the combined model can be viewed in the following way. At each of the treatment combinations among the z_l's let us assume the model in the x_i's is fitted so that the $b_i, b_{ij}, \ldots, b_{12\cdots q}$ in each model are calculated like the b_{Sr} of Eq. (2.37) and their variances are $\tilde{g}(r)\sigma^2$. Now, having obtained the 2^n models in the mixture components, (one at each of the 2^n points in the 2^n design), let us fix Sr and choose the b_{Sr} from each of the 2^n models. For fixed Sr, estimates $g_{Sr}^0, g_{Sr}^l, g_{Sr}^{lm}, \ldots, g_{Sr}^{12\cdots n}$ are obtained using linear combinations of the b_{Sr} and since the b_{Sr} are independent,

$$\text{var}(g_{Sr}^0) = \text{var}(g_{Sr}^{12\cdots n}) = \tilde{g}(r)\sigma^2/2^n \tag{5.55}$$

Let us illustrate the calculations that are used for obtaining the values of the estimates g_{12}^0 and g_{12}^1 where $Sr = 12$ for the case $q = 3$ and $n = 3$ (see Figure 5.6*b*). Assume that at each of the eight process variable combinations the estimate b_{12} is calculated. Setting up linear combinations of the eight b_{12}'s the parameter estimates g_{12}^0, g_{12}^l, and g_{12}^{lm}, $l, m = 1, 2, 3, l < m$, in Eq. (5.54) are obtained. Each estimate, g_{12}^0, g_{12}^l, and g_{12}^{lm}, $l, m = 1, 2, 3, l < m$, has a variance of $\text{var}(g_{12}^0) = 24\sigma^2/8 = 3\sigma^2$ according to Eq. (5.55), and the covariances between the estimates are

$$\begin{aligned} \text{cov}(g_{Sr}^{\lambda}, g_{Sr'}^{\lambda'}) &= 0 \qquad \text{if } \lambda \neq \lambda' \\ &= rr' \frac{\{f(r+r', h)\}\sigma^2}{2^n} \qquad \text{if } \lambda = \lambda' \end{aligned} \tag{5.56}$$

where h is the number of elements common to Sr and Sr'. In particular, the linear combinations of the b_{12}'s that are used for calculating g_{12}^0 and g_{12}^1 are

$$g_{12}^0 = \tfrac{1}{8}[\text{sum of the eight } b_{12}\text{'s}] = \text{average of the } b_{12}\text{'s}$$

$$g_{12}^1 = \tfrac{1}{2}[(\text{average of the } b_{12}\text{'s where } z_1 = +1) - (\text{average of the } b_{12}\text{'s where } z_1 = -1)]$$

We now present some experimental results of a fish patty experiment which was set up and performed by the Department of Food Science and Human Nutrition located at the University of Florida in Gainesville, Florida.

5.12. PREPARING FISH PATTIES USING THREE FISH SPECIES AND THREE PROCESSING FACTORS

Three fish species (mullet, sheepshead, and croaker) were blended together to form sandwich patties. The fish comprised 98.8% of the total mixture with the remaining 1.2% being alginate (0.4%), TPP (0.3%), and salt (0.5%). Seven fish combinations were studied. They were 100% mullet, 100% sheepshead, and 100% croaker; mullet and sheepshead (50% : 50%), mullet and croaker (50% : 50%), and sheepshead and croaker (50% : 50%); and mullet, sheepshead, and croaker (33% : 33% : 33%).

The effects of three process variables on the blending behavior of the three fish were also to be studied during the experimentation. One factor was the *time of deep fat frying*, where the patties were immersed either for 25 seconds or 40 seconds in a bath at a temperature of 400°F. After deep fat frying, the patties were cooked at one of two different oven temperatures, 375° or 425°F, for one of two different cooking times, 25 or 40 minutes. *Cooking oven temperature* and *cooking time* were the other process variables. It was assumed that the range of cooking conditions from 375°F for 25 minutes to 425°F for 40 minutes was sufficiently wide to produce detectable linear effects of the process variables. In Table 5.6, z_1 represents the coded level of oven cooking temperature, z_2 the coded

TABLE 5.6. Average texture reading in grams of force $\times 10^{-3}$ taken from twin patties

Scaled Settings of the Process Variables			Mixture Composition						
z_1	z_2	z_3	(1, 0, 0)	(0, 1, 0)	(0, 0, 1)	$(\frac{1}{2}, \frac{1}{2}, 0)$	$(\frac{1}{2}, 0, \frac{1}{2})$	$(0, \frac{1}{2}, \frac{1}{2})$	$(\frac{1}{3}, \frac{1}{3}, \frac{1}{3})$
−1	−1	−1	1.84	0.67	1.51	1.29	1.42	1.16	1.59
1	−1	−1	2.86	1.10	1.60	1.53	1.81	1.50	1.68
−1	1	−1	3.01	1.21	2.32	1.93	2.57	1.83	1.94
1	1	−1	4.13	1.67	2.57	2.26	3.15	2.22	2.60
−1	−1	1	1.65	0.58	1.21	1.18	1.45	1.07	1.41
1	−1	1	2.32	0.97	2.12	1.45	1.93	1.28	1.54
−1	1	1	3.04	1.16	2.00	1.85	2.39	1.60	2.05
1	1	1	4.13[a]	1.30	2.75	2.06	2.82	2.10	2.32

[a] The observed texture value was actually higher than 4.13 but it was thought that the patties were burnt and so the value 4.13 was substituted.

level of oven cooking time, and z_3 the coded level of deep fat frying time. The values of z_1, z_2, and z_3 were computed using $z_1 = [(\text{oven temperature}) - 400°\text{F}]/25°$, $z_2 = [(\text{oven time}) - 32.5]/7.5$ and $z_3 = [(\text{frying time}) - 32.5]/7.5$.

Several properties of the patties were studied. One was the texture of the patties. The texture was measured by a compression test reading in grams of force required to initially puncture the surface of a patty. The test was conducted with an Instron Model TM. The data in Table 5.6 represents the average force in grams of six readings (three measurements taken on each of two patties) for each of the 56 combinations of the mixture components and the process variables. An estimate of the variance for each average of six readings is $s_{\bar{y}}^2 = (s_p^2 + s_e^2/3)/2 = 0.0184$ where $s_p^2 = 0.0153$ is an estimate of the between-patty variance component and $s_e^2 = 0.0644$ is an estimate of the within-patty variance component.

In preparing for the analysis of the experimental data, the kinds of questions to be answered are as follows:

1. Are there combinations of mullet, sheepshead, and croaker which are as acceptable from a texture standpoint as the single-fish patties? An acceptable texture range is somewhere between 2000 and 3500 grams of force (or in the scaled units, between 2.0 and 3.50 units).
2. How do oven cooking temperature and cooking time affect the texture of the patties? Does raising the temperature to 425°F and increasing the cooking time to 40 minutes burn the patties? Are the fish blends affected by the oven temperature differently than the pure fish patties? Do oven cooking temperature and oven cooking time act independently?
3. Does the length of time of deep fat frying affect the patty texture? Are combination-fish patties affected more or less than the pure fish patties by time of deep fat frying?
4. What are the optimal settings of oven cooking temperature–cooking time and deep fat frying time with respect to the texture of the patties? How do these optimal settings change, if at all, with the various fish combinations?

To the data in Table 5.6, a complete 56-term model of the form presented in Eq. (5.54) with $q = 3$ and $n = 3$ was fitted. The parameter estimates are presented in Table 5.7 along with their standard errors (s.e.); the interpretations of the individual estimates thought to be significantly different from zero are discussed now.

In each row of Table 5.7 estimates of the parameters thought to be nonzero are indicated by a superscript [a]. The nonzero parameters help

TABLE 5.7. Individual parameter estimates g_S^λ, and their standard errors

	Mean	z_1	z_2	z_3	z_1z_2	z_1z_3	z_2z_3	$z_1z_2z_3$	s.e.
x_1	2.87[a]	0.49[a]	0.71[a]	−0.09	0.07	−0.05	0.10	0.04	0.05
x_2	1.08[a]	0.18[a]	0.25[a]	−0.08	−0.03	−0.05	−0.03	−0.04	0.05
x_3	2.01[a]	0.25[a]	0.40[a]	0.01	0.00	0.17[a]	0.05	0.04	0.05
x_1x_2	−1.14[a]	−0.81[a]	−0.59	0.10	−0.06	0.14	−0.19	−0.09	0.23
x_1x_3	−1.00[a]	−0.54	−0.01	−0.03	−0.06	−0.27	−0.43	−0.12	0.23
x_2x_3	0.20	−0.14	0.07	−0.19	0.23	−0.25	0.12	0.27	0.23
$x_1x_2x_3$	3.18	0.07	−1.41	0.11	1.74	−0.71	1.77	−1.33	1.65

[a] Individual estimates thought to be significantly different from zero ($|t_{\text{calculated}}| > |t_{0.005,56}| = 2.67$).

to describe the effects present among the process variables at the various locations (blends) in the composition space. For example, in the row for x_2, the presence of the estimates $g_2^1 = 0.18$, $g_2^2 = 0.25$, and $g_2^3 = -0.08$ tells us that each process variable influences in some way the blending effect of component 2. The positive estimates g_2^1 and g_2^2 indicate that higher cooking temperature and higher cooking time respectively increase the firmness of the pure sheepshead patties. The negative value of $g_2^3 = -0.08$ implies that the longer time of deep fat frying (40 seconds) resulted in pure sheepshead patties with a lower average texture than the patties deep fat fried for only 25 seconds.

In the x_1x_2 row, the estimate $g_{12}^0 = -1.14$ says that a lower patty texture results when components 1 and 2 are mixed compared to the single-fish patties, where the mixed and pure patties are averaged over all combinations of the process variables. In other words, the patties formed by blending mullet with sheepshead had on the average a lower texture than did the average of the pure mullet patties with the pure sheepshead patties over all combinations of the process variables. The estimate $g_{12}^1 = -0.81$ indicates that the lower texture of the combination mullet and sheepshead patties is significantly greater at the high level of z_1 (temperature at 425°F) than at the lower level of z_1 (temperature at 375°F).

The significance of the coefficient estimate $g_3^{13} = 0.17$ indicates that an interaction is present between z_1 and z_3 with respect to the average texture of the pure croaker patties. In the pure croaker patties two-way table is presented the average texture (averaged over both levels of z_2) of the pure croaker patties along with the simple effects of the oven cooking temperature and deep fat frying time, respectively.

Pure croaker patties

		z_3 = −1	z_3 = +1	Simple effect of z_3
z_1	−1	1.91	1.60	−0.31
	+1	2.08	2.44	0.36
Simple effect of z_1		0.17	0.84	

Defining the interaction effect between z_1 and z_3 to be one-half the difference between the simple effects of z_1, the magnitude of the interaction effect is $(0.84-0.17)/2=0.34$, which is two times g_3^{13}. One interpretation of the interaction is that the difference in the average texture of the pure croaker patties at the two cooking temperatures is significantly greater with the patties that were deep fat fried for 40 seconds than with the patties that were deep fat fried for 25 seconds. (A greater cooking temperature effect was observed with the pure croaker patties deep fat fried for 40 seconds than was found with the croaker patties deep fat fried for only 25 seconds.) Or, with the pure croaker patties, the simple effect due to the time of deep fat frying was significantly different when prepared at the two cooking temperatures.

From the standpoint of the three fish types, one notices from Table 5.6 that the lowest texture value in every row is found in the sheepshead column (0, 1, 0). Furthermore, every entry in the $(\frac{1}{2}, \frac{1}{2}, 0)$ mullet : sheepshead column is lower than the corresponding entry in the mullet column (1, 0, 0), and every entry in the $(0, \frac{1}{2}, \frac{1}{2})$ sheepshead : croaker column is lower than the corresponding entry in the pure croaker column (0, 0, 1). The conclusion therefore is that the addition of sheepshead with mullet or with croaker lowers the texture of the patties regardless of the process conditions.

In an experiment of this size, so often it happens that a considerable number of effects (main effects, interactions, synergisms, etc.) are present. Any attempt to interpret the results of the overall experiment without going into great detail as we have done thus far is almost impossible. However, another way of studying how the process variables affect the blending characteristics of the components is to isolate the individual blending surfaces at each of the eight process variable treatment combinations, and then compare the contours of these surfaces. (In a similar manner we might wish to study the effects of the process variables at each point in the composition space.) Such an attempt was made with the data of Table 5.6 and the eight texture surfaces are presented in Figure 5.7. In Figure 5.7, the shaded regions in the triangles indicate blends which are

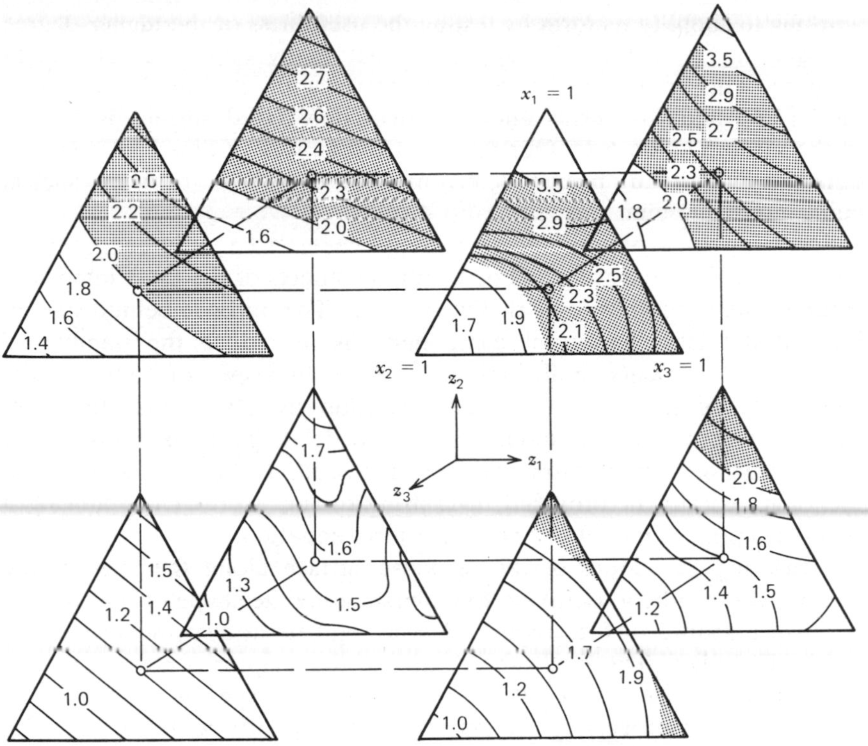

FIGURE 5.7. Contour plots of the texture at the eight process variable combinations. The shaded regions represent blends estimated to produce patties with acceptable texture values in the 2.0–3.5 × 10^3 grams range.

estimated to produce patties with acceptable texture values in the 2.00–3.50 × 10^3 grams range. A breakdown of the first-order, second-order, and special cubic surfaces is as follows: First order at $(-1,-1,1)$ and $(-1,1,-1)$; second order at $(-1,1,1)$, $(1,-1,-1)$, $(1,-1,1)$, $(1,1,-1)$ and $(1,1,1)$; and special cubic at $(-1,-1,-1)$.

5.13. SUMMARY OF CHAPTER 5

In this chapter, several techniques that are used in the analysis of mixture data were presented. In Section 5.1, three data sets were presented and served to help us initiate our model-fitting scheme. The modeling procedure that was suggested was the fitting of the complete Scheffé special cubic model and the subsequent testing of the usefulness of the terms in the model. The strategy of starting with the special cubic model and working to simplify its form by testing the usefulness of the higher-degree terms is a reversal of the approach that many experimenters might propose. This is because usually one starts with the simplest form of model and adds on terms until an adequately fitted surface is accomplished (see Section 6.1). Our reverse strategy was outlined because the data sets had already been collected and this enabled us to fit the special cubic model initially.

Screening out the unimportant components from a large group of components by measuring and detecting the effects of the most important components is another topic that we present. The purpose behind screening out the least important components is to reduce the number of essential components in the mixture system in hopes of reducing the complexity of the system. Model reduction by eliminating the nonsignificant terms in the model either by deleting components or by combining component proportions is discussed. Data from a gasoline-blending experiment provided the setting for an exercise in reducing a seven-component model to only a five-term equation.

Measuring the slopes of the response surface along the axes of the components can be a useful aid to understanding the shape of the surface. A flavor response surface arising from tasting beef and soymeal sandwich patties is studied. The final two sections of the chapter contain considerable discussion of the results of a six-variable experiment where the effects of three processing factors are studied jointly with the blending characteristics of three mixture components (salt water fish species).

5.14. REFERENCES AND RECOMMENDED READING

Becker, N. G. (1978). Models and designs for experiments with mixtures. *Austral. J. Stat.*, **20**, 195–208.

Cornell, J. A. (1975). Some comments on designs for Cox's mixture polynomial. *Technometrics*, **17**, 25–35.

Cornell, J. A. and L. Ott (1975). The use of gradients to aid in the interpretation of mixture response surfaces. *Technometrics*, **17**, 409–424.

Cox, D. R. (1971). A note on polynomial response functions for mixtures. *Biometrika*, **58**, 155–159.

Marquardt, D. W. and R. D. Snee (1974). Test statistics for mixture models. *Technometrics*, **16**, 533–537.

Mallows, C. L. (1973). Some comments on C_p. *Technometrics*, **15**, 661–675.

Park, S. H. (1978). Selecting contrasts among parameters in Scheffé's mixture models: Screening components and model reduction. *Technometrics*, **20**, 273–279.

Scheffé, H. (1963). The simplex-centroid design for experiments with mixtures. *J. R. Stat. Soc., B*, **25**, 235–263.

Snee, R. D. (1973). Techniques for the analysis of mixture data. *Technometrics*, **15**, 517–528.

Snee, R. D. and D. W. Marquardt (1976). Screening concepts and designs for experiments with mixtures. *Technometrics*, **18**, 19–29.

Wagner, T. O. and J. W. Gorman (1962). The lattice method for design of experiments with fuels and lubricants. *Applications of Statistics and Computers to Fuel and Lubricant Research Problems*, 123–145. Proceedings of a symposium sponsored by the Office of the Chief of Ordinance, Department of the Army, San Antonio, Texas.

QUESTIONS FOR CHAPTER 5

5.1. In a three component system where data is collected at the seven points of a simplex-centroid design, a decision is to be made on whether or not to fit the special cubic model to the data values collected at the points. List some advantages to fitting the complete model as well as some advantages to fitting a reduced model form.

5.2. To the data of set III in Table 5.1, fit the first-degree model $y = \beta_1 x_1 + \beta_2 x_2 + \beta_3 x_3 + \epsilon$ and test the hypothesis H_0: $\beta_1 = \beta_2 = \beta_3 = \beta_0$. Are you satisfied with a planar model fitted to this data set? If not, explain.

5.3. The following data set was collected in an experiment involving the blending of a salt water fish (x_1) with two brands (A and B) of soy protein supplement filler. Fish patties were formulated and the scores represent the sum of flavor plus texture.

$x_1 = 100\% F$	$x_2 = 70\% F : 30\% S_A$	$x_3 = 70\% F : 30\% S_B$	Replicated Scores
1.0	0	0	6.8, 6.2
0	1.0	0	3.7, 5.7
0.5	0.5	0	9.5, 10.6, 10.2
0	0	1.0	2.9, 3.5
0	0.5	0.5	8.2, 8.0, 6.2
0.5	0	0.5	11.6, 11.9, 14.3

a. Fit a first-degree model to the scores and compute the analysis of variance quantities including R_A^2. Test H_0: $\beta_1 = \beta_2 = \beta_3 =$ constant.

b. Fit a second-degree model to the scores and compute the increase in the regression sum of squares with this model over the model in **a**. Is the increase significant? What is the value of R_A^2 with the second-degree model? Calculate the C_p values associated with the first- and second-degree models using $s^2 = 1.09$.

5.4. With the three component system of 5.3, firmness measurements were recorded on patties made from the ten blends listed below.

blend:	1	2	3	4	5	6	7	8	9	10
x_1	1	0	0	$\frac{1}{2}$	$\frac{1}{2}$	0	$\frac{2}{3}$	$\frac{1}{6}$	$\frac{1}{6}$	$\frac{1}{3}$
x_2	0	1	0	$\frac{1}{2}$	0	$\frac{1}{2}$	$\frac{1}{6}$	$\frac{2}{3}$	$\frac{1}{6}$	$\frac{1}{3}$
x_3	0	0	1	0	$\frac{1}{2}$	$\frac{1}{2}$	$\frac{1}{6}$	$\frac{1}{6}$	$\frac{2}{3}$	$\frac{1}{3}$
$y_u =$	5.00	3.25	6.38	2.00	6.38	4.00	5.00	3.38	5.88	4.75

a. A special cubic model was fitted to all 10 values and the $(\mathbf{X}'\mathbf{X})^{-1}$ matrix was, along with the vector **b** of parameter estimates

$$(\mathbf{X}'\mathbf{X})^{-1} = \begin{bmatrix} 0.934 & 0.025 & 0.025 & -2.08 & -2.08 & 0.10 & 2.12 \\ & 0.934 & 0.025 & -2.08 & 0.10 & -2.08 & 2.12 \\ & & 0.934 & 0.10 & -2.08 & -2.08 & 2.12 \\ & & & 23.68 & 4.04 & 4.04 & -63.53 \\ \text{(symmetric)} & & & & 23.68 & 4.04 & -63.53 \\ & & & & & 23.68 & -63.53 \\ & & & & & & 1029.18 \end{bmatrix}$$

$$\mathbf{b} = \begin{bmatrix} b_1 \\ b_2 \\ b_3 \\ b_{12} \\ b_{13} \\ b_{23} \\ b_{123} \end{bmatrix} = \begin{bmatrix} 5.06 \\ 3.24 \\ 6.37 \\ -8.39 \\ 2.86 \\ -3.27 \\ 26.60 \end{bmatrix}$$

Complete the analysis and set up the analysis of variance table. Do you feel the special cubic model is necessary to explain the shape of the surface?

b. To the blends numbered 1, 2, 3, 4, 5, 6, and 10, a second-degree model was fitted producing the estimates, along with the $(\mathbf{X}'\mathbf{X})^{-1}$ matrix,

$$\mathbf{b} = \begin{bmatrix} 4.94 \\ 3.19 \\ 6.32 \\ -7.31 \\ 3.94 \\ -2.06 \end{bmatrix}$$

$$
(\mathbf{X}'\mathbf{X})^{-1} = \begin{bmatrix} 0.992 & -0.008 & -0.008 & -1.848 & -1.848 & 0.152 \\ & 0.992 & -0.008 & -1.848 & 0.152 & -1.848 \\ & & 0.992 & 0.152 & -1.848 & -1.848 \\ & \text{(symmetric)} & & 20.969 & 0.969 & 0.969 \\ & & & & 20.969 & 0.969 \\ & & & & & 20.969 \end{bmatrix}
$$

How does the fit of this model compare with the fit of the model in **a**?

c. In **a** the points listed as blends 7, 8, and 9 (as well as 10) might serve as check points against which to test the adequacy of the fitted model in **b**. How might we proceed to use the data at these points for a check on the fitted model in **a** and in **b**?

5.5. For the restricted estimator $\tilde{y}(\mathbf{x})$ presented in Eq. (5.25), determine the values of the estimated standard errors of the restricted coefficient estimates.

5.6. The integrated difference between the mean square errors of $\tilde{y}(\mathbf{x})$ and $\hat{y}(\mathbf{x})$ is expressed as

$$
J_D = \frac{K}{\sigma^2} \int_R \{\text{MSE}[\hat{y}(\mathbf{x})] - \text{MSE}[\tilde{y}(\mathbf{x})]\} \, dW(\mathbf{x})
$$

Let the weight function $W(\mathbf{x}) = 1/N$ at the design points and $W(\mathbf{x}) = 0$ elsewhere. Show that

$$
\hat{J}_D = \text{trace}[\mathbf{LC}(\mathbf{X}'\mathbf{X})^{-1}\mathbf{C}'\mathbf{L}'\mathbf{M}] - \mathbf{b}'\mathbf{C}'\mathbf{L}'\mathbf{MLCb}/s^2
$$

where $\mathbf{M} = (\mathbf{X}'\mathbf{X})/N$, can be written as a function of the F-statistic in Eq. (5.17) where $\mathbf{C}$ is the $r \times p$ contrast matrix. Given the formula for $\hat{J}_D$ as a function of the F-statistic, a large value of F would imply what action should be taken with regards to using $\tilde{y}(\mathbf{x})$ or $\hat{y}(\mathbf{x})$? Also, since the C_p statistic, discussed by Mallows (1973), is defined as $C_p = [\text{RSS}_p/s^2] - N + 2p$ where RSS_p is the residual sum of squares, then $\hat{J}_D$ is related to C_p. Write the formula expressing $\hat{J}_D$ as a function of C_p.

5.7. The prediction equation for the average quality rating of six blends of yarn was given by Eq. (5.13) as

$$
\hat{y}(\mathbf{x}) = 5.69x_1 + 5.82x_2 + 4.46x_3 - 0.24x_1x_2 + 3.72x_1x_3 + 3.44x_2x_3
$$

with $s^2 = 0.04$. Calculate the estimated slopes of the surface along the x_1, x_2, and x_3 axes and describe the surface shape from the slope plots. The model above was simplified to read as

$$\hat{y}(\mathbf{x}) = 5.74(x_1 + x_2) + 4.46x_3 + 3.62(x_1 + x_2)x_3$$

With this model calculate the estimate of the slope of the surface along the axis of $(x_1 + x_2)$ and compare this slope estimate with the slope estimates for x_1 and x_2 computed previously.

5.8. Continue the discussion of the results of the fish patty experiment by explaining the significance of the parameter estimates g_1^{23}, g_{13}^{23}, and g_{123}^{0}, denoted by a superscript ^{a} in Table 5.7. Based on the analysis of the combined sets of data, can you make recommendations about which fish blends to try at which processing conditions in a follow-up experiment so as to maintain the texture around 3.00×10^3 grams of force?

5.9. Fit a combined model of the form

$$\eta = \sum_{i=1}^{2} \left[\gamma_i^0 + \sum_{l=1}^{2} \gamma_i^l z_l \right] x_i + \epsilon$$

to the data in the top four rows of Table 5.6 under the column headings (1, 0, 0), (0, 1, 0) and $(\frac{1}{2}, \frac{1}{2}, 0)$. Set up an analysis of variance table and comment on your findings.

5.10. A $\{q, 2\}$ simplex-lattice is set up in q components. Let $\bar{y}_i$ represent the average of r_1 observations collected at the vertex where $x_i = 1$ for $i = 1, 2, \ldots, q$, and let $\bar{y}_{ij}$ be the average of r_2 observations collected at the binary blend $x_i = x_j = \frac{1}{2}$ for all $i, j = 1, 2, \ldots, q$, $i < j$.

a. Derive the expressions for the coefficients a, b, c, and d, in the slope formula for the second-degree model where

$$\widehat{\text{slope}}(\text{at } x_i) = a\bar{y}_i + b\sum_{j \neq i}^{q} \bar{y}_j + c\left[\sum_{l=1}^{i-1} \bar{y}_{li} + \sum_{j=i+1}^{q} \bar{y}_{ij}\right] + d \sum_{\substack{j<k \\ j,k \neq i}}^{q} \sum \bar{y}_{jk}$$

b. If the errors are independent and identically distributed $(0, \sigma^2)$, show that the variance of the slope estimate when $q = 4$, where $\widehat{\text{slope}}(\text{at } x_i) = g_0 + g_1 x_i$, is

$$\text{var}[\widehat{\text{slope}}(\text{at } x_i)] = \frac{\sigma^2}{27}\left\{\left(\frac{28}{r_1} + \frac{208}{r_2}\right) - \left(\frac{224}{r_1} + \frac{704}{r_2}\right)x_i + \left(\frac{448}{r_1} + \frac{640}{r_2}\right)x_i^2\right\}$$

5.11. An extruded product was formed by blending wheat flour (W), peanut flour (P), and corn meal (C). The product was to be a candidate for a high protein cereal or snack. The wheat flour resulted in a puffiness upon cooling after extrusion but the wheat contained only approximately 10% protein. The peanut flour was high in protein while the corn meal served mainly as a filler ingredient.

The data represent diameter measurements in millimeters of extruded and chopped two-inch rods. Twelve combinations of W–P–C were performed. The pairs of diameter values represent data collected on each of two days.

a. Fit a model to all 24 diameter readings assuming that the difference in readings from day to day is random variation. The objective is to model the rod diameter surface over the triangle.

b. Include a term in the model to represent the difference between the diameter values taken on the two days. Is the difference in the average diameter values for the two days significantly greater than zero?

c. If only rods having a diameter of at least 7.5 mm are desirable, suggest mixtures that probably yield an acceptable product.

Ingredient Proportions			Diameter Measurements	
W	P	C	Day 1	Day 2
1.00	0	0	9.40	9.18
0	1.00	0	9.00	8.91
0.50	0.50	0	7.75	8.25
0	0	1.00	6.17	5.36
0	0.50	0.50	5.54	5.40
0.50	0	0.50	7.69	7.21
0.20	0.30	0.50	6.44	6.36
0.30	0.50	0.20	7.42	6.84
0.50	0.40	0.10	7.90	8.00
0.50	0.30	0.20	7.35	7.27
0.50	0.20	0.30	7.60	6.72
0.50	0.10	0.40	7.50	7.45

APPENDIX 5A. THE DERIVATION OF THE MOMENTS OF THE SIMPLEX REGION

The region moments of the simplex factor space are used in the average variance formula when a comparison is made between the mean square error of the restricted least squares estimator $\tilde{y}(\mathbf{x}) = \mathbf{x}'\tilde{\mathbf{b}}$ and the mean square error of the standard least squares estimator $\hat{y}(\mathbf{x}) = \mathbf{x}'\mathbf{b}$. The region moments appear in Eq. (5.23) as the elements of the matrix $\mathbf{M}$. In this appendix we derive the formulas for the elements of $\mathbf{M}$.

Let us set the weighting function $W(\mathbf{x})$ to be uniform over the simplex region of interest, which we denote by R, so that

$$\mathbf{M} = K \int_R x_1^{c_1} x_2^{c_2} \cdots x_q^{c_q}\, dx_1\, dx_2 \cdots dx_q$$

where $K^{-1} = \int_R dx_1\, dx_2 \cdots dx_q$. For computational reasons, we rewrite R in terms of the first $q-1$ components as $R^* = \{(x_1, x_2, \ldots, x_{q-1}): x_i \geq 0$ and $x_1 + \cdots + x_{q-1} \leq 1\}$ where $x_q = 1 - (x_1 + \cdots + x_{q-1})$. Then

$$\int_{R^*} x_1^{c_1} x_2^{c_2} \cdots x_{q-1}^{c_{q-1}}\, dx_1\, dx_2 \cdots dx_{q-1} = \prod_{i=1}^{q-1} \frac{c_i!}{\left(\sum_{i=1}^{q-1} c_i + q - 2\right)!} \int_0^1 y^{\{\sum_{i=1}^{q-1} c_i + q - 2\}}\, dy \qquad (5A.1)$$

where the symbol ! denotes factorial.

The formulas for the elements of the matrix $\mathbf{M}$ are derived as follows. The constant $K = (q-1)!$ since in the formula for K^{-1}, all the $c_i = 0$, $i = 1, 2, \ldots, q-1$ and $0! = 1$. For the element $m_{11} = K \int_{R^*} x_1^2\, dx$, in $\int_{R^*} x_1^2\, dx$, $c_1 = 2$, $c_2 = c_3 = \cdots = c_{q-1} = 0$ so that $\int_{R^*} x_1^2\, d\mathbf{x} = (2!/q!) \int_0^1 y^q\, dy = 2/(q+1)!$ Therefore $m_{11} = 2(q-1)!/(q+1)!$ In a similar manner, for $i \neq j$,

$$m_{ij} = K \int_{R^*} x_i x_j\, d\mathbf{x} = \frac{(q-1)!}{(q+1)!} \qquad m_{iijj} = K \int_{R^*} x_i^2 x_j^2\, d\mathbf{x} = \frac{4(q-1)!}{(q+3)!}$$

$$m_{iij} = K \int_{R^*} x_i^2 x_j\, d\mathbf{x} = \frac{2(q-1)!}{(q+2)!} \qquad m_{iiij} = K \int_{R^*} x_i^3 x_j\, d\mathbf{x} = \frac{6(q-1)!}{(q+3)!} \qquad (5A.2)$$

The form of the region moment matrix $\mathbf{M}$ in Section 5.5 is

$$\mathbf{M} = K \int_{R^*} \mathbf{x}\mathbf{x}'\, d\mathbf{x} = K \int_{R^*} \begin{bmatrix} x_1^2 & x_1x_2 & x_1x_3 & x_1^2x_2 & x_1^2x_3 & x_1x_2x_3 \\ & x_2^2 & x_2x_3 & x_1x_2^2 & x_1x_2x_3 & x_2^2x_3 \\ & & x_3^2 & x_1x_2x_3 & x_1x_3^2 & x_2x_3^2 \\ & & & x_1^2x_2^2 & x_1^2x_2x_3 & x_1x_2^2x_3 \\ & \text{(symmetric)} & & & x_1^2x_3^2 & x_1x_2x_3^2 \\ & & & & & x_2^2x_3^2 \end{bmatrix} d\mathbf{x}$$

$$= \frac{1}{180}\begin{bmatrix} 30 & 15 & 15 & 6 & 6 & 3 \\ & 30 & 15 & 6 & 3 & 6 \\ & & 30 & 3 & 6 & 6 \\ & & & 2 & 1 & 1 \\ & \text{(symmetric)} & & & 2 & 1 \\ & & & & & 2 \end{bmatrix}$$

where $\mathbf{x} = (x_1, x_2, x_3, x_1x_2, x_1x_3, x_2x_3)'$ and in the products $x_1x_2x_3$, $x_1^2x_2x_3$, $x_1x_2^2x_3$ and $x_1x_2x_3^2$, the component x_3 is replaced by $1 - x_1 - x_2$.

CHAPTER 6

Other Mixture Model Forms

In Chapter 5, nearly all of the techniques that were suggested for analyzing data from mixture experiments were developed around the fitting of the Scheffé polynomials. The use of polynomial models was possible because the data evolved from (or at least was assumed to evolve from) systems that were well behaved. Well-behaved systems have surfaces that are expressible with a functional form that is continuous in all of the mixture variables and within the ranges of the experimental data, the degree of the functional form is less than or equal to three. For modeling well-behaved systems, most of the times the Scheffé polynomials are adequate.

In this chapter we investigate other types of systems for which functional forms, other than the Scheffé polynomials, are more appropriate than the Scheffé models. We begin in the next section with a slight modification made to the Scheffé polynomials, namely the inclusion of terms which are reciprocals (or inverse terms) x_i^{-1} of the component proportions. Such terms were introduced in mixture models by Draper and St. John (1977).

6.1. THE INCLUSION OF INVERSE TERMS IN THE SCHEFFÉ POLYNOMIALS

To model an extreme change in the response behavior as the value of one or more components tends to a boundary of the simplex region (i.e., where one or more $x_i \rightarrow 0$), the following equations containing inverse terms have been proposed:

$$\eta = \sum_{i=1}^{q} \beta_i x_i + \sum_{i=1}^{q} \beta_{-i} x_i^{-1} \tag{6.1}$$

$$\eta = \sum_{i=1}^{q} \beta_i x_i + \sum_{i<j}^{q} \sum \beta_{ij} x_i x_j + \sum_{i=1}^{q} \beta_{-i} x_i^{-1} \tag{6.2}$$

$$\eta = \sum_{i=1}^{q} \beta_i x_i + \sum_{i<j}^{q} \sum \beta_{ij} x_i x_j + \sum_{i<j<k}^{q} \sum \sum \beta_{ijk} x_i x_j x_k + \sum_{i=1}^{q} \beta_{-i} x_i^{-1} \tag{6.3}$$

$$\eta = \sum_{i=1}^{q} \beta_i x_i + \sum_{i<j}^{q} \sum \beta_{ij} x_i x_j + \sum_{i<j<k}^{q} \sum \sum \beta_{ijk} x_i x_j x_k + \sum_{i<j}^{q} \sum \gamma_{ij} x_i x_j (x_i - x_j) + \sum_{i=1}^{q} \beta_{-i} x_i^{-1} \tag{6.4}$$

The models presented as Eqs. (6.1)–(6.4) are extensions of the Scheffé polynomials with the additional terms of the form x_i^{-1} included to account for the possible extreme change in the response as x_i approaches zero. It is assumed that the value of x_i *never* reaches zero, but that the value could be very close to zero, that is, $x_i \to \epsilon_i > 0$ where ϵ_i is some small quantity that is defined for each application of these models. Also, although the models presented as Eqs. (6.1)–(6.4) contain x_i^{-1} terms for all $i = 1, 2, \ldots, q$, if only a couple of components are likely to produce extreme changes in the response as $x_i \to \epsilon_i$, then only these terms are included in the model.

When lower bound contraints of the form shown below are considered (as in Section 4.2),

$$0 < a_i \le x_i, \qquad \text{for all } i, \sum_{i=1}^{q} a_i < 1$$

pseudocomponents $x_i' = (x_i - a_i)/(1 - \Sigma_{i=1}^{q} a_i)$ are suggested for use in place of the x_i. If an extreme change in the response occurs as x_i approaches the boundary a_i, to model this change we define $a_i' = a_i - \epsilon_i$ where $a_i > \epsilon_i > 0$ so that instead of the previous pseudocomponent definition, the pseudocomponent is now

$$x_i' = \frac{x_i - a_i'}{1 - \sum_{i=1}^{q} a_i'} \tag{6.5}$$

When $x_i = a_i$, then $x_i' = [\epsilon_i/(1 - \Sigma_{i=1}^{q} a_i')] > 0$, so that the boundary has been transformed to the zero boundary. Terms such as $(x_i')^{-1}$ are included in the pseudocomponent model to represent extreme changes near (but not on) the zero boundary.

When, according to the design, some of the blends have $x_i = 0$, then in order to include a term like x_i^{-1} in the model, we can add a small positive amount, say c_i, to each x_i. This is equivalent to working again with the pseudocomponents x_i' where this time they are defined as

$$x_i = \frac{x_i' - c_i}{1 - \sum_{i=1}^{q} c_i} \qquad \text{or} \qquad x_i' = x_i\left(1 - \sum_{i=1}^{q} c_i\right) + c_i \tag{6.6}$$

that is, the c_i is the lower bound for x_i'. Draper and St. John (1977) suggest that when working with the pseudocomponent model, a rule of thumb for values of the c_i would be somewhere between 0.02 and 0.05, and ideally to let $c_1 = c_2 = \cdots = c_q = c$. These c_i values are suitable probably for most problems.

The need to include terms like x_i^{-1} in the models is most likely discovered during a program of sequential model fitting. Such a program might resemble the "family tree of mixture models" presented in Figure 6.1. Each box in the figure contains the types of terms contained in the

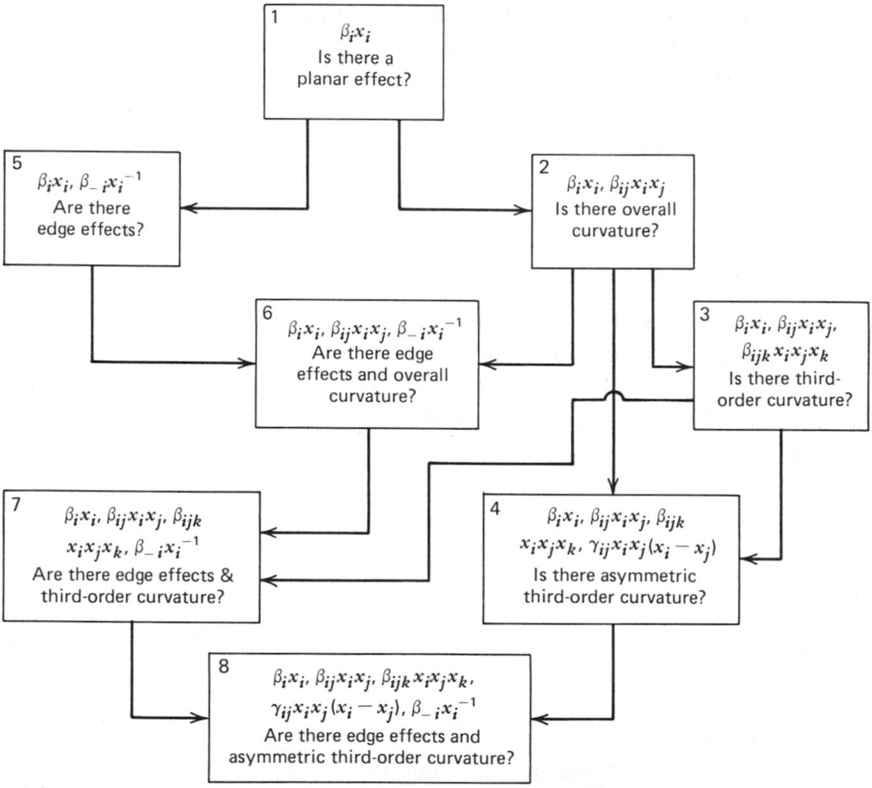

FIGURE 6.1. Sequential model fitting. Reproduced from Draper and St. John (1977), p. 40, Figure 1, with permission of the authors.

various models and the arrows indicate the possible paths that are taken when going from one model to the next. The sequential program of building up a model beginning with the Scheffé first-degree polynomial and advancing all the way if necessary to a cubic polynomial can be staged as stated by Draper and St. John (© 1977, American Statistical Association, p. 39, used by permission of the authors):

We assume that an experimenter would often follow a sequential approach when fitting Scheffé's polynomial. That is, an experimenter would first fit the linear canonical polynomial and make a judgement about the adequacy of this model to fit his data. If he deemed the model inadequate, he would then fit the second order polynomial. If he also considered this model inadequate he would then fit either the cubic canonical polynomial or the special cubic canonical polynomial.

We now illustrate the sequential model fitting approach along with the use of inverse terms in the fitted model using data from a gasoline-blending experiment.

6.2. FITTING GASOLINE OCTANE NUMBERS USING INVERSE TERMS IN THE MODEL

Research octane numbers were recorded for nine blends of an olefin (x_1), an aromatic (x_2) and a saturate (x_3) at 1.5 milliliters of lead per gallon. The octane numbers and the blending compositions, along with the pseudocomponent settings, are presented in Table 6.1.

TABLE 6.1. Research octane numbers

Blending Components			Pseudocomponents			Research Octane No. at 1.5 ml Pb/gal
x_1	x_2	x_3	x'_1	x'_2	x'_3	
0.010	0.870	0.120	0.029	0.838	0.133	111.5
0.541	0.000	0.459	0.529	0.020	0.451	101.3
0.427	0.061	0.512	0.421	0.077	0.501	80.6
0.022	0.464	0.514	0.041	0.456	0.503	91.0
0.007	0.957	0.036	0.027	0.920	0.054	107.0
0.414	0.278	0.308	0.409	0.281	0.310	97.0
0.648	0.030	0.322	0.629	0.048	0.323	98.6
0.162	0.514	0.324	0.172	0.503	0.325	92.2
0.008	0.068	0.924	0.028	0.084	0.889	77.8

Particularly noticeable are the three octane numbers (111.5, 101.3, and 107.0) that exceeded 100.0 and the respective component proportions of the blends that yielded the high octane numbers. In particular, $x_1 = 0.010$ with 111.5, $x_2 = 0.000$ with 101.3, and $x_3 = 0.030$ with 107.0. We choose to work with pseudocomponents x'_i whose levels are slightly higher than the corresponding levels $x_1 = 0.010$, $x_2 = 0.000$, and $x_3 = 0.030$ for the three blends so that the inverse terms $(x'_i)^{-1}$ could be included in the model. The lower bound for the x'_i is chosen to be 0.02, that is, $c_1 = c_2 = c_3 = c = 0.02$, $1 - \Sigma_{i=1}^{3} c_i = 0.94$, and

$$x'_i = x_i(0.94) + 0.02 \tag{6.7}$$

For example, the pseudocomponent proportions corresponding to the blend $x_1, x_2, x_3 = 0.01, 0.87, 0.12$, which produced $y_1 = 111.5$, are

$$x'_1 = 0.010(0.94) + 0.02 = 0.029$$

$$x'_2 = 0.870(0.94) + 0.02 = 0.838$$

$$x'_3 = 0.120(0.94) + 0.02 = 0.133$$

The results of the sequential buildup of the fitted Scheffé pseudocomponent polynomial with and without inverse terms is displayed in Table 6.2. In terms of producing a high value of R_A^2 or a low value for the estimate of the error variance σ^2, the linear plus complete inverse terms model, model 7, appears to perform best. Models 4 and 6 are next in the

TABLE 6.2. Values of R_A^2 and s^2 corresponding to the different models fitted to the octane numbers

Model	Fitted Model*	R_A^2	s^2
1	$\hat{y}(\mathbf{x}') = 110.0x'_1 + 110.8x'_2 + 71.1x'_3$	0.663	42.1
2	$\hat{y}(\mathbf{x}') = 145.8x'_1 + 114.1x'_2 + 73.9x'_3 - 96.1x'_1x'_2 - 77.5x'_1x'_3 - 0.5x'_2x'_3$	0.530	58.7
3	$\hat{y}(\mathbf{x}') = 120.8x'_1 + 98.3x'_2 + 59.1x'_3 + 0.4(x'_1)^{-1}$	0.712	36.0
4	$\hat{y}(\mathbf{x}') = 96.8x'_1 + 113.2x'_2 + 66.9x'_3 + 0.4(x'_2)^{-1}$	0.798	25.2
5	$\hat{y}(\mathbf{x}') = 110.1x'_1 + 111.2x'_2 + 71.0x'_3 - 0.03(x'_3)^{-1}$	0.597	50.3
6	$\hat{y}(\mathbf{x}') = 105.4x'_1 + 105.1x'_2 + 60.4x'_3 + 0.2(x'_1)^{-1} + 0.3(x'_2)^{-1}$	0.793	25.9
7	$\hat{y}(\mathbf{x}') = 117.2x'_1 + 111.2x'_2 + 48.5x'_3 + 0.5(x'_1)^{-1} + 0.3(x'_2)^{-1} - 1.1(x'_3)^{-1}$	0.888	14.0

*Fitted model obtained using the values of x'_i, $i = 1, 2, 3$, from Table 6.1.

hierarchy of the best prediction equations with the remaining models following.

The model with the single inverse term $(x_2')^{-1}$, model 4, appears to be an improvement over the use of the simple linear model, model 1. This improvement is present in the amount of increase in the R_A^2 value of 0.135, that is, from 0.663 to 0.798 (or similarly a decrease in the value of s^2 of $42.1 - 25.2 = 16.9$ or of approximately 40%). The single inverse term model, model 4, also appears to be a better fitted model than the Scheffé second-degree polynomial, model 2. Thus with this set of data, we have a good example of observing values of a surface where a second-degree equation does not improve on the fit of the surface obtained with the first-degree model but the addition of inverse terms in the first-degree model does improve on the fit.

Exercise 6.1 Calculate the value of the C_p statistic (see Question 2.12) for each of the models 1–7 in Table 6.2 using s^2 from each model. Are the results the same with C_p as we found with R_A^2? Are these statistics related? Show.

The modeling of the data using the equation of model 7 in Table 6.2 is an example of the use of an equation other than a polynomial. Another phenomena that cannot be modeled very well by polynomial equations is a system where one or more components exhibits linear blending (i.e., their effects are additive) in an otherwise nonlinear (or nonplanar surface) system. Such nonpolynomial forms can be discovered in a typical modeling program, as we see now.

6.3. AN ALTERNATIVE MODEL FORM FOR MODELING THE ADDITIVE BLENDING EFFECT OF ONE COMPONENT IN A MULTICOMPONENT SYSTEM

The occurrence of a single component exhibiting linear blending might be expected in systems where the component i serves as a diluent. As a diluent, component i has the effect of diluting the mixture in the sense that, as the proportion x_i increases, the effect of the remaining components on the response diminishes in proportion. Thus the mean or average response changes linearly as x_i is varied between 0 and 1.0 along every line in the simplex region, while fixing or holding constant the relative proportions of the remaining components.

An example of a three-component system where component 3 blends linearly, but synergism is present between components 1 and 2, is shown in Figure 6.2, where the response drops off linearly along the edges

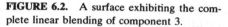

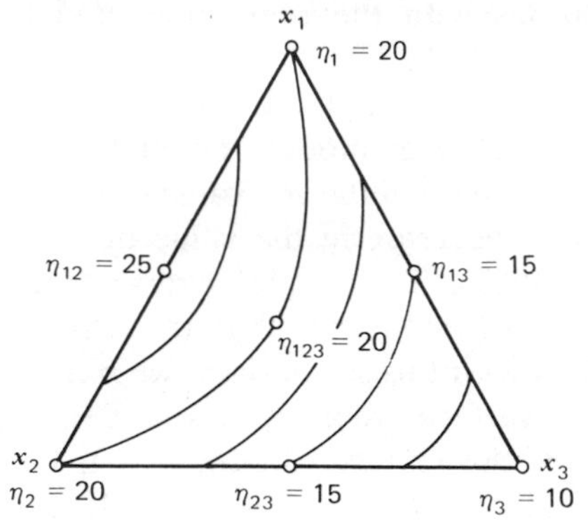

FIGURE 6.2. A surface exhibiting the complete linear blending of component 3.

connecting components 1 and 3, and components 2 and 3. On first sight one might expect that such a system could be represented by

$$\eta = \beta_1 x_1 + \beta_2 x_2 + \beta_3 x_3 + \beta_{12} x_1 x_2$$

which, according to the numbers in Figure 6.2, would be

$$\eta = 20x_1 + 20x_2 + 10x_3 + 20x_1 x_2 \tag{6.8}$$

Although Eq. (6.8) does suffice for the true binaries involving x_1 and x_3 and, x_2 with x_3, as seen by substituting $x_1 = x_3 = \frac{1}{2}$ into Eq. (6.8) to get $\eta = 15 = \eta_{13}$, the ternary blends involving component three with mixtures of components one and two, are not modeled correctly. A more correct model for the system is a nonpolynomial model of the form

$$\eta = \beta_1 x_1 + \beta_2 x_2 + \beta_3 x_3 + \delta_{12} \frac{x_1 x_2}{x_1 + x_2} \tag{6.9}$$

where the term $x_1 x_2/(x_1 + x_2)$ takes the value of zero whenever $x_1 + x_2 = 0$ such as at the vertex where $x_3 = 1$.

To understand why Eq. (6.9) is better than the special cubic model in an additive environment, we show first how the special cubic model fails to depict the linear blending of one of the components in an otherwise nonlinear or curvilinear system.

Let us assume that a three-component system has complete linear blending with respect to component 3, accompanied by the curvilinear blending of components 1 and 2. Initially, the system is modeled with the special cubic model

$$\eta = \beta_1 x_1 + \beta_2 x_2 + \beta_3 x_3 + \beta_{12} x_1 x_2 + \beta_{13} x_1 x_3 + \beta_{23} x_2 x_3 + \beta_{123} x_1 x_2 x_3$$

Since component 3 only blends linearly with components 1 and 2 (this means that component 3 has an additive effect when in combination with either component 1 or 2), then $\beta_{13}=0$ and $\beta_{23}=0$. However, components 1 and 2 blend curvilinearly, that is, $\beta_{12}\neq 0$, and so the model becomes

$$\eta=\beta_1x_1+\beta_2x_2+\beta_3x_3+\beta_{12}x_1x_2+\beta_{123}x_1x_2x_3 \tag{6.10}$$

Next, let us impose the condition that component three blends linearly with any mixture containing *both* components 1 and 2. Then along any arbitrary ray such as defined by *ab* in Figure 6.3, the response is expressible as a linear equation in x_3,

$$\eta=(1-x_3)\eta'+\beta_3x_3 \tag{6.11}$$

where η' is the response to the blend (x_1', x_2', $x_3=0$). However, curvilinear blending is present with components 1 and 2, so that

$$\eta'=\beta_1x_1'+\beta_2x_2'+\beta_{12}x_1'x_2' \tag{6.12}$$

and η' is constant along ray *ab*. A decreasing in the value of η along ray *ab* as x_3 goes from 0 to 1 is shown in Figure 6.4.

To express Eq. (6.10) in the form of Eq. (6.11), we note that along ray *ab*, the ratio x_1/x_2 is a constant, that is, the relative proportions of the components other than 3 remain fixed. If along the ray, $(x_1/x_2)=c$, then in terms of x_3

$$x_1=\frac{c(1-x_3)}{1+c}, \qquad x_2=\frac{(1-x_3)}{1+c} \tag{6.13}$$

Upon substituting these expressions for x_1 and x_2 into Eq. (6.10), a modified form of the special cubic model, which is applicable along ray *ab*, is

$$\eta=(1-x_3)\left\{\frac{\beta_1c}{1+c}+\frac{\beta_2}{1+c}+\frac{\beta_{12}(1-x_3)c}{(1+c)^2}+\frac{\beta_{123}x_3(1-x_3)c}{(1+c)^2}\right\}+\beta_3x_3 \tag{6.14}$$

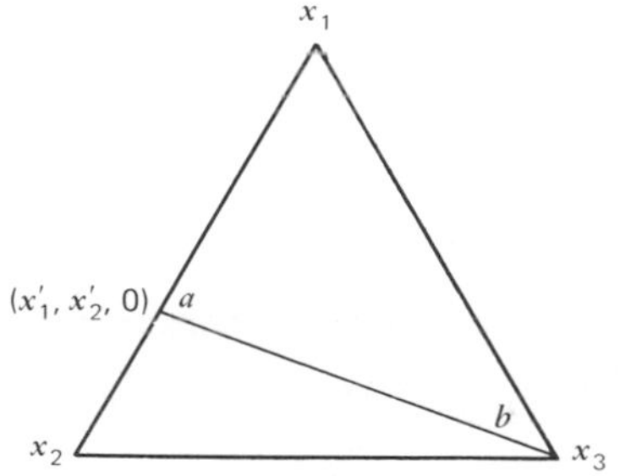

FIGURE 6.3. Ray *ab* from $x_3=0$ to $x_3=1$. Reproduced from Cornell and Gorman (1978), with permission of the Biometric Society.

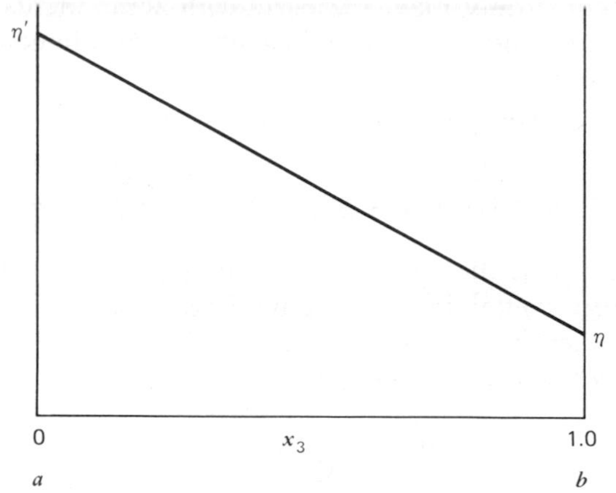

FIGURE 6.4. A decreasing linear response along ray *ab*.

In Eq. (6.14) the multiplier of $(1-x_3)$ in curly brackets corresponds to η' in Eq. (6.11), and η' is required to be constant along ray *ab*. However, the quantity in curly brackets is constant only when both β_{12} and β_{123} are zero, since both are multipliers of x_3 and x_3 varies from 0 to 1 along the ray. Furthermore, since curvilinear blending is assumed to exist between components 1 and 2, that is $\beta_{12} \neq 0$, this means the reduced special cubic polynomial equation (6.10) cannot satisfy the complete linear blending or additive blending of component 3 and at the same time account for curvilinear blending of components 1 and 2.

To derive the particular model form presented in Eq. (6.9), we notice from Eq. (6.13) that at the point a, $x_1' = c/(c+1)$. Thus η' in Eq. (6.11) can be expressed as $\eta' = (c\beta_1 + \beta_2 + c\beta_{12}/(c+1))/(c+1)$ and at any point along ray *ab*,

$$x_1' = \frac{c}{c+1} = \frac{x_1/x_2}{(x_1/x_2)+1} = \frac{x_1}{x_1+x_2}, \qquad x_2' = \frac{1}{c+1} = \frac{1}{(x_1/x_2)+1} = \frac{x_2}{x_1+x_2}$$

Upon substituting these expressions for x_1' and x_2' into Eq. (6.12) we have for η'

$$\eta' = \frac{\beta_1 x_1}{x_1+x_2} + \frac{\beta_2 x_1}{x_1+x_2} + \frac{\beta_{12} x_1 x_2}{(x_1+x_2)^2}$$

so that the linear blending of component three along the ray is expressed as

$$\begin{aligned}\eta &= (1-x_3)\eta' + \beta_3 x_3 \\ &= (1-x_3)\left\{\frac{\beta_1 x_1}{x_1+x_2} + \frac{\beta_2 x_2}{x_1+x_2} + \frac{\beta_{12} x_1 x_2}{(x_1+x_2)^2}\right\} + \beta_3 x_3\end{aligned}$$

Since $x_1 + x_2 + x_3 = 1$ everywhere on the simplex, then $(1 - x_3) = x_1 + x_2$, and therefore the nonpolynomial form in Eq. (6.9) is

$$\eta = \beta_1 x_1 + \beta_2 x_2 + \delta_{12} \frac{x_1 x_2}{x_1 + x_2} + \beta_3 x_3 \tag{6.15}$$

Equation (6.15) is used to express the complete linear blending (or additive effect) of component 3 with all blends of components 1 and 2.

We ask ourselves, "Are there clues, in terms of the coefficients of the special cubic model, that tell us the simpler model of Eq. (6.15) is the more likely alternative model?" The answer is "yes," particularly when the special cubic model for $q = 3$ is fitted to the responses at the points of the lattice design. In this case, the additive blending of component 3, particularly at the points $x_1 = x_3 = \frac{1}{2}$, $x_2 = x_3 = \frac{1}{2}$, for example, forces $\beta_{13} = \beta_{23} = 0$. Furthermore, if the expected responses at the design points are denoted by η_i, η_{ij}, and η_{ijk}, as presented in Figure 6.2, then it can be shown (see, for example, Cornell and Gorman (1978), Appendix C) that the expected response at the centroid η_{123} should be related to η_3 and η_{12} by the relation

$$\eta_{123} = \tfrac{2}{3}\eta_{12} + \tfrac{1}{3}\eta_3 \tag{6.16}$$

Also, the condition presented in Eq. (6.16) would require that the coefficients in the special cubic model be related by

$$\beta_{123} = \tfrac{3}{2}\beta_{12} \tag{6.17}$$

Thus, in the analysis of a special cubic design when $q = 3$, if two of the three b_{ij} are zero and b_{123} is $\frac{3}{2}$ times the nonzero b_{ij}, an equation of the form presented in Eq. (6.15) is a possible model. Of course, when the observations are subject to error, the relations presented in Eqs. (6.16) and (6.17) are approximate, holding only in expectation.

For illustrative purposes, let us refer to Figure 6.2 and fit the special cubic model to the data at the lattice points. We have

$$\beta_1 = \eta_1 = 20, \qquad \beta_2 = \eta_2 = 20, \qquad \beta_3 = \eta_3 = 10$$
$$\beta_{12} = 4\eta_{12} - 2(\eta_1 + \eta_2) = 4(25) - 2(20 + 20) = 20$$
$$\beta_{13} = 4(15) - 2(20 + 10) = 0,$$
$$\beta_{123} = 27\eta_{123} - 12(\eta_{12} + \eta_{13} + \eta_{23}) + 3(\eta_1 + \eta_2 + \eta_3)$$
$$\beta_{123} = 27(20) - 12(25 + 15 + 15) + 3(20 + 20 + 10) = 30$$

and the special cubic model, which fits the seven data points exactly is

$$\eta = 20x_1 + 20x_2 + 10x_3 + 20x_1x_2 + 30x_1x_2x_3 \tag{6.18}$$

Unfortunately, Eq. (6.18) produces an artificial buckling in all of the ternary blends involving component three as shown by the curved line in Figure 6.7. This buckling effect is a maximum along the ray known as the x_3-axis which goes from $x_3 = 1$, through the centroid design point, to $x_3 = 0$ at the midpoint of the opposite side. Hence we are prompted to try the model form

$$\eta = 20x_1 + 20x_2 + 10x_3 + 20\left(\frac{x_1 x_2}{x_1 + x_2}\right) \tag{6.19}$$

since this model does not produce the buckling effect shown in Figure 6.7. Also, if a single observation only is collected at each of the seven points in the special cubic design, an estimate of the error variance with 3 degrees of freedom is available if needed since the model presented in Eq. (6.19) contains only four terms.

Although we have chosen to discuss the $q = 3$ case in detail, the ideas presented in this section are extended easily to q components. With q components, if one component, say $x_{i'}$, exhibits complete linear blending, the nonpolynomial form becomes

$$\eta = \sum_{i=1}^{q} \beta_i x_i + \frac{\sum\sum \beta_{ij} x_i x_j}{\sum_{\substack{i=1 \\ i \neq i'}}^{q} x_i}$$

If in addition the components which blend nonlinearly require cubic or quartic models, the divisors $D = \Sigma_{i=1}^{q} x_i$, $i \neq i'$, of the polynomial numerators are raised to a power one less than the degree of the polynomial term in the numerator. An example is the full cubic model in $q = 4$ components with the blending of the single component x_1 being completely additive with the other components so that the model is

$$\eta = \sum_{i=1}^{4} \beta_i x_i + \sum_{2 \le i < j}^{4}\sum \frac{\beta_{ij} x_i x_j}{1 - x_1} + \sum_{2 \le i < j}^{4}\sum \frac{\gamma_{ij} x_i x_j (x_i - x_j)}{(1 - x_1)^2} + \frac{\beta_{234} x_2 x_3 x_4}{(1 - x_1)^2}$$

We now present a three-component example where two components represent liquid pesticides and the third component represents a wettable powder pesticide. The pesticides are used to control mite numbers on plants.

6.4. A BIOLOGICAL EXAMPLE ON THE LINEAR EFFECT OF A POWDER PESTICIDE IN COMBINATION WITH TWO LIQUID PESTICIDES USED FOR SUPPRESSING MITE POPULATION NUMBERS

An experiment was designed to measure the suppression of the population numbers of mites on strawberry plants. During the experimental program, three pesticides were applied biweekly to the strawberry plants which had been infested with the mites prior to the start of the experiment. The data in Table 6.3 represent the seasonal average mite population numbers determined on a per leaf basis where the averages were calculated using data recorded from five plants sampled from each plot and taken at six biweekly dates. Prior to spraying the plants at the beginning of the experiment, 10 plants were positioned in every experimental plot and 50 mites were placed on five leaves of each of the 10 plants.

Initially the plan was to observe the average mite numbers corresponding to the first seven mixtures in Table 6.3 only. However, because component 3 is a wettable powder and it was suspected that the effect of the powder would be additive when used in combination with the two liquids whose proportions are represented by x_1 and x_2, eight additional

TABLE 6.3. Strawberry mite experimental data

	Component Proportions			Observed Numbers	Predicted Numbers Using Special Cubic Eq. (6.20)	Predicted Numbers Using Nonpolynomial Eq. (6.21)
Run	x_1	x_2	x_3	y_u		
1	1.00	0	0	49.8	49.7	49.2
2	0.50	0.50	0	35.8	36.8	35.4
3	0	1.00	0	84.2	83.4	83.5
4	0	0.50	0.50	52.4	52.1	52.3
5	0	0	1.00	20.1	20.2	21.0
6	0.50	0	0.50	34.7	34.3	35.1
7	0.20	0.20	0.60	26.1	28.2	26.8
8	0	0.75	0.25	66.0	67.9	67.9
9	0	0.25	0.75	39.4	36.2	36.6
10	0.25	0	0.75	28.8	27.1	28.0
11	0.75	0	0.25	41.3	41.8	42.1
12	0.40	0.40	0.20	32.7	30.7	32.5
13	0.30	0.30	0.40	29.6	28.8	29.6
14	0.25	0.25	0.50	27.9	28.5	28.2
15	0.10	0.10	0.80	23.3	26.3	23.9

spray combinations, listed as run numbers 8–15 in Table 6.3, were prepared. The design and observed mite numbers are presented in Figure 6.5.

A special cubic model was fitted to all 15 values resulting in

$$\begin{aligned}\hat{y}(\mathbf{x}) = {} & \underset{(2.03)}{49.70x_1} + \underset{(2.03)}{83.42x_2} + \underset{(1.80)}{20.21x_3} - \underset{(10.14)}{119.14x_1x_2} - \underset{(9.02)}{2.55x_1x_3} + \underset{(9.02)}{1.24x_2x_3} \\ & - \underset{(54.30)}{232.54x_1x_2x_3} \end{aligned} \tag{6.20}$$

The quantities in parentheses below the coefficient estimates are the estimated standard errors of the estimates and the analysis of variance is presented in Table 6.4.

At first glance Eq. (6.20) appears to be as good as any other model form because with this Eq. (6.20), $R_A^2 = 0.985$. From Table 6.4, an estimate of the error variance is $s^2 = 4.65$. However, particularly noticeable in Eq. (6.20) are the extreme differences in the magnitudes of the estimates of the binary parameters β_{12}, β_{13} and β_{23} where $b_{12} = -119.14$, $b_{13} = -2.55$, and $b_{23} = 1.24$. Also, the magnitude of the estimate $b_{123} = -232.54$, which is significantly less than zero, is approximately twice the magnitude of b_{12}. The relative sizes of b_{123} and b_{12} are a little disturbing, since it suggests the

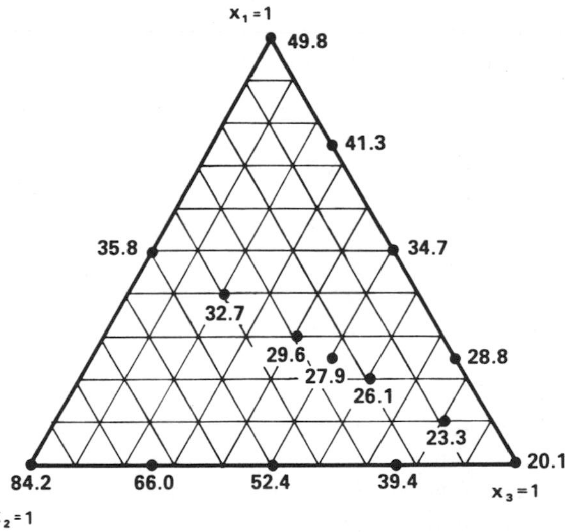

FIGURE 6.5. Observed seasonal average mite population numbers. Reproduced from Cornell and Gorman (1978), with permission of the Biometric Society.

TABLE 6.4. The analysis of variance for the mite data with the special cubic equation (6.20)

Source of Variation	Degrees of Freedom	Sum of Squares	Mean Square	R_A^2
Regression	6	4187.23	697.87	
Residual	8	37.24	4.65	0.985
Total	14	4224.47		

existence of a ternary antagonistic effect twice the size of the largest binary effect and we are nearly sure there should not be much of a ternary effect at all. Hence, since b_{13} and b_{23} are both nearly zero and $b_{123} \approx 2b_{12}$, which approximates the relation shown in Eq. (6.17), it was decided to fit the nonpolynomial model in Eq. (6.9). Fitting the nonpolynomial equation (6.9) to the 15 observation values in Table 6.3 produced

$$\hat{y}(\mathbf{x}) = \underset{(0.89)}{49.18x_1} + \underset{(0.89)}{83.54x_2} + \underset{(0.65)}{20.98x_3} - \underset{(3.90)}{123.82}\frac{x_1x_2}{x_1+x_2}, \qquad (6.21)$$

With this fitted model, the error variance estimate is $s^2 = 1.40$. The adjusted R_A^2 value with Eq. (6.21) is $R_A^2 = 0.995$ and a contour plot of the estimated surface using predicted values from Eq. (6.21) is presented in Figure 6.6.

The linear terms for the models of Eqs. (6.20) and (6.21) are similar in size, and so are the coefficients for the x_1x_2 term of the special cubic model and the $x_1x_2/(x_1+x_2)$ term of the nonpolynomial form. Thus, calculated values (predicted response values) from each model form agree reasonably well for binary mixtures but differ considerably for three-component blends at points along the x_3-axis as shown on Figure 6.7. On the x_3-axis the special cubic model tends to fit the points near $x_3 = 0, \frac{1}{2}$ and 1.0, but the model exhibits a cubic buckling between these points. Therefore to model the additive effect of a component, a model other than the special cubic equation is recommended.

Several other types of model forms exist that would fit the data just as well as the model presented in Eq. (6.21). Two equation forms which come to mind are models that are homogeneous of degree one where in Eq. (6.21) the term $\delta_{12}x_1x_2/(x_1+x_2)$ is replaced by $\delta'_{12}\min(x_1, x_2)$ or by $\delta'_{12}(x_1x_2)^{1/2}$. Each of these models would produce a different set of surface contours for Figure 6.6 as we show in the next section while providing

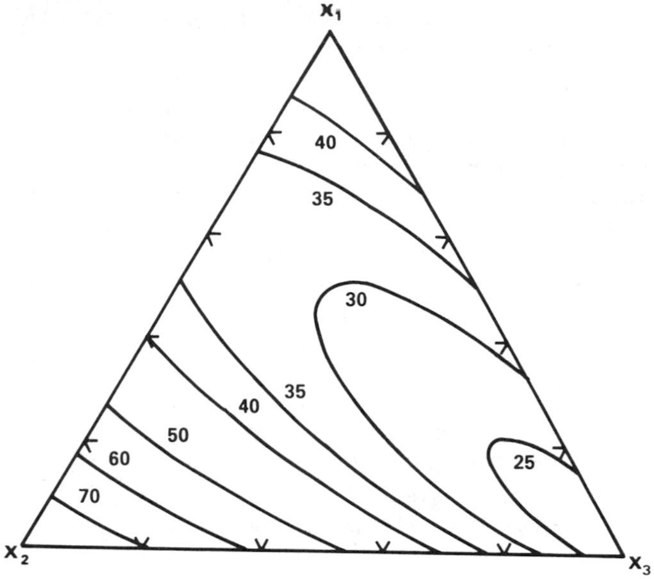

FIGURE 6.6. Contours of constant estimated seasonal average mite population numbers. Reproduced from Cornell and Gorman (1978), with permission of the Biometric Society.

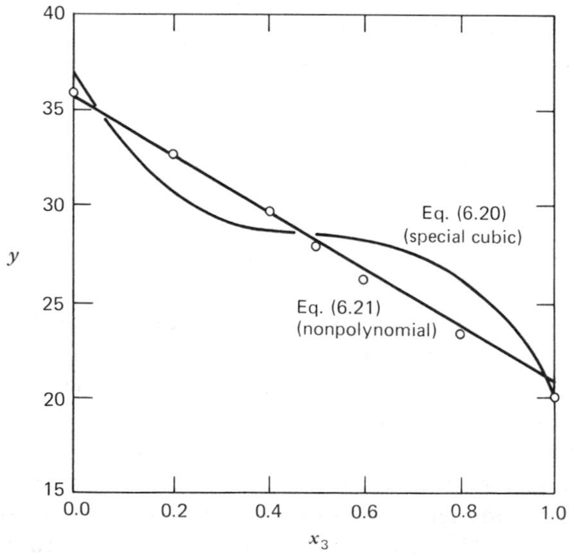

FIGURE 6.7. The fit of the models in Eqs. (6.20) and (6.21) along the x_3-axis. Reproduced from Cornell and Gorman (1978), with permission of the Biometric Society.

approximately equivalent fits to the data, however. The discrepancies in the contours arise because data in the example were taken along the three rays emanating from the $x_3 = 1$ vertex only. If additional data had been taken on the $x_3 = 0$ side resulting in different values to be realized by the terms x_1x_2, $\min(x_1, x_2)$ and $(x_1x_2)^{1/2}$, then some differences among the models might be observed also. In fact, data at these points could be used to determine which of the three models most closely resembles the true surface.

6.5. OTHER MODELS THAT ARE HOMOGENEOUS OF DEGREE ONE

We were able to observe with the three-component example of the last section, how the polynomial model fails to satisfy the modeling of the additive effect of one component and at the same time accommodate the curvilinear blending effects of the remaining two components. To model these effects jointly, a composite equation was required consisting of the first-degree terms from a polynomial with an additional term that is homogeneous of degree one. Since quite often the need to use composite model forms is the rule rather than the exception, we pursue this topic further by discussing some additional models that are homogeneous of degree one. The interested reader is directed to the papers by Becker (1968) and Snee (1973), listed at the end of this chapter, for additional discussion on fitting models of the homogeneous type.

In order to express functional forms that are capable of modeling systems containing components with additive effects, we begin by assuming the system consists only of $q-1$ components and write the function

$$\eta_{q-1} = f(x_1, \ldots, x_{q-1}) \tag{6.22}$$

to represent the response to mixtures of the $q-1$ components. An additional component, represented by x_q, is said to have an additive effect with respect to any mixture of the first $q-1$ components if the response to the q-component system is given by

$$\eta_q = \beta_q x_q + (1 - x_q) f\left\{\frac{x_1}{1-x_q}, \frac{x_2}{1-x_q}, \ldots, \frac{x_{q-1}}{1-x_q}\right\} \tag{6.23}$$

The particular form of Eq. (6.23) suggests that the interpretation of the coefficients of the terms in the functional form $f\{\ \}$ must be invariant under the addition to the mixture of the additive component, component q.

To illustrate the invariance principle for interpreting the coefficients, let $q-1=3$ and suppose the fourth component whose proportion is denoted by x_4 is thought to have an additive effect. Let us write the functional form $f(x_1, x_2, x_3)$ in Eq. (6.22) as a second-degree model. Then the polynomial term x_1x_2 in $f(x_1, x_2, x_3)$ of Eq. (6.22) becomes $x_1x_2/(1-x_4)$ in Eq. (6.23) and the interpretation of this term may change with changing values of x_3 even though values of x_1 and x_2 are fixed. However, to ensure that the interpretation of the coefficients of the terms in $f\{\ \}$ in Eq. (6.23) be invariant under the introduction of the fourth component with an additive effect, we require the equality hold

$$f(x_1, x_2, x_3) = (1-x_4)f\left\{\frac{x_1}{1-x_4}, \frac{x_2}{1-x_4}, \frac{x_3}{1-x_4}\right\} \tag{6.24}$$

With q-components, the equality is written as

$$f(x_1, \ldots, x_{q-1}) = \left(\sum_{i=1}^{q-1} x_i\right) f\left\{\frac{x_1}{\sum_{i=1}^{q-1} x_i}, \frac{x_2}{\sum_{i=1}^{q-1} x_i}, \ldots, \frac{x_{q-1}}{\sum_{i=1}^{q-1} x_i}\right\} \tag{6.25}$$

The expression in Eq. (6.24) imposes no restriction on $f\{\ \}$ over the triangle defined by $x_1+x_2+x_3=1$. However, when $x_4>0$, what is required in $f\{\ \}$ is the property that for any t

$$f(tx_1, tx_2, tx_3) = \left(\sum_{i=1}^{3} tx_i\right) f\left\{\frac{tx_1}{\sum_{i=1}^{3} tx_i}, \frac{tx_2}{\sum_{i=1}^{3} tx_i}, \frac{tx_3}{\sum_{i=1}^{3} tx_i}\right\} = tf\{x_1, x_2, x_3\} \tag{6.26}$$

The equality in Eq. (6.26) indicates simply that $f(x_1, x_2, x_3)$ should be *homogeneous of degree one*.

There are several model forms each homogeneous of degree one which satisfy the invariance criterion of Eq. (6.26). Three models which were suggested by Becker (1968) are

$$\text{H1:} \quad \eta = \sum_{i=1}^{q} \beta_i x_i + \sum_{i<j}^{q}\sum \beta_{ij} \min(x_i, x_j) + \cdots + \beta_{12\cdots q} \min(x_1, x_2, \ldots, x_q) \tag{6.27}$$

$$\text{H2:} \quad \eta = \sum_{i=1}^{q} \beta_i x_i + \sum_{i<j}^{q}\sum \frac{\beta_{ij}x_ix_j}{(x_i+x_j)} + \cdots + \frac{\beta_{12\cdots q}x_1x_2\cdots x_q}{(x_1+\cdots+x_q)^{q-1}} \tag{6.28}$$

$$\text{H3:} \quad \eta = \sum_{i=1}^{q} \beta_i x_i + \sum_{i<j}^{q}\sum \beta_{ij}(x_ix_j)^{1/2} + \cdots + \beta_{12\cdots q}(x_1x_2\cdots x_q)^{1/q} \tag{6.29}$$

H1, H2, and H3 are three different model forms that are homogeneous of degree one. Equations (6.27), (6.28), and (6.29) will be designated as models H1, H2, and H3, respectively. In H2, if the denominator of a term is zero, the value of the term is defined to be zero.

The nature of the additivity in these models is seen by observing that terms consisting of more than a single component can be reduced to terms involving only one component. For example, for $q \geq 3$, and if $x_1 = rx_2$, then in H1, H2, and H3, respectively we have

$$\text{H1:}\quad \min(x_1, x_2) = \min(r, 1)x_2, \qquad \text{H2:}\quad \frac{x_1x_2}{x_1 + x_2} = \frac{rx_2}{1+r}$$

$$\text{H3:}\quad (x_1x_2)^{1/2} = r^{1/2}x_2 \tag{6.30}$$

compared with $x_1x_2 = rx_2^2$ which arises with the second-degree polynomial. Thus by the reduction of a crossproduct term to one involving only the single component 2 in Eq. (6.30), the response is linearly expressible in x_2. Similarly, when $q \geq 4$ and $x_i = rx_j$ and $x_k = sx_j$, terms consisting of products are linearly expressed in x_j as

$$\text{H1:}\quad \min(x_i, x_j, x_k) = \min(r, 1, s)x_j, \qquad \text{H2:}\quad \frac{x_ix_jx_k}{(x_i + x_j + x_k)^2} = \frac{rsx_j}{(1+r+s)^2}$$

$$\text{H3:}\quad (x_ix_jx_k)^{1/3} = (rs)^{1/3}x_j$$

compared with rsx_j^3 for the polynomial and therefore the response is easily depicted as behaving linearly over the range of the single component whose proportion is x_j.

Implicit in the forms of the models H1, H2, and H3 is the assumption that the surface attains a maximum (or minimum) at the centroid of the simplex. To overcome this restriction, however, a "decentralized" form of the model in Eq. (6.27) for example is defined by replacing $\min(x_1, x_2, \ldots, x_q)$ with $\min(x_1/r_1, x_2/r_2, \ldots, x_q/r_q)$ to reflect the possible noncentrality, where the r_i are additional parameters whose values are constrained by the relations $r_i \geq 0$ and $\Sigma_{i=1}^{q} r_i = 1$. The decentralized form of H1 is

$$\text{H1}':\quad \eta = \sum_{i=1}^{q} \beta_i x_i + \sum_{i<j}^{q}\sum \beta_{ij} \min\left(\frac{x_i}{r_i}, \frac{x_j}{r_j}\right) + \cdots + \beta_{12\cdots q} \min\left(\frac{x_1}{r_1}, \frac{x_2}{r_2}, \ldots, \frac{x_q}{r_q}\right) \tag{6.31}$$

The point whose coordinates are expressible as $(r_1, r_2, \ldots, r_q)$ is the location on the simplex at which the model expressed by Eq. (6.31) is centered. If the coordinates $(r_1, r_2, \ldots, r_q)$ are unknown, one suggestion might be to use the coordinates of that mixture which, during the experimentation, yielded the maximum response.

Let us compare the models defined as H1, H2, and H3, by fitting each of them to the strawberry mite data which was presented in Table 6.3. For convenience we shall use matrices to illustrate the least squares fitting of the H1 model minus the $\beta_{123}\min(x_1, x_2, x_3)$ term. For $q=3$, the model is

$$\eta = \sum_{i=1}^{3} \beta_i x_i + \beta_{12}\min(x_1, x_2) + \beta_{13}\min(x_1, x_3) + \beta_{23}\min(x_2, x_3)$$

Over the $N=15$ observations presented in Table 6.3, the matrix form of the model is

$$\mathbf{y} = \mathbf{X}\boldsymbol{\beta} + \boldsymbol{\epsilon}$$

where

$$\mathbf{Y} = \begin{bmatrix} 49.8 \\ 35.8 \\ 84.2 \\ 52.4 \\ 20.1 \\ 34.7 \\ 26.1 \\ 66.0 \\ 39.4 \\ 28.8 \\ 41.3 \\ 32.7 \\ 29.6 \\ 27.9 \\ 23.3 \end{bmatrix}, \quad \mathbf{X} = \begin{bmatrix} 1.0 & 0 & 0 & 0 & 0 & 0 \\ 0.5 & 0.5 & 0 & 0.5 & 0 & 0 \\ 0 & 1.0 & 0 & 0 & 0 & 0 \\ 0 & 0.5 & 0.5 & 0 & 0 & 0.5 \\ 0 & 0 & 1.0 & 0 & 0 & 0 \\ 0.5 & 0 & 0.5 & 0 & 0.5 & 0 \\ 0.2 & 0.2 & 0.6 & 0.2 & 0.2 & 0.2 \\ 0 & 0.75 & 0.25 & 0 & 0 & 0.25 \\ 0 & 0.25 & 0.75 & 0 & 0 & 0.25 \\ 0.25 & 0 & 0.75 & 0 & 0.25 & 0 \\ 0.75 & 0 & 0.25 & 0 & 0.25 & 0 \\ 0.4 & 0.4 & 0.2 & 0.4 & 0.2 & 0.2 \\ 0.3 & 0.3 & 0.4 & 0.3 & 0.3 & 0.3 \\ 0.25 & 0.25 & 0.5 & 0.25 & 0.25 & 0.25 \\ 0.1 & 0.1 & 0.8 & 0.1 & 0.1 & 0.1 \end{bmatrix} \quad \boldsymbol{\beta} = \begin{bmatrix} \beta_1 \\ \beta_2 \\ \beta_3 \\ \beta_{12} \\ \beta_{13} \\ \beta_{23} \end{bmatrix}$$

and $\boldsymbol{\epsilon}$ is the 15×1 vector of errors. The least squares estimates of the elements of $\boldsymbol{\beta}$ are

$$\mathbf{b} = (\mathbf{X}'\mathbf{X})^{-1}\mathbf{X}'\mathbf{y} = \begin{bmatrix} 49.56 \\ 83.38 \\ 21.24 \\ -61.68 \\ -1.75 \\ 0.30 \end{bmatrix}$$

and the H1 prediction equation is of the form

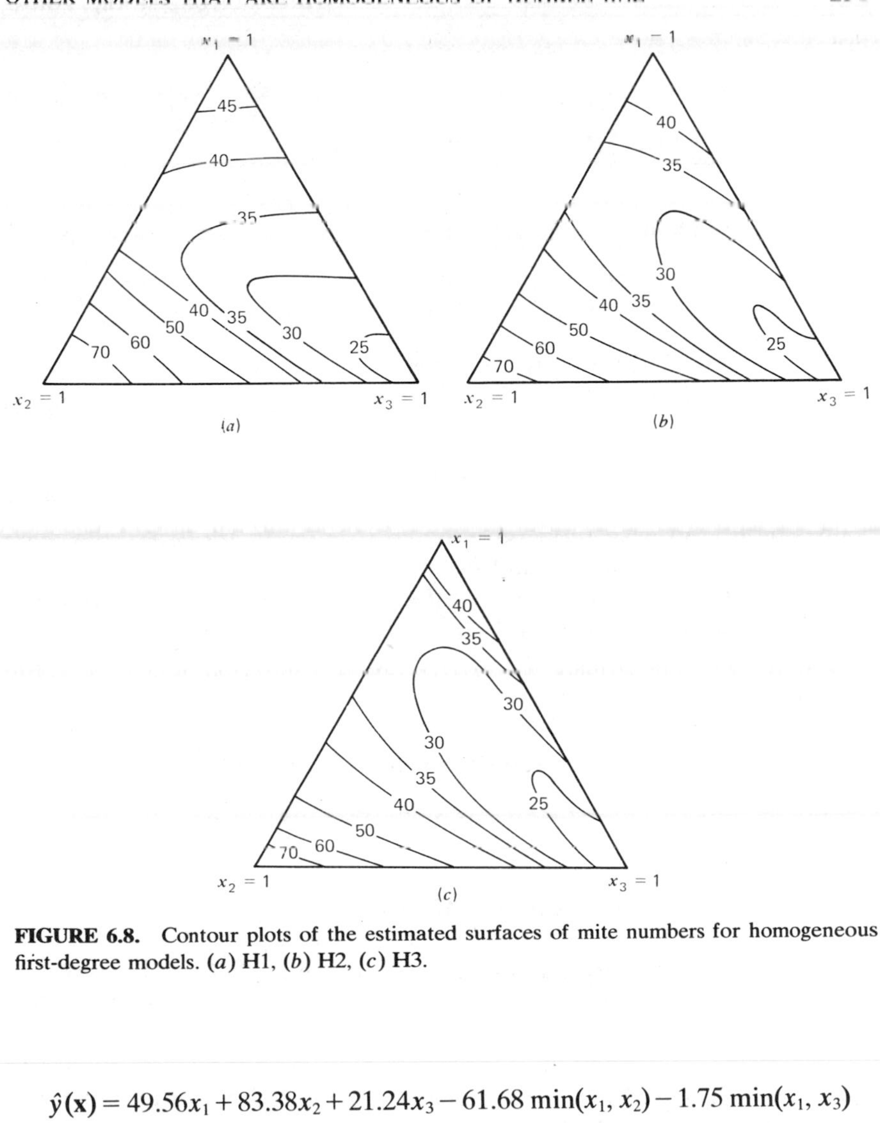

FIGURE 6.8. Contour plots of the estimated surfaces of mite numbers for homogeneous first-degree models. (*a*) H1, (*b*) H2, (*c*) H3.

$$\begin{aligned}\hat{y}(\mathbf{x}) = \underset{(1.11)}{49.56x_1} + \underset{(1.11)}{83.38x_2} + \underset{(0.92)}{21.24x_3} - \underset{(2.17)}{61.68\min(x_1, x_2)} - \underset{(2.53)}{1.75\min(x_1, x_3)} \\ + \underset{(2.53)}{0.30\min(x_2, x_3)} \qquad (6.32)\end{aligned}$$

The value of R_A^2 with Eq. (6.32) is $R_A^2 = 1 - (1.62/301.75) = 0.9946$ where $s^2 = 1.62$.

Using similar matrix manipulations to obtain the parameter estimates for the H2 and H3 models, these models and the corresponding R_A^2 values, are

$$\text{H2:}\quad \underset{}{\hat{y}(\mathbf{x})} = \underset{(1.12)}{49.63x_1} + \underset{(1.12)}{83.36x_2} + \underset{(0.99)}{21.34x_3} - \underset{(4.39)}{123.14x_1x_2/(x_1+x_2)}$$

$$\underset{(4.36)}{-3.40x_1x_3/(x_1+x_3)} + \underset{(4.36)}{0.39x_2x_3/(x_2+x_3)} \quad R_A^2 = 0.9947 \qquad (6.33)$$

$$\text{H3:}\quad \hat{y}(\mathbf{x}) = \underset{(1.11)}{49.66x_1} + \underset{(1.11)}{83.37x_2} + \underset{(1.04)}{21.45x_3} - \underset{(2.18)}{61.50(x_1x_2)^{1/2}} - \underset{(2.03)}{1.74(x_1x_3)^{1/2}}$$

$$\underset{(2.03)}{+0.03(x_2x_3)^{1/2}} \quad R_A^2 = 0.9948 \qquad (6.34)$$

All three models are similar, particularly in the magnitudes of their parameter estimates b_1, b_2, and b_3 (and standard errors) and in their values of R_A^2. In each form represented by Eqs. (6.32)–(6.34), the magnitudes of the estimates b_{13} and b_{23} are small relative to the estimate b_{12} indicating that binary curvilinear effects involving x_3 are probably not important (perhaps nonexistent) and that the surface is mainly linear with respect to changes in the value of x_3. We also note the similarity in the magnitudes of the b_{12} estimate with the H1 and H3 models plus the fact that these b_{12} estimates are approximately one-half the size of b_{12} in the H2 model. Contour plots of constant response for the H1, H2, and H3 systems are presented in Figure 6.8.

6.6. THE USE OF RATIOS OF COMPONENTS

In some areas of mixture experimentation, one is likely to be interested in one or more of the components not so much from the standpoint of their proportions in the mixtures, but from their relationship to the other components in the mixtures in the form of ratios of their proportions. In the manufacturing of a particular type of porcelain glass, for example, it is sometimes more meaningful to think of the ratio of silica to soda rather than to look at the actual proportions of each in a blend. A third ingredient, lime, might also be studied by looking at the ratio of silica to lime, or by looking at the ratio of soda to lime.

The use of ratios of components can be handled in a variety of ways. With three components whose proportions are denoted by x_1, x_2, and x_3, several possible sets of transformations from the x_i to the ratio variables r_1 and r_2 are

Set I

$$r_1 = \frac{x_2}{x_1}, \qquad r_2 = \frac{x_2}{x_3}$$

Set II

$$r_1 = \frac{x_1}{x_2}, \qquad r_2 = \frac{x_2}{x_3} \tag{6.35}$$

Set III

$$r_1 = \frac{x_1}{x_2 + x_3}, \qquad r_2 = \frac{x_2}{x_3}$$

In each set of ratios, each ratio r_i, $i = 1$ or 2, contains at least one of the components used in the other ratio of the same set. The number of ratios r_i in each set should be less than the number of components in the system. If the number of ratios in a set equals q, the ratios form a redundant set because the sum of the component proportions is unity. Note that any type of ratio can be used in a set as long as there is a tie-in with a component in one of the other ratios in the same set.

A simple first-degree model in r_1 and r_2 is

$$\eta = \alpha_0 + \alpha_1 r_1 + \alpha_2 r_2$$

which becomes in x_1, x_2, and x_3, with the first set in Eqs. (6.35):

$$\eta = \alpha_0 + \alpha_1\left(\frac{x_2}{x_1}\right) + \alpha_2\left(\frac{x_2}{x_3}\right)$$

The model is fitted to data collected at design point settings chosen in the r_i's. The corresponding settings of the x_i in terms of r_1 and r_2 are then

$$x_1 = \frac{r_2}{r_1 + r_1 r_2 + r_2}$$

$$x_2 = \frac{r_1 r_2}{r_1 + r_1 r_2 + r_2}$$

$$x_3 = \frac{r_1}{r_1 + r_1 r_2 + r_2}$$

Let us illustrate with a numerical example the construction of a factorial design in the ratios r_1 and r_2 and the subsequent fitting of a second-degree model in r_1 and r_2 to data collected at the design points. Let us assume that a three-component system currently operates at the composition $\mathbf{x}_0 = (x_{10}, x_{20}, x_{30})' = (0.40, 0.20, 0.40)'$ but that the region in which the response surface is to be studied is defined by the constraints

$$0.2 \leq x_1 \leq 0.6, \qquad 0.1 \leq x_2 \leq 0.4, \qquad 0.2 \leq x_3 \leq 0.6$$

The constrained region is drawn in Figure 6.9.

The ratios of interest are defined as

$$r_1 = \frac{x_3}{x_1}, \qquad r_2 = \frac{x_3}{x_2}$$

Specifically, we shall want to look at values of r_1 of 0.5, 1.0, and 1.5 in combination with values of r_2 of 1.0, 2.0, and 3.0. A second-degree model of the form

$$y = \alpha_0 + \alpha_1 r_1 + \alpha_2 r_2 + \alpha_{11} r_1^2 + \alpha_{22} r_2^2 + \alpha_{12} r_1 r_2 + \epsilon \tag{6.36}$$

is to be fitted to observations collected at the $3^2 = 9$ design points.

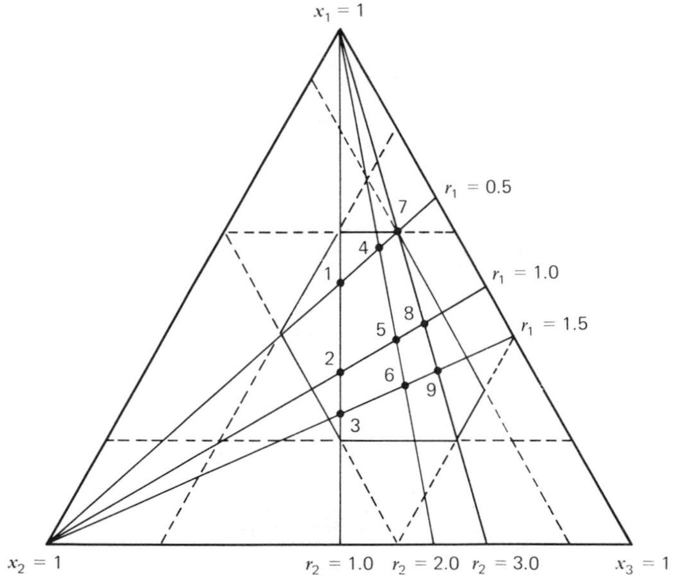

FIGURE 6.9. Design points in the constrained region obtained by using equally spaced values for the ratios $r_1 = x_3/x_1$ and $r_2 = x_3/x_2$.

The settings (proportions) in the mixture components corresponding to the nine factorial settings in r_1 and r_2 are determined as follows. Corresponding to the combination $(r_1, r_2) = (0.5, 1.0)$ we have only to solve the equations

$$r_1 = \frac{x_3}{x_1} = 0.5, \qquad r_2 = \frac{x_3}{x_2} = 1.0, \qquad x_1 + x_2 + x_3 = 1.0$$

to get $x_1 = 0.5$, $x_2 = 0.25$, and $x_3 = 0.25$. In a similar manner, corresponding to $(r_1, r_2) = (1.5, 2.0)$, we can set up the equations

$$\frac{x_3}{x_1} = 1.5, \qquad \frac{x_3}{x_2} = 2.0, \qquad x_1 + x_2 + x_3 = 1.0$$

to get $x_1 = \frac{4}{13}$, $x_2 = \frac{3}{13}$, and $x_3 = \frac{6}{13}$. The seven compositions corresponding to the remaining seven (r_1, r_2) combinations are presented in Table 6.5 along with response values observed from the ratios. The nine compositions are noted in Figure 6.9.

The second-degree model presented as Eq. (6.36) can be fitted to the data of Table 6.5 using the values of r_1 and r_2 as listed, or, the values of r_1 and r_2 can be scaled to enable us to utilize orthogonal design settings. Since the values of r_1 are equally spaced, as are the values of r_2, we may rewrite Eq. (6.36) as

$$y = \gamma_0 + \gamma_1 z_1 + \gamma_2 z_2 + \gamma_{11} z_1^2 + \gamma_{22} z_2^2 + \gamma_{12} z_1 z_2 + \epsilon \tag{6.37}$$

TABLE 6.5. Octane numbers for ratios of components

Design Point	Ratio Variables: $r_1 = x_3/x_1$	$r_2 = x_3/x_2$	Component Proportions: x_1	x_2	x_3	Octane Number (y)
1	0.5	1.0	0.50	0.25	0.25	82.8
2	1.0	1.0	0.33	0.33	0.33	87.3
3	1.5	1.0	0.25	0.37	0.37	84.9
4	0.5	2.0	0.57	0.14	0.29	85.6
5	1.0	2.0	0.40	0.20	0.40	93.0
6	1.5	2.0	0.31	0.23	0.46	89.3
7	0.5	3.0	0.60	0.10	0.30	87.2
8	1.0	3.0	0.43	0.14	0.43	89.6
9	1.5	3.0	0.33	0.17	0.50	90.6

where $z_1 = (r_1 - 1.0)/0.5$ and $z_2 = (r_2 - 2.0)/1.0$. The second-degree equation (6.37) was fitted to the nine octane values resulting in the prediction model

$$\begin{aligned} \hat{y}(\mathbf{z}) = 91.46 &+ 1.53z_1 + 2.07z_2 - 3.23z_1^2 - 2.23z_2^2 + 0.33z_1z_2 \\ (1.23)\ &(0.67)\quad (0.67)\quad (1.17)\quad (1.17)\quad (0.83) \end{aligned}$$

The value of $R_A^2 = 0.72$ and the estimate of the error variance is $s^2 = 2.73$.

The corresponding model in the ratios r_1 and r_2 is

$$\begin{aligned} \hat{y}(\mathbf{r}) = 63.69 &+ 27.63r_1 + 10.35r_2 - 12.93r_1^2 - 2.23r_2^2 + 0.65r_1r_2 \\ (6.72)\ &(10.00)\quad (5.00)\quad (4.67)\quad (1.17)\quad (1.65) \end{aligned} \tag{6.38}$$

which when written in the mixture components is

$$\begin{aligned} \hat{y}(\mathbf{x}) = 63.69 &+ 27.63\left(\frac{x_3}{x_1}\right) + 10.35\left(\frac{x_3}{x_2}\right) - 12.93\left(\frac{x_3}{x_1}\right)^2 - 2.23\left(\frac{x_3}{x_2}\right)^2 + 0.65\left(\frac{x_3^2}{x_1x_2}\right) \\ (6.72)\quad &(10.00)\qquad (5.00)\qquad (4.67)\qquad (1.17)\qquad (1.65) \end{aligned}$$

From this latter model [as well as with Eq. (6.38)], we see that increasing the value of x_3 relative to both x_1 and x_2 up to a certain level produces higher estimated octane numbers (since both $\hat{\gamma}_1$ and $\hat{\gamma}_2$ are positive). However, above certain values of r_1 and r_2, increasing the value of x_3 results in a reduction (both $\hat{\gamma}_{11}$ and $\hat{\gamma}_{22}$ are negative) in the value of the estimate of the octane number.

Problems where the number of components is greater than three can be handled by the use of additional ratios. As a reminder, the number of ratios in a set should be kept equal to $q-1$. We also remember that each ratio in the set should contain at least one of the components used in at least one of the other ratios belonging to the set.

6.7. COX'S MIXTURE POLYNOMIALS: MEASURING COMPONENT EFFECTS

Let us express the first- and second-degree polynomial models in the more general forms

First-degree model

$$\eta(\mathbf{x}) = \beta_0 + \sum_{i=1}^{q} \beta_i x_i \tag{6.39}$$

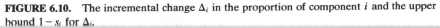

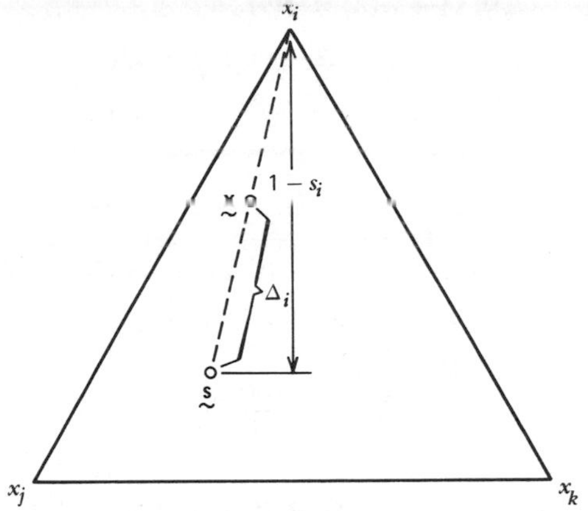

FIGURE 6.10. The incremental change Δ_i in the proportion of component i and the upper bound $1-s_i$ for Δ_i.

Second-degree model

$$\eta(\mathbf{x}) = \beta_0 + \sum_{i=1}^{q} \beta_i x_i + \sum_{i=1}^{q}\sum_{j=1}^{q} \beta_{ij} x_i x_j, \quad \beta_{ij} = \beta_{ji} \tag{6.40}$$

where $\eta(\mathbf{x})$ is written to mean the expected response at the point $\mathbf{x}$. The individual parameters in Eqs. (6.39) and (6.40) can be given different interpretations than with the Scheffé polynomials by imposing constraints on their values. This is done in the paper by Cox (1971), where the parameters are made to represent relative changes in the measured response which is accomplished by comparing the value of the response at points in the simplex against the value of the response taken at a standard mixture.

The parameter β_i in Eq. (6.39) can be made to represent the effect on the response of changing the proportion of component i. To show this, let us select some standard mixture and denote it by $\mathbf{s}$, and let us select another point $\mathbf{x}$ in the simplex, different from $\mathbf{s}$, where at $\mathbf{x}$ the proportion of component i is $x_i = s_i + \Delta_i$, and s_i is the proportion of component i at $\mathbf{s}$. If at $\mathbf{x}$, the contributions of the other $q-1$ components are in the same relative proportions as at $\mathbf{s}$, these contributions are $x_j = s_j - \Delta_i s_j/(1-s_i) = s_j(1-x_i)/(1-s_i)$, $j \neq i$; see Figure 6.10.

Let us write the expected response using the first-degree equation (6.39). Then the change in the expected response, which is measured by taking the difference $\eta(\mathbf{x}) - \eta(\mathbf{s})$, is

$$\begin{aligned}\Delta\eta(\mathbf{x}) &= \eta(\mathbf{x}) - \eta(\mathbf{s})\\ &= \beta_0 + \sum_{i=1}^{q}\beta_i x_i - \beta_0 - \sum_{i=1}^{q}\beta_i s_i\\ &= \beta_i\Delta_i + \sum_{j\neq i}^{q}\beta_j(x_j - s_j)\end{aligned}$$

If we force the constraint $\Sigma_{i=1}^{q}\beta_i s_i = 0$ on the values of the β_i, then the change in the expected response is expressed as

$$\Delta\eta(\mathbf{x}) = \frac{\beta_i\Delta_i}{(1 - s_i)} \tag{6.41}$$

Rewriting the equality (6.41) so that β_i is expressed as a function of the change $\Delta\eta(\mathbf{x})$, we have $\beta_i = \Delta\eta(\mathbf{x})(1 - s_i)/\Delta_i$ and an estimate of β_i which we call an estimate of the effect of component i is

$$b_i = \frac{(1 - s_i)}{\Delta_i}\{y(\mathbf{x}) - y(\mathbf{s})\}, \qquad i = 1, 2, \ldots, q \tag{6.42}$$

where $y(\mathbf{x})$ and $y(\mathbf{s})$ are values of the response observed at $\mathbf{x}$ and at $\mathbf{s}$.

In the formula for b_i, the difference in the heights of the surface at $\mathbf{x}$ and at $\mathbf{s}$ is weighted by the incremental change Δ_i made in the proportion of component i, relative to the amount $(1 - s_i)$ that is an upper bound for the value of Δ_i. The weight $(1 - s_i)/\Delta_i$ is greater than or equal to unity. An example of a weight of approximately 2 units is shown in Figure 6.10 for the system consisting of components i, j, and k.

When the surface is more appropriately represented by the second-degree polynomial presented in Eq. (6.40), the change $\Delta\eta(\mathbf{x})$ in the expected response contains an additional quantity which is $\beta_{ii}\Delta_i^2/(1 - s_i)^2$. In other words, if $\eta(\mathbf{x})$ is more correctly written as Eq. (6.40), the change expressed in Eq. (6.41) would read as $\Delta\eta(\mathbf{x}) = \eta(\mathbf{x}) - \eta(\mathbf{s}) = \beta_i\Delta_i/(1 - s_i) + \beta_{ii}\Delta_i^2/(1 - s_i)^2$, since in addition to placing the constraint $\Sigma_{i=1}^{q}\beta_i s_i = 0$ on the coefficients, the q constraints $\Sigma_{k=1}^{q}\beta_{jk}s_k = 0$, for $j = 1, 2, \ldots, q$ are imposed also.

Let us now consider estimating the parameters associated with all of the q components simultaneously with either the first- or the second-degree model, subject to the constraints being placed on the parameter estimates. A general formula can be written for estimating the restricted parameters in Cox's models since the Cox estimates can be written as functions of the parameter estimates in the corresponding Scheffé polynomial model. To show this we shall use matrix notation.

Let us write the matrix formula for estimating the coefficients in Scheffé's model to be $\mathbf{g}_l = (\mathbf{X}_l'\mathbf{X}_l)^{-1}\mathbf{X}_l'\mathbf{y}$ where the subscript l denotes the degree of the model, $l = 1$ or 2. When $l = 1$, then $\mathbf{y} = \mathbf{X}_1\boldsymbol{\gamma}_1 + \boldsymbol{\epsilon}$ and $\mathbf{y}$ is an $N \times 1$ vector of observations, $\mathbf{X}_1$ is an $N \times q$ matrix of component proportions, $\boldsymbol{\gamma}_1$ is a $q \times 1$ vector of unknown parameters, $\mathbf{g}_1$ is the estimator of $\boldsymbol{\gamma}_1$, and $\boldsymbol{\epsilon}$ is an $N \times 1$ vector of random errors where $E(\boldsymbol{\epsilon}) = \mathbf{0}$ and $E(\boldsymbol{\epsilon}\boldsymbol{\epsilon}') = \sigma^2\mathbf{I}$. The vector of estimates $\mathbf{b}_l$ associated with the lth-degree Cox model is

$$\mathbf{b}_l = \mathbf{B}_l\begin{bmatrix}\mathbf{g}_l\\ \mathbf{0}\end{bmatrix} = \mathbf{B}_l\begin{bmatrix}(\mathbf{X}_l'\mathbf{X}_l)^{-1}\mathbf{X}_l'\mathbf{y}\\ \mathbf{0}\end{bmatrix}, \qquad l = 1, 2 \tag{6.43}$$

where if $l = 1$, the matrix $\mathbf{B}_1$ is $(q+1)\times(q+1)$ and $\mathbf{0}$ is 1×1, and if $l = 2$, the matrix $\mathbf{B}_2$ is $(q+1)(q+2)/2 \times (q+1)(q+2)/2$ and $\mathbf{0}$ is $(q+1)\times 1$.

The elements of the matrices $\mathbf{B}_1$ and $\mathbf{B}_2$ in Eq. (6.43) are the coefficients in the linear relations of the β_i's to the γ_i's. For example, when $l = 1$, the β_i's in terms of the γ_i's, subject to the constraint $\Sigma_{i=1}^{q} \beta_i s_i = 0$, are

$$\beta_0 = \sum_{i=1}^{q} \gamma_i s_i, \qquad \beta_i = \gamma_i(1 - s_i) - \sum_{j\neq i}^{q} \gamma_j s_j, \qquad i = 1, 2, \ldots, q$$

The form of the matrix $\mathbf{B}_1$ is

$$\mathbf{B}_1 = \left[\begin{array}{cccc|c} s_1 & s_2 & \cdots & s_q & 0\\ \hline & & \mathbf{I} - \mathbf{1s}' & & \mathbf{0}\end{array}\right] \tag{6.44}$$

where $\mathbf{1}$ and $\mathbf{0}$ are $q \times 1$ vectors of ones and zeros, respectively, and $\mathbf{s} = (s_1, s_2, \ldots, s_q)'$. In a similar way, the elements of the matrix $\mathbf{B}_2$ can be found. A special case is presented in Appendix 6A.

The variance–covariance matrix of the restricted estimates of Eq. (6.43) is

$$\text{var}(\mathbf{b}_l) = \mathbf{B}_l\begin{bmatrix}(\mathbf{X}_l'\mathbf{X}_l)^{-1} & \mathbf{0}\\ \mathbf{0} & \mathbf{0}\end{bmatrix}\mathbf{B}_l'\sigma^2 \tag{6.45}$$

The elements of the matrix $\text{var}(\mathbf{b}_l)$ are functions of the component proportions in $\mathbf{X}_l$. Of particular interest to an experimenter are the magnitudes of the diagonal elements of the matrix $\text{var}(\mathbf{b}_l)$ which are the variances of the individual b_i.

In choosing a design configuration for fitting the first-degree polynomial presented in Eq. (6.39), several suggestions are made in the paper by Cornell (1975). Specifically, axial designs of the type introduced in Section

2.13 are recommended. Two approaches were taken by Cornell in searching for a design, but we discuss briefly only the case where the standard mixture **s** is positioned at the center of the simplex and changes in the sizes of the elements of var($\mathbf{b}_1$) are brought about by varying the distances to the design points, that is, by varying the elements in $\mathbf{X}_1$.

Let us assume one is interested in estimating the effects of each of the q components simultaneously. An axial design is proposed, and the points are to be positioned on the axes a distance $\boldsymbol{\Delta}$ from the centroid **s** either toward each of the q vertices [max $\Delta = (q-1)/q$] or away from (max $\Delta = 1/q$) each of the q vertices. If r observations are taken at each point in the design (but none at **s**), then the form of matrix $(\mathbf{X}_1'\mathbf{X}_1)$ in Eq. (6.43) is

$$\frac{1}{r}(\mathbf{X}_1'\mathbf{X}_1) = b\mathbf{I} + c\mathbf{J} \qquad \text{where } b = \frac{\Delta^2 q^2}{(q-1)^2}, \qquad c = \frac{q^2 - \Delta^2 q^2 - 2q + 1}{q(q-1)^2}$$

where the matrix $\mathbf{I}$ is an identity matrix of order q and the matrix $\mathbf{J}$ is a $q \times q$ matrix of ones. The inverse $(\mathbf{X}_1'\mathbf{X}_1)^{-1}$ is

$$r(\mathbf{X}_1'\mathbf{X}_1)^{-1} = d\mathbf{I} + e\mathbf{J}$$

where $d = 1/b$, $e = -c/b(b + cq)$, and the form of the variance matrix of the estimates is

$$\text{var}(\mathbf{b}_1) = \frac{\sigma^2}{rq}\begin{bmatrix} 1 & \mathbf{0} \\ \mathbf{0} & \dfrac{(q-1)^2}{\Delta^2 q}\left\{\mathbf{I} - \dfrac{1}{q}\mathbf{J}\right\} \end{bmatrix} \tag{6.46}$$

The form of the matrix var($\mathbf{b}_1$) in Eq. (6.46) suggests that as the magnitude of Δ increases, the variances of the individual b_i, $i = 1, 2, \ldots, q$, decrease. In fact, if the points are placed at the vertices of the simplex resulting in the Scheffé $\{q, 1\}$ lattice, then $\Delta = (q-1)/q$ and $\text{var}(b_i) = \sigma^2(q-1)/qr$, $i = 1, 2, \ldots, q$. Also, we notice that even though the Scheffé estimates are uncorrelated when the extreme vertices design is used, because of the restriction $\Sigma_{i=1}^{q} b_i s_i = \Sigma_{i=1}^{q} b_i/q = 0$, the estimates b_i and b_j in the Cox polynomial are dependent, that is, the covariance between b_i and b_j is $\text{cov}(b_i, b_j) = -\sigma^2/qr$.

Exercise 6.2 Write the formula for var(b_i) when the design points are placed at the bases of the q component axes so that $\Delta = 1/q$. Compare the magnitudes of the variances of these estimates to the variances of the estimates of the same coefficients obtained with the $\{3, 1\}$ simplex-lattice.

6.8. AN EXAMPLE ILLUSTRATING THE FITS OF COX'S MODEL AND SCHEFFÉ'S POLYNOMIAL

In Section 2.4 we discussed briefly the experimental results of blending the components polyethylene (x_1), polystyrene (x_2), and polypropylene (x_3) on the response, thread elongation of spun yarn. Again we shall discuss the blending of these same ingredients, but this time the response is the knot strength of the yarn. We assume the components polystyrene and polypropylene are approximately the same in terms of cost and availability, and we choose to measure the effect of the first component, polyethylene, relative to the others owing to the high availability and low cost of polyethylene. It is assumed that the importance of polyethylene can be measured by fitting a second-degree polynomial, since the knot strength values are suspected of behaving in a curvilinear fashion for increasing percentages of polyethylene. However, for purposes of illustrating the fitting of the complete Cox polynomial as well as for discussing and comparing the interpretations given to the parameter estimates in the Cox and Scheffé polynomials initially we shall fit the first-degree models after which the discussion is extended to the second-degree models.

Let us assume that the cost of polyethylene is less than one-half the cost of either polystyrene or polypropylene so that only mixtures are considered where the polyethylene proportion is greater than or equal to 0.50. Also, since only blends consisting of all three components simultaneously (interior points of the simplex) are practical and each component proportion must be present by at least $x_i \geq 0.02$, the original components are transformed to pseudocomponents (see Section 4.1.1). The feasible blends comprise the upper interior triangular region (whose sides are the dashed lines) displayed in Figure 6.11.

Since the purpose of this example is the comparison of the fits of the Cox and Scheffé models, the position of the standard mixture will be the centroid of the triangular factor space as in our discussion in the previous section. The coordinates of the standard mixture in the pseudocomponent system is $(s_1', s_2', s_3') = (\frac{1}{3}, \frac{1}{3}, \frac{1}{3})'$.

The design in the pseudocomponent system at whose points the data is collected in order to estimate the parameters in the first-degree models is the $\{3, 1\}$ simplex-lattice. The choice of the $\{3, 1\}$ lattice is made because the points are spread out and thus the parameter estimates are minimum variance estimates. Furthermore, the $\{3, 1\}$ lattice can be augmented easily to the $\{3, 2\}$ simplex-lattice for use in fitting the second-degree models.

The proportions for pseudocomponents 1, 2, and 3 are found using

$$x_1' = \frac{x_1 - 0.50}{1 - 0.54}, \qquad x_2' = \frac{x_2 - 0.02}{1 - 0.54}, \qquad x_3' = \frac{x_3 - 0.02}{1 - 0.54} \qquad (6.47)$$

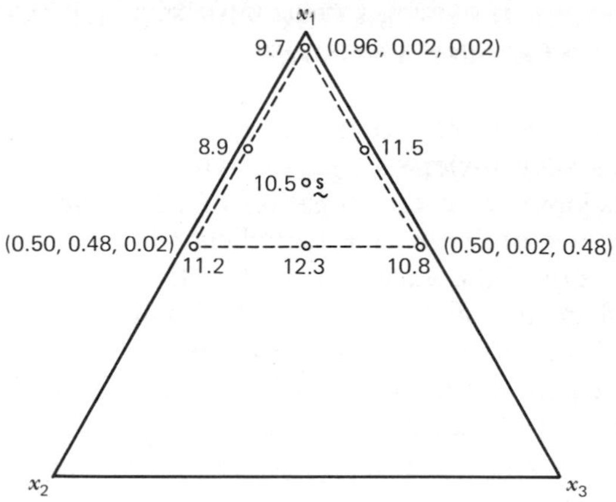

FIGURE 6.11. The design of the three-component yarn experiment with the observed knot strength values at the design points.

where $0.54 = 0.50 + 0.02 + 0.02$ is the sum of the lower bounds for x_1, x_2, and x_3. Corresponding to the pseudocomponent proportions $x_1' = 1$, $x_2' = x_3' = 0$, and $x_1' = 0$, $x_2' = 1$, $x_3' = 0$, and $x_1' = 0$, $x_2' = 0$, $x_3' = 1$, which are the vertices of the pseudocomponent triangle, the original component proportions are $x_1 = 0.96$, $x_2 = x_3 = 0.02$, and $x_1 = 0.50$, $x_2 = 0.48$, $x_3 = 0.02$, and, $x_1 = 0.50$, $x_2 = 0.02$, $x_3 = 0.48$, respectively. For the {3, 2} lattice later, in addition we will have the pseudocomponent values $x_i' = x_j' = 0.5$, $x_k' = 0$, $k \neq j \neq i$. The mixture component settings of the x_i' and the x_i at the six design points of the {3, 2} lattice plus at the standard mixture **s** are listed in Table 6.6 where the first three points, 1, 2, and 3, represent the {3, 1} lattice.

TABLE 6.6. Spun yarn example

	Pseudocomponents			Original Components			Observed Knot Strength (lbs force)
Point	x_1'	x_2'	x_3'	x_1	x_2	x_3	
1	1	0	0	0.96	0.02	0.02	9.7
2	0	1	0	0.50	0.48	0.02	11.2
3	0	0	1	0.50	0.02	0.48	10.8
4	0.5	0.5	0	0.73	0.25	0.02	8.9
5	0	0.5	0.5	0.50	0.25	0.25	12.3
6	0.5	0	0.5	0.73	0.02	0.25	11.5
7	0.33	0.33	0.33	0.653	0.173	0.173	10.5

Also listed in Table 6.6 are the knot strength values at the compositions where the strength values are measured in pounds of force exerted on the knots before rupturing the yarn threads. To enable us to calculate the standard errors of the parameter estimates, an independent estimate of the observation variance is given to be $s^2 = 0.30$.

Corresponding to the knot strength values at the three vertices which are listed as points 1, 2, and 3, the fitted response equations using the Cox polynomial as well as the Scheffé polynomials in the original components x_i and in the pseudocomponents x'_i, are

Cox polynomial

$$\hat{y}(\mathbf{x}) = b_0 + \sum_{i=1}^{3} b_i x_i = \underset{(0.32)}{10.56} - \underset{(0.50)}{0.97}x_1 + \underset{(1.28)}{2.29}x_2 + \underset{(1.28)}{1.42}x_3 \tag{6.48}$$

Scheffé polynomial

$$\hat{y}(\mathbf{x}) = \sum_{i=1}^{3} g_i x_i = \underset{(0.60)}{9.59}x_1 + \underset{(1.31)}{12.85}x_2 + \underset{(1.31)}{11.98}x_3 \tag{6.49}$$

$$\hat{y}(\mathbf{x}') = \sum_{i=1}^{3} g'_i x'_i = \underset{(0.55)}{9.7}x'_1 + \underset{(0.55)}{11.2}x'_2 + \underset{(0.55)}{10.8}x'_3 \tag{6.50}$$

In the pseudocomponent polynomial of Eq. (6.50) the coefficient estimates g'_i represent the heights of the knot strength planar surface at the vertices of the pseudocomponent triangle inside the original triangle. The coefficient estimate g'_i is the knot strength value observed at the vertex of the pseudocomponent triangle, that is, $g'_i = y_u$ at $x'_i = 1$, $x'_j = 0$, $j \neq i$, $i = 1$, 2 and 3.

For the Scheffé polynomial in the original components, the coefficient estimate g_i represents the difference between the height of the surface at the ith pseudocomponent vertex and the weighted sum of the heights at the three vertices where the weights are the lower bounds of the original components. The deviation arises as a result of placing constraints on the size of the factor space of interest. Computationally,

$$g_i = \frac{g'_i - [0.50g'_1 + 0.02g'_2 + 0.02g'_3]}{1 - 0.54} \tag{6.51}$$

where the denominator $1 - 0.54 = 0.46$, represents the length of the pseudocomponent axes in the original units.

With the Cox polynomial represented by Eq. (6.48), the coefficient estimate b_0 represents the weighted mean of the differences g_i in Eq. (6.49) where the weights are the coordinates of the standard mixture s in the original component system [see the top row of the matrix $\mathbf{B}_1$ in Eq. (6.44)],

$$b_0 = \sum_{i=1}^{3} s_i g_i = 0.653(9.59) + 0.173(12.85) + 0.173(11.98)$$
$$= 10.56$$

The estimate b_i in Eq. (6.48) represents a contrast among the parameter estimates in the Scheffé's model of Eq. (6.49)

$$b_i = g_i - \sum_{i=1}^{3} s_i g_i = (1 - s_i)g_i - \sum_{j \neq i}^{3} s_j g_j$$

For example, $b_2 = -0.653(9.59) + (1 - 0.173)(12.85) - 0.173(11.98) = 2.29$. The matrix formulas for obtaining these estimates are shown by Eqs. (6.43) and (6.44). A check on the b_i values is $b_1 s_1 + b_2 s_2 + b_3 s_3 = 0$.

Fitting all seven response values in Table 6.6 with a second-degree model, the estimated response equation using Cox's restrictions as well as the Scheffé polynomial in the original components and the pseudocomponents, respectively, are

Cox Polynomial

$$\begin{aligned}\hat{y}(\mathbf{x}) = {} & \underset{(0.34)}{10.81} - \underset{(0.68)}{2.41x_1} + \underset{(1.75)}{0.91x_2} + \underset{(1.75)}{8.25x_3} + \underset{(1.06)}{1.35x_1^2} + \underset{(6.78)}{16.88x_2^2} - \underset{(6.78)}{17.80x_3^2} \\ & + \underset{(10.86)}{20.45x_2x_3} + \underset{(4.91)}{3.99x_1x_3} - \underset{(4.91)}{14.27x_1x_2}\end{aligned} \tag{6.52}$$

Scheffé Polynomial

$$\hat{y}(\mathbf{x}) = \underset{(0.70)}{9.75x_1} + \underset{(6.21)}{28.59x_2} + \underset{(6.21)}{1.26x_3} + \underset{(11.86)}{21.37x_2x_3} + \underset{(11.86)}{20.43x_1x_3} - \underset{(11.86)}{32.50x_1x_2} \tag{6.53}$$

$$\hat{y}(\mathbf{x}') = \underset{(0.54)}{9.7x_1'} + \underset{(0.54)}{11.2x_2'} + \underset{(0.54)}{10.8x_3'} + \underset{(2.51)}{4.5x_2'x_3'} + \underset{(2.51)}{4.3x_1'x_3'} - \underset{(2.51)}{6.9x_1'x_2'} \tag{6.54}$$

The estimated coefficients in Eq. (6.52) were obtained using a value of $a = 3.8$ in the matrix $\mathbf{B}_2$ of Appendix 6A. Since an observation value at s was used in obtaining the parameter estimates in the models shown in

Eqs. (6.52)–(6.54), it is difficult to give precise interpretations to the coefficient estimates g'_i, g'_{ij}, $i<j$ and g_i, g_{ij}, $i<j$, in the Scheffé Eqs. (6.54) and (6.53) respectively. If exterior points only had been used for obtaining the estimates in Eq. (6.54), the g'_i would be estimates of the heights of the surface at the three vertices $x'_i=1$, $x'_j=0$, $j\neq i$, and the g'_{ij} would represent estimated departures of the surface from the plane defined by $g'_1x'_1+g'_2x'_2+g'_3x'_3$. Since an observation was taken at **s** and used in the analysis, however, these interpretations are relaxed slightly, particularly if the observation at **s** does not fall on the plane specified by $g'_1x'_1+g'_2x'_2+g'_3x'_3$. Nevertheless, one often chooses to interpret the parameter estimates as if interior points had not been included for reasons of simplicity in the interpretation.

The parameter estimates in the original mixture component model of Eq. (6.53) are linear functions of the parameter estimates in the pseudocomponent model. Since the simplex space is restricted by lower bounds on the x_i, the coefficient estimate g_i, $i=1$, 2, and 3 approximately represents the difference between the height of the surface at the pseudocomponent vertex $x'_i=1$, $x'_j=0$, $j\neq i$, and the weighted mean of the heights at all of the vertices as in Eq. (6.51) plus some measure of departure from the plane. The g_{ij} represent only departure from the plane scaled by the square of the length of the pseudocomponent axes in the original units.

With the Cox polynomial, the estimates b_0 and b_i, $i=1$, 2, and 3, represent linear combinations of the g_i and the g_{ij} according to the top four rows of the matrix $\mathbf{B}_2$ in Appendix 6A. The estimates b_{ij}, $i\leq j=1$, 2, and 3 are linear functions only of the g_{ij} (and hence only of the g'_{ij}) and thus represent departure from the plane. For example, the estimate b_{11} is calculated approximately as $b_{11}=[g_{23}-2(g_{13}+g_{12})]/(a+2)^2$ and is a comparison of the binary blending of components 2 and 3 versus the binary blending of polyethylene with the other components. Note that for increasing values of a (where $s_1=as_2=as_3$), while the contrast inside the brackets is unchanged, less weight is assigned to the contrast owing to the reduction in the value of $1/(a+2)$. Similar contrasts can be written for b_{22} and b_{33}.

A final note on the comparison of Cox's model and Scheffé's model concerns the estimated response contours corresponding to both models. While Cox's model is not designed for empirically fitting a response surface above the simplex, nevertheless the coefficient estimates in Cox's model are simple linear functions of the parameter estimates in Scheffé's model and therefore both estimated response functions are the same and must have identical response contours. For example, the predicted value of the response at $(x_1, x_2, x_3)=(0.5, 0.5, 0)$ with the models of Eqs. (6.52) and (6.53) is $\hat{y}(\mathbf{x})=11.05$, and, $\widehat{\text{var}}[\hat{y}(\mathbf{x})]=0.39$.

Exercise 6.3 An additional knot strength value of 9.2(lbs force) is observed at the blend $x_1 = 0.80$, $x_2 = 0.10$ and $x_3 = 0.10$. Use this knot strength value and calculate an estimate of the effect of polyethylene using Eq. (6.42) and compare this estimate against the estimate of the effect of polyethylene using the knot strength value of 9.7 at $x_1 = 0.96$, $x_2 = 0.02$ and $x_3 = 0.02$ from Table 6.6. Are the estimates the same? Should the form of the model be revised? Comment. Use $s^2 = 0.30$.

6.9. OCTANE BLENDING MODELS

When gasoline components are mixed, there are two important properties of the blend which are used by refiners to evaluate the component blending. These properties are the research octane rating and the motor octane rating of the blend. Most often, the higher the octane rating the more desirable the blend. This is true of the research and motor octane ratings.

To determine the research and motor octane ratings (or octane numbers) of a blend, certain basic properties of the individual components must be known. To show this, let us define the following properties for component i, along with the (probable range for the value of the property)

$$\begin{aligned}
x_i &= \text{volume fraction } (0 \le x_i \le 1.0)\\
r_i &= \text{research octane number } (50 \le r_i \le 110)\\
m_i &= \text{motor octane number } (50 \le m_i \le 110)\\
s_i &= \text{sensitivity of component } i,\ s_i = r_i - m_i \ (0 \le s_i \le 60)\\
O_i &= \text{olefin content } (0 \le O_i \le 100)\\
A_i &= \text{aromatic content } (0 \le A_i \le 100)
\end{aligned}$$

The volumetric averages of each of the above properties are computed over the q components along with the averages of cross products between some of the properties. These averages are

$$\bar{r} = \sum r_i x_i, \quad \bar{m} = \sum m_i x_i, \quad \bar{s} = \sum s_i x_i, \quad \bar{O} = \sum O_i x_i, \quad \bar{A} = \sum A_i x_i \tag{6.55}$$

$$\overline{rs} = \sum r_i s_i x_i, \quad \overline{ms} = \sum m_i s_i x_i, \quad \overline{O^2} = \sum O_i^2 x_i, \quad \overline{A^2} = \sum A_i^2 x_i$$

where all of the summations (Σ_i^q) are over $i = 1, 2, \ldots, q$.

The *prediction equation* for the *research octane number* of the blend is formed by using some of the averages in (6.55). For some blend l say, the equation is

$$\hat{y}_{R_l} = \bar{r}_l + a_1(\overline{rs} - \bar{r}\bar{s})_l + a_2(\overline{O^2} - \bar{O}^2)_l + a_3(\overline{A^2} - \bar{A}^2)_l \tag{6.56}$$

where $\bar{r}_l$ is the calculated average research number across the q components in the blend l and a_1, a_2, and a_3 are coefficients that are to be estimated. The term $(\overline{rs} - \bar{r}\bar{s})$ in Eq. (6.56) is an estimate of the covariance between the research and the sensitivity values, the term $(\overline{O^2} - \bar{O}^2)$ is the estimated variance of the olefin contents of the components and $(\overline{A^2} - \bar{A}^2)$ is the variance of the aromatic contents. In a similar manner, the prediction equation for the *motor octane number* of a blend l is

$$\hat{y}_{M_l} = \bar{m}_l + c_1(\overline{ms} - \bar{m}\bar{s})_l + c_2(\overline{O^2} - \bar{O}^2)_l + c_3\left(\frac{\overline{A^2} - \bar{A}^2}{100}\right)_l^2 \tag{6.57}$$

where c_1, c_2, and c_3 are coefficients to be estimated.

Let us illustrate the calculations that are necessary in obtaining the prediction equations (6.56) and (6.57) for some blend l. Listed in Table 6.7 are four components that make up the blend l along with the research and octane numbers of each of the components. The components and their volume fractions are catalytically cracked naphtha $(x_1 = 0.69)$, reformate $(x_2 = 0.12)$, C_5-isomer $(x_3 = 0.14)$ and C_6-isomer $(x_4 = 0.05)$. Also listed in Table 6.7 is the sensitivity $s_i = r_i - m_i$, the olefin content and the aromatic content for each component. The research (r_i) and motor octane numbers (m_i) were taken at the lead level of 3.0 milliliters per gallon.

For the particular blend, the volumetric averages and crossproducts of the properties of the four components are

$$\bar{r} = 0.69(98.1) + 0.12(102.2) + 0.14(84.7) + 0.05(104.4) = 97.0$$

$$\bar{m} = 0.69(84.3) + 0.12(90.9) + 0.14(83.7) + 0.05(104.4) = 86.0$$

$$\bar{s} = 0.69(13.8) + 0.12(11.3) + 0.14(1.0) + 0.05(0.00) = 11.0$$

$$\bar{O} = 0.69(43.2) + 0.12(2.1) + 0.14(0.0) + 0.05(0.00) = 30.1$$

$$\bar{A} = 0.69(22.1) + 0.12(61.6) + 0.14(0.0) + 0.05(0.00) = 22.6$$

$$\begin{aligned}\overline{rs} &= 0.69(98.1)(13.8) + 0.12(102.2)(11.3) + 0.14(84.7)(1.0) \\ &\quad + 0.05(104.4)(0.00) \\ &= 1084.6\end{aligned}$$

$$\overline{ms} = 937.7, \qquad \overline{O^2} = 1288.2, \qquad \overline{A^2} = 792.3$$

TABLE 6.7. Sample data for calculating research and motor octane prediction equations for a particular blend

Component	Volume Fraction (x_i)	r_i	m_i	s_i	O_i	A_i
Catalytically Cracked	0.69	98.1	84.3	13.8	43.2	22.1
Reformate	0.12	102.2	90.9	11.3	2.1	61.6
C_5-isomer	0.14	84.7	83.7	1.0	0.0	0.0
C_6-isomer	0.05	104.4	104.4	0.0	0.0	0.0

Substituting these average numbers so as to form the terms in the prediction equations, the research and motor octane number equations for the particular blend of four components at lead level 3.0 milliliters per gallon are

$$\begin{aligned}\hat{y}_{R_l} &= 97.0 + a_1[1084.6 - (97.0)(11.0)] + a_2[1288.2 - (30.1)^2] \\ &\quad + a_3[792.3 - (22.6)^2] \\ &= 97.0 + a_1[17.6] + a_2[382.2] + a_3[281.5] \end{aligned} \tag{6.58}$$

and

$$\begin{aligned}\hat{y}_{M_l} &= 86.0 + c_1[937.7 - (86.0)(11.0)] + c_2[1288.2 - (30.1)^2] \\ &\quad + c_3\left[\frac{792.3 - (22.6)^2}{100}\right]^2 \\ &= 86.0 + c_1[-8.3] + c_2[382.2] + c_3[7.9] \end{aligned} \tag{6.59}$$

where a_1, a_2, and a_3 and c_1, c_2, and c_3 are coefficients to be estimated.

To obtain the prediction equations for the research and motor octane numbers of a *group of blends*, the coefficients a_1, a_2, and a_3 in Eq. (6.58) have to be estimated, and the coefficients c_1, c_2, and c_3 in Eq. (6.59) have to be estimated. The estimation is accomplished by first obtaining the research and octane numbers for three or more blends in the group. The estimates of the a_i, and of the c_i, $i = 1, 2, 3$ are then found in separate analyses using least squares.

An alternative scheme is to use only the research (r_i) and motor (m_i) octane ratings along with the volume fractions (x_i) of each of the components in forming the prediction equations. The models which are presented in Eqs. (6.56) and (6.57) are rewritten to include the average research rating or average motor rating and crossproducts between the component fractions in the forms as

$$\hat{y}_{R_l} = \bar{r}_l + a_1 \sum_{i<j}\sum^{q} (r_i - r_j)(s_i - s_j)x_i x_j + a_2 \sum_{i<j}\sum^{q} (O_i - O_j)^2 x_i x_j$$
$$+ a_3 \sum_{i<j}\sum^{q} (A_i - A_j)^2 x_i x_j$$
$$= \bar{r}_l + \sum_{i<j}\sum^{q} g_{ij} x_i x_j \qquad (6.60)$$

and

$$\hat{y}_{M_l} = \bar{m}_l + c_1 \sum_{i<j}\sum^{q} (m_i - m_j)(s_i - s_j)x_i x_j + c_2 \sum_{i<j}\sum^{q} (O_i - O_j)^2 x_i x_j$$
$$+ c_3 \sum_{i<j}\sum^{q} (A_i - A_j)^2 x_i x_j$$
$$= \bar{m}_l + \sum_{i<j}\sum^{g} h_{ij} x_i x_j \qquad (6.61)$$

The models presented by Eqs. (6.60) and (6.61) are called "interaction models" (see Morris (1975)). The blending models of Eqs. (6.56) and (6.57) can be shown to be special cases of the interaction models.

The major difference between the use of the blending models in Eqs. (6.56) and (6.57) and the interaction models in Eqs. (6.60) and (6.61) is that in the blending models, the quantities $\overline{ms}$, $\overline{O^2}$, $\overline{A^2}$, and $\overline{rs}$, are computed from the component properties (component research octane numbers, sensitivities, olefin content, etc.), while the values of the terms $x_i x_j$ in Eqs. (6.60) and (6.61) are calculated directly from the component proportions. The values of the variables $(\overline{rs} - \bar{r}\bar{s})$, $(\overline{O^2} - \bar{O}^2)$, $(\overline{A^2} - \bar{A}^2)$ and $(\overline{ms} - \bar{m}\bar{s})$ in the blending models are subject to error in measurement and if a considerable amount of measurement error is present when measuring these variables, the estimation of the coefficients a_i, $i = 1, 2, 3$, in Eq. (6.56) and of the c_i, $i = 1, 2, 3$, in Eq. (6.57) is performed with additional measures of error; more perhaps than may have been anticipated.

6.10. A NUMERICAL EXAMPLE ILLUSTRATING THE CALCULATIONS REQUIRED FOR OBTAINING THE RESEARCH AND MOTOR OCTANE PREDICTION EQUATIONS FOR A GROUP OF BLENDS

In this example, we work through the steps and calculations that are required to obtain the blending models and the interaction models for predicting the research and motor octane rating of a blend. To aid us in our calculations, we refer to Table 6.8 which contains the component proportions for catalytically cracked (x_1), reformate (x_2), C_5-isomer (x_3), and C_6-isomer (x_4), along with the other measured properties of each component for a group of thirteen blends. The values of the entries in Table 6.8 are artificial and are provided simply for illustrative purposes.

To obtain an equation of the form shown in Eq. (6.56) to use for predicting the research octane number of a blend, the average research rating $\bar{r}_l$ for each of the $l = 1, 2, \ldots, 13$ blends is subtracted from the corresponding observed research octane number y_{R_l} and the coefficients a_1, a_2, and a_3 are estimated using the formula

$$\begin{bmatrix} a_1 \\ a_2 \\ a_3 \end{bmatrix} = (\mathbf{R}'\mathbf{R})^{-1}\mathbf{R}'(\mathbf{y}_R - \bar{\mathbf{r}})$$

where the matrix $\mathbf{R}$ and the vector $(\mathbf{y}_R - \bar{\mathbf{r}})$ are

$$\mathbf{R} = [(\overline{rs} - \bar{r}\bar{s}), \overline{O^2} - \bar{O}^2, \overline{A^2} - \bar{A}^2] = \begin{bmatrix} 17.6 & 382.2 & 281.6 \\ 10.3 & 107.9 & 92.7 \\ 0.4 & 408.5 & 335.3 \\ \vdots & \vdots & \vdots \\ 18.4 & 555.5 & 109.7 \end{bmatrix}$$

$$(\mathbf{y}_R - \bar{\mathbf{r}}) = \begin{bmatrix} 98.4 - 96.0 \\ 94.8 - 94.1 \\ 89.3 - 86.2 \\ \vdots \\ 93.3 - 93.6 \end{bmatrix} = \begin{bmatrix} 2.4 \\ 0.7 \\ 3.1 \\ \vdots \\ -0.3 \end{bmatrix}$$

The estimates are $a_1 = -0.080$, $a_2 = 0.004$ and $a_3 = 0.004$ and the prediction equation for the research octane number for some blend l, where the component proportions are within the limits

$$0.20 \le x_1 \le 0.72, \quad 0.04 \le x_2 \le 0.41, \quad 0.01 \le x_3 \le 0.42, \quad 0.02 \le x_4 \le 0.27 \tag{6.62}$$

TABLE 6.8. Illustrative data for calculating blending models

Blend	Components x_1	x_2	x_3	x_4	Observed Research Octane No. (y_R)	$\bar{r}$	Observed Motor Octane No. (y_M)	$\bar{m}$	$(\overline{rs} - \bar{r}\bar{s})$	$(\overline{ms} - \bar{m}\bar{s})$	$(\overline{O^2} - \bar{O}^2)$	$(\overline{A^2} - \bar{A}^2)$	$[(\overline{A^2} - \bar{A}^2)/100]^2$
1	0.69	0.12	0.14	0.05	98.4	96.0	84.2	81.1	17.6	−8.3	382.2	281.6	7.9
2	0.58	0.17	0.15	0.10	94.8	94.1	85.2	83.7	10.3	3.9	107.9	92.7	0.9
3	0.72	0.05	0.04	0.19	89.3	86.2	82.0	81.2	0.4	−3.6	408.5	335.3	11.2
4	0.42	0.38	0.15	0.05	88.1	85.5	85.2	82.8	−5.1	−18.4	85.1	62.6	0.4
5	0.48	0.39	0.01	0.12	94.1	94.5	87.3	86.1	2.4	−0.7	97.2	172.2	3.0
6	0.33	0.41	0.12	0.14	93.5	91.7	86.0	84.3	5.3	−2.6	181.9	400.7	16.1
7	0.52	0.23	0.18	0.07	94.8	92.6	83.2	84.6	−3.5	−7.7	238.9	378.5	14.3
8	0.52	0.26	0.02	0.20	94.6	95.1	84.2	83.4	−1.2	−5.2	95.0	24.9	0.1
9	0.43	0.34	0.21	0.02	91.6	88.5	82.1	80.8	9.4	−2.1	707.2	168.6	2.8
10	0.44	0.41	0.06	0.09	86.0	83.3	78.6	76.3	−5.4	−8.3	522.8	78.3	0.6
11	0.20	0.16	0.42	0.22	83.1	80.3	74.9	76.0	1.8	−0.2	223.1	445.5	19.8
12	0.29	0.23	0.21	0.27	91.6	90.1	86.9	85.3	3.9	−1.1	303.3	382.1	14.6
13	0.46	0.04	0.33	0.17	93.3	93.6	86.1	84.2	18.4	12.9	555.5	109.7	1.2

is

$$\hat{y}_{R_l} = \bar{r}_l - \underset{(0.041)}{0.080}(\overline{rs} - \bar{r}\bar{s})_l + \underset{(0.001)}{0.004}(\overline{O^2} - \bar{O}^2)_l + \underset{(0.002)}{0.004}(\overline{A^2} - \bar{A}^2)_l \qquad (6.63)$$

The number in parenthesis below each coefficient estimate is the estimated standard error of the estimate. The standard errors are the square roots of the diagonal elements of $(\mathbf{R}'\mathbf{R})^{-1}s^2$, where $s^2 = 1.07$ is the residual mean square taken from Table 6.9.

The prediction equation (6.57) for the motor octane number is obtained in a similar manner. The estimated coefficients c_1, c_2, and c_3, are found using

$$\begin{bmatrix} c_1 \\ c_2 \\ c_3 \end{bmatrix} = (\mathbf{M}'\mathbf{M})^{-1}\mathbf{M}'(\mathbf{y}_M - \bar{\mathbf{m}})$$

where

$$\mathbf{M} = [(\overline{ms} - \bar{m}\bar{s}), (\overline{O^2} - \bar{O}^2), (\overline{A^2} - \bar{A}^2)^2/100^2] = \begin{bmatrix} -8.3 & 382.2 & 7.9 \\ 3.9 & 107.9 & 0.9 \\ -3.6 & 408.5 & 11.2 \\ \vdots & \vdots & \vdots \\ 12.9 & 555.5 & 1.2 \end{bmatrix}$$

$$(\mathbf{y}_M - \bar{\mathbf{m}}) = \begin{bmatrix} 84.2 - 81.1 \\ 85.2 - 83.7 \\ 82.0 - 81.2 \\ \vdots \\ 86.1 - 84.2 \end{bmatrix} = \begin{bmatrix} 3.1 \\ 1.5 \\ 0.8 \\ \vdots \\ 1.9 \end{bmatrix}$$

TABLE 6.9. Analysis of variance of research octane data with blending model

Source	Degrees of Freedom	Sum of Squares	Mean Square	R_A^2
Regression	2	11.28	5.64	
Residual	10	10.69	1.07	0.416
Total	12	21.97		

The values of the estimates c_1, c_2, and c_3 are substituted into an equation of the form (6.57) and the motor octane prediction equation for blend l whose component proportions are inside the ranges shown in Eq. (6.62) is

$$\hat{y}_{M_l} = \bar{m}_l - \underset{(0.050)}{0.063}(\overline{ms} - \bar{m}\bar{s})_l + \underset{(0.001)}{0.004}(\overline{O^2} - \bar{O}^2)_l - \underset{(0.050)}{0.047}\left[\frac{\overline{A^2} - \bar{A}^2}{100}\right]_l^2 \qquad (6.64)$$

where $s^2 = 1.77$ is estimated from the data values.

The parameters in the "interaction" models of Eqs. (6.60) and (6.61) are estimated using only the measured octane numbers (y_R and y_M) and the values of the x_i in Table 6.8. For example, to estimate the coefficients in the research octane equation (6.60), again the $\bar{r}_l$ value for each blend is subtracted from the observed research value and

$$\begin{bmatrix} g_{12} \\ g_{13} \\ \vdots \\ g_{34} \end{bmatrix} = (\mathbf{X}'\mathbf{X})^{-1}\mathbf{X}'(\mathbf{y}_R - \bar{\mathbf{r}})$$

where

$$\mathbf{X} = \begin{bmatrix} 0.08 & 0.10 & \cdots & 0.01 \\ 0.10 & 0.09 & & 0.02 \\ 0.04 & 0.03 & & 0.01 \\ \vdots & \vdots & & \vdots \\ 0.02 & 0.15 & \cdots & 0.06 \end{bmatrix}, \qquad (\mathbf{y}_R - \bar{\mathbf{r}}) = \begin{bmatrix} 2.4 \\ 0.7 \\ 3.1 \\ \vdots \\ -0.3 \end{bmatrix}$$

(columns of $\mathbf{X}$: x_1x_2, x_1x_3, …, x_3x_4)

The "interaction" prediction equation for the research octane number is

$$\hat{y}_{R_l} = \bar{r}_l + \underset{(9.81)}{5.77}x_1x_2 - \underset{(12.32)}{12.67}x_1x_3 + \underset{(9.98)}{20.94}x_1x_4 + \underset{(20.35)}{59.93}x_2x_3 - \underset{(31.87)}{48.85}x_2x_4 + \underset{(27.75)}{2.83}x_3x_4 \qquad (6.65)$$

The numbers in parentheses are the estimated standard errors of the coefficient estimates and are the square roots of the diagonal elements of $(\mathbf{X}'\mathbf{X})^{-1}s^2$, where $s^2 = 1.31$.

The precision of Eqs. (6.63)–(6.65) for predicting research and motor octane numbers is much less than one would normally expect of these models in practice. The loss of precision of the models is due to the introduction of a considerable amount of variation in the research octane numbers (y_R) and in the motor octane numbers (y_M) listed in Table 6.8. The introduction of the large amount of variation, estimated by $s^2 = 1.07$ and $s^2 = 1.77$ was intentional, and is designed to serve as a reminder to us that the data in Table 6.8 is artificial and is to be used only for purposes of illustrating the calculation formulas. By contrast, octane blending models of the form presented by Eqs. (6.63) and (6.64) that are used in practice are reported in the literature to possess model standard errors of magnitude less than one (1.0) octane number. Additional discussion on the use of blending models was presented by Snee (1974).

6.11. SUMMARY OF CHAPTER 6

The use of model forms other than the standard polynomials in the x_i is the main theme of Chapter 6. Inverse terms or reciprocals (x_i^{-1}) of the component proportions are suggested for modeling extreme changes in the response behavior as the value of x_i becomes very close to zero. Extreme changes in the response behavior are very real occurrences in many areas of chemical experimentation especially when certain component proportions approach boundary conditions. The modeling sequence of including inverse terms in the Scheffé polynomials starts with the simplest model form and grows with the addition of extra terms. This sequence is the reverse of that in Chapter 5.

When a component, say i, serves as a diluent in the sense that as the proportion x_i increases the effect of the remaining components on the response diminishes in proportion, then component i is said to blend with the other components in an additive manner. To model such an effect of one component in an otherwise nonlinear blending system, models that are homogeneous of degree one are superior to the Scheffé polynomials. It is shown, using the analysis of a three-component special cubic model, how to discover when one of the components blends additively with the other two components. Three types of homogeneous models are listed.

In some experiments ratios of the component proportions are more meaningful than the proportions by themselves. By defining $q - 1$ ratios as variables, standard polynomial models can be written in the ratio-variables and the models can also be fitted to data collected at the points of standard factorial arrangements. This is illustrated by the construction of a 3^2 factorial arrangement in a constrained region of a three-component triangle where the factorial arrangement uses equally spaced levels for the ratios $r_1 = x_3/x_1$ and $r_2 = x_3/x_2$.

When certain restrictions are placed on the parameter values in the standard polynomial models, the interpretation of the parameter estimates are closer in meaning to the treatment effects (or to the component effects defined in Section 5.7) in an analysis of variance setting of a comparative-type experiment. Cox's mixture polynomials with restrictions imposed on the parameter values are compared to the Scheffé polynomials in terms of the meanings attached to the parameters in the respective models and some comments are made concerning the best designs to use for fitting Cox's polynomials. Gasoline-blending models are presented in the final two sections of the chapter. Artificial data values are used to illustrate the necessary calculations for constructing the research octane and motor octane prediction equations for a group of gasoline blends.

6.12. TOPICS THAT WERE OMITTED BUT WHICH WILL BE COVERED IN FUTURE EDITIONS

As mentioned in the preface, in this first edition on experiments with mixtures, it has been necessary to exercise considerable selectivity in the choice of topics covered. The need to be selective in the topics discussed has been caused by the brevity of time set aside in the preparation of this work. It would have been nice indeed to have been able to include additional work by A. K. Nigam, by D. P. Lambrakis, by N. G. Becker, and by myself with various colleagues. Nevertheless, as a reminder and promise to you, the reader, that an effort will be made to cover these works in any future rewriting by this writer, the following list of authors and paper titles is presented with a key word or key words italicized in the paper title informing us of the importance of the paper. Further details of the particular source of each paper can be found in the bibliography.

1. N. G. Becker. *Regression problems* when the predictor variables are proportions.
2. N. G. Becker. *Models* and designs for experiments with mixtures.
3. J. A. Cornell. *Weighted versus unweighted estimates* using Scheffé's mixture model for symmetrical error variance patterns.
4. J. A. Cornell and A. I. Khuri. *Obtaining constant prediction variance* on concentric triangles for ternary mixture systems.
5. D. P. Lambrakis. Experiments with *p-component* mixtures.
6. A. K. Nigam. Some designs and models for mixture experiments for the *sequential exploration* of response surfaces.
7. S. K. Saxena and A. K. Nigam. Symmetric-simplex *block designs* for mixtures.
8. R. D. Snee. Experimental designs for mixture systems with *multicomponent constraints.*

6.13. REFERENCES AND RECOMMENDED READING

Becker, N. G. (1968). Models for the response of a mixture. *J. R. Stat. Soc., B*, **30**, 349–358.

Becker, N. G. (1978). Models and designs for experiments with mixtures. *Austral. J. Stat.*, **20**, No. 3, 195–208.

Cornell, J. A. (1975). Some comments on designs for Cox's mixture polynomial. *Technometrics*, **17**, 25–35.

Cornell, J. A. and J. W. Gorman (1978). On the detection of an additive blending component in multicomponent mixtures. *Biometrics*, **34**, No. 2, 251–263.

Cox, D. R. (1971). A note on polynomial response functions for mixtures. *Biometrika*, **58**, 155–159.

Draper, N. R. and R. C. St. John (1977). A mixtures model with inverse terms. *Technometrics*, **19**, 37–46.

Healy, W. C. Jr., C. W. Maassen, and R. T. Peterson (1959). A new approach to blending octanes. Paper presented at Midyear Meeting of American Petroleum Institute. Division of Refining, New York, NY.

Kenworthy, O. O. (1963). Factorial experiments with mixtures using ratios. *Ind. Qual. Control*, **19**, 24–26.

Leonpacker, R. M. (1978). The Ethyl technique of Octane prediction. Ethyle Corporation Petroleum Chemicals Division Report No. RTM-400, Ethyl Corporation, Baton Rouge, LA.

Marquardt, D. W. and R. D. Snee (1974). Test statistics for mixture models. *Technometrics*, **16**, 533–537.

Morris, W. E. (1975). The interaction approach to gasoline blending. E. I. duPont de Nemours and Co., Inc., Petroleum Chemicals Division Report AM-75-30.

Scheffé, H. (1958). Experiments with mixtures. *J. R. Stat. Soc., B*, **20**, No. 2, 344–360.

Scheffé, H. (1963). Simplex-centroid design for experiments with mixtures. *J. R. Stat. Soc., B*, **25**, No. 2, 235–263.

Snee, R. D. (1971). Design and analysis of mixture experiments. *J. Qual. Technol.*, **3**, 159–169.

Snee, R. D. (1973). Techniques for the analysis of mixture data. *Technometrics*, **15**, No. 3, 517–528.

Snee, R. D. (1974). Techniques for developing blending models. Paper presented at Annual Meeting of American Statistical Association, St. Louis, MO.

QUESTIONS FOR CHAPTER 6

6.1. The model in the ratios r_1 and r_2 shown by Eq. (6.38) was fitted to the octane data in Table 6.5 and produced a value of $R_A^2 = 0.72$. Fit a linear plus inverse term model with the two terms (x_1^{-1}) and (x_3^{-1}) to the data and comment on the use of these terms in the model versus the use of second-degree terms. Calculate the C_p values using $s^2 = 2.50$.

6.2. In Table 6.3 are listed predicted mite numbers obtained with the special cubic Eq. (6.20) and the nonpolynomial Eq. (6.21). Using the observed mite numbers at the 15 blends, compute the residuals (r_u) where $r_u = y_u - \hat{y}_u$ with each of the models. Comment on the use of the residuals for testing the performance of each of the fitted model forms as well as for comparing the fitted model forms.

6.3. For the three component system where data is collected at the points of a simplex-centroid design, the conditions listed in Eq. (6.16) and (6.17) combined with a fitted model of the form in Eq. (6.15) imply the additive blending effect of component three. A similar procedure would be test the adequacy of the model

$$E[y_{lu}] = \mu + \beta_l(1 - x_3)$$

where μ is some overall mean value and β_l is the slope of the surface along the lth ray extending from $x_3 = 0$ to $x_3 = 1$ where we might have a set of k rays (i.e., $l = 1, 2, \ldots, k$). Comment on this latter testing procedure and explain the similarity of it to the use of Eqs. (6.15)–(6.17).

6.4. Becker (1978) suggests a model of the form

$$E[y_{lu}] = \mu_l + \beta_l(1 - x_3) + \gamma_l(1 - x_3)^2$$

can be fitted to the data collected on a set of rays ($1 \leq l \leq k$) extending from $x_3 = 0$ to $x_3 = 1$. A test of the hypothesis

$$H_0:\quad \beta_l = \gamma_l = 0$$

provides a test of whether component 3 is inactive (that is, changes in x_3 do not alter the value of the response) while a test of the hypothesis

$$h_0:\quad \mu_l = \mu_0,\ \gamma_l = 0$$

provides a test of whether or not component 3 has an additive effect. Use the data of Table 6.3 to perform such tests. (There are three rays $l = 1, 2, 3$, as shown in Figure 6.5.)

6.5. In a three component system, the ratio variables $r_1 = x_1/x_2$ and $r_2 = x_2/x_3$ are to be studied. The design variables $z_1 = (r_1 - 1.0)/0.5$, $z_2 = (r_2 - 2.0)/1.0$ are used to define a second-degree rotatable design in the (z_1, z_2) space. Determine the mixture blends in x_1, x_2, and x_3 corresponding to the levels of z_1 and z_2 and with the response values y_u, obtain a prediction model in the r's and the x's.

z_1:	-1	1	-1	1	$-\sqrt{2}$	$\sqrt{2}$	0	0	0	0	0
z_2:	-1	-1	1	1	0	0	$-\sqrt{2}$	$\sqrt{2}$	0	0	0
y_u:	4.8	8.4	10.2	6.0	3.7	4.8	4.9	8.8	14.3	16.2	14.4

6.6. The interaction models presented in Eqs. (6.59) and (6.60) are said to be special cases of the blending models of Eqs. (6.55) and (6.56), respectively. Show algebraically that this is true.

6.7. On the axes of each component in a five-component blending system, the following response values were collected:

x_1	x_2	x_3	x_4	x_5	Response Values (y_u)
0.80	0.05	0.05	0.05	0.05	24.3, 28.5
0.05	0.80	0.05	0.05	0.05	14.0, 15.2
0.05	0.05	0.80	0.05	0.05	9.2, 10.6, 7.2
0.05	0.05	0.05	0.80	0.05	5.8, 4.8, 4.4
0.05	0.05	0.05	0.05	0.80	22.3, 24.5, 25.2

$\mathbf{s} = (0.20, 0.20, 0.20, 0.20, 0.20)'$ 8.0

a. Obtain an estimate of the effect of each component using Eq. (6.42). Use the replicates to obtain an estimate of the error variance and test $H_0: \beta_i = 0$ against $H_a: \beta_i \neq 0$ for each $i = 1, 2, 3, 4, 5$.

b. Estimate the parameters β_i, in the model $y = \beta_1 x_1 + \beta_2 x_2 + \beta_3 x_3 + \beta_4 x_4 + \beta_5 x_5 + \epsilon$. How do these estimates b_i differ from the b_i in **a**? With the model parameters estimates, compute the component effects using Eq. (5.29) or Eq. (5.30) from Section 5.6. How do these estimates of the component effects differ from the estimates in **a**? Which set of estimates, the set from Section 5.6 or the set in Section 6.7, are most meaningful to you, and why?

APPENDIX 6A. THE FORM OF THE MULTIPLIER MATRIX $\mathbf{B}_2$ FOR EXPRESSING THE PARAMETERS IN COX'S QUADRATIC MODEL AS FUNCTIONS OF THE PARAMETERS IN SCHEFFÉ'S MODEL

In Chapter 2, the form of the second degree Scheffé polynomial was derived by applying the constraint $x_1 + x_2 + \cdots + x_q = 1$ to the terms in the general second-degree polynomial resulting in

$$\eta(\mathbf{x}) = \beta_0 + \sum_{i=1}^{q} \beta_i x_i + \sum_{i=1}^{q}\sum_{j=1}^{q} \beta_{ij} x_i x_j, \qquad \beta_{ij} = \beta_{ji}$$

$$\eta(\mathbf{x}) = \sum_{i=1}^{q} \gamma_i x_i + \sum_{i<j}\sum^{q} \gamma_{ij} x_i x_j \tag{6A.1}$$

where in (6A.1)

$$\gamma_i = \beta_0 + \beta_i + \beta_{ii}, \; \gamma_{ij} = 2\beta_{ij} - \beta_{ii} - \beta_{jj}, \; i = 1, 2, \ldots, q \text{ and } i<j \tag{6A.2}$$

Now, suppose with the model in Eq. (6A.1) the coefficients β_0, β_i, and β_{ij}, $i \le j$, are redefined in terms of the γ_i and γ_{ij}, $i<j$, subject to the constraints,

$$\sum_{i=1}^{q} \beta_i s_i = 0, \qquad \sum_{j=1}^{q} \beta_{ij} s_j = 0 \qquad \text{for all } i \tag{6A.3}$$

where s_i is the proportion of component i in some standard mixture. The result would be a quadratic model of the form suggested by Cox (1971).

To develop a system of equations relating the parameters in Cox's quadratic model to the parameters in Scheffé's quadratic model, let us position the standard mixture $\mathbf{s}$ on the axes of component i and let $s_i > 0$. Then

$$s_i = a s_j, \qquad a \neq 0, \qquad j = 1, 2, \ldots, q, \qquad j \neq i \tag{6A.4}$$

and $s_i = a/(a+q-1)$ so that $s_j = 1/(a+q-1)$, $j \neq i$. If we set $i = 1$ and use equations (6A.2), (6A.3), and (6A.4), then in the formula of Eq. (6.43), or, $\mathbf{b}_2 = \mathbf{B}_2\begin{bmatrix}\mathbf{g}_2\\ \mathbf{0}\end{bmatrix}$, the elements of the matrix $\mathbf{B}_2$ are found by solving the following equations,

$$\sum_{j=2}^{q} g_{1j} = -(a+q-1)b_{11} - \sum_{j=2}^{q} b_{jj}$$

$$\sum_{i=2}^{q}\sum_{j\neq i}^{q} g_{ij} + (a-1)\sum_{j=2}^{q} g_{1j} = -a(q-1)b_{11} - (a+q+1)\sum_{j=2}^{q} b_{jj}$$

$$a g_1 + \sum_{j=2}^{q} g_j = (a+q-1)b_0 + a b_{11} + \sum_{j=2}^{q} b_{jj}$$

For example, in the case where $q = 3$, the values of the restricted Cox estimates $\mathbf{b}_2 = (b_0, b_1, b_2, b_3, b_{11}, b_{22}, b_{33}, b_{23}, b_{13}, b_{12})'$ in terms of the values of the unrestricted Scheffé estimates $\mathbf{g}_2 = (g_1, g_2, g_3, g_{23}, g_{13}, g_{12})'$ are found using the following matrix $\mathbf{B}_2$ where $D = (a+2)^2$:

$$
\mathbf{B}_2 = \left[\begin{array}{ccc|ccc|cccc}
\frac{a}{a+2} & \frac{1}{a+2} & \frac{1}{a+2} & \frac{1}{D} & \frac{a}{D} & \frac{a}{D} & 0 & 0 & 0 & 0 \\
\frac{2}{a+2} & \frac{-1}{a+2} & \frac{-1}{a+2} & \frac{-2}{D} & \frac{-(a-2)}{D} & \frac{-(a-2)}{D} & 0 & 0 & 0 & 0 \\
\frac{-a}{a+2} & \frac{a+1}{a+2} & \frac{-1}{a+2} & \frac{a}{D} & \frac{-2a}{D} & \frac{a^2}{D} & 0 & 0 & 0 & 0 \\
\frac{-a}{a+2} & \frac{-1}{a+2} & \frac{a+1}{a+2} & \frac{a}{D} & \frac{a^2}{D} & \frac{-2a}{D} & 0 & 0 & 0 & 0 \\
\hline
0 & 0 & 0 & \frac{1}{D} & \frac{-2}{D} & \frac{-2}{D} & 0 & 0 & 0 & 0 \\
0 & 0 & 0 & \frac{-(a+1)}{D} & \frac{a}{D} & \frac{-a(a+1)}{D} & 0 & 0 & 0 & 0 \\
0 & 0 & 0 & \frac{-(a+1)}{D} & \frac{-a(a+1)}{D} & \frac{a}{D} & 0 & 0 & 0 & 0 \\
\hline
0 & 0 & 0 & \frac{a^2+2a+2}{D} & \frac{-a^2}{D} & \frac{-a^2}{D} & 0 & 0 & 0 & 0 \\
0 & 0 & 0 & \frac{-a}{D} & \frac{3a+2}{D} & \frac{a-2}{D} & 0 & 0 & 0 & 0 \\
0 & 0 & 0 & \frac{-a}{D} & \frac{a-2}{D} & \frac{3a+2}{D} & 0 & 0 & 0 & 0
\end{array}\right]
$$

In the appendix of Cornell (1975), the number 2 was erroneously entered in the denominators of the elements in the lower center partition of the matrix $\mathbf{B}_2$. The matrix $\mathbf{B}_2$ above is correct.

CHAPTER 7

Matrix Algebra, Least Squares, and the Analysis of Variance

In this chapter, we present some background ideas on the use of matrices and vectors. The use of matrices and vectors is fundamental for describing and constructing some of the mixture designs as well as necessary to some procedures for analyzing mixture data. We present only the basic concepts of matrix algebra which are pertinent to mixture experiments and we refer the reader to the books listed at the end of this chapter for a more complete coverage.

7.1. MATRIX ALGEBRA

Let us consider a linear transformation from the q variables, $x_1, x_2, \ldots, x_q$ to the N variables $y_1, y_2, \ldots, y_N$ expresssed as

$$\begin{aligned}
y_1 &= m_{11}x_1 + m_{12}x_2 + \cdots + m_{1q}x_q \\
y_2 &= m_{21}x_1 + m_{22}x_2 + \cdots + m_{2q}x_q \\
&\vdots \\
y_N &= m_{N1}x_1 + m_{N2}x_2 + \cdots + m_{Nq}x_q
\end{aligned}$$

The rectangular (or square if $N = q$) array of coefficients $[m_{ij}]$ of the transformation is called the *matrix* of the linear transformation. The matrix **M** is

$$\underset{N\times q}{\mathbf{M}} = [m_{ij}] = \begin{bmatrix} m_{11} & m_{12} & \cdots & m_{1q} \\ m_{21} & m_{22} & \cdots & m_{2q} \\ \vdots & \vdots & \ddots & \vdots \\ m_{N1} & m_{N2} & \cdots & m_{Nq} \end{bmatrix} \tag{7.1}$$

and $\mathbf{M}$ has N rows and q columns. The rows are of equal length, as are the columns and the matrix $\mathbf{M}$ is said to have dimensions $N \times q$. The matrix $\mathbf{M}$ will be denoted by the use of square brackets $[m_{ij}]$ where the element m_{ij} is positioned in the ith row and jth column.

A matrix consisting of only a single column is called a *column vector.* A *row vector* is a matrix with just a single row. Sometimes a vector will be denoted by a set of parentheses. Notationally a matrix is denoted by an uppercase boldface letter and vectors are denoted by lowercase boldface letters. For example, the $N \times 3$ matrix $\mathbf{M}$ consists of the three $N \times 1$ column vectors $\mathbf{m}_1$, $\mathbf{m}_2$, and $\mathbf{m}_3$

$$\mathbf{M} = [\mathbf{m}_1, \mathbf{m}_2, \mathbf{m}_3] = \begin{bmatrix} m_{11} & m_{12} & m_{13} \\ m_{21} & m_{22} & m_{23} \\ \vdots & \vdots & \vdots \\ m_{N1} & m_{N2} & m_{N3} \end{bmatrix} \tag{7.2}$$

7.2. SOME FUNDAMENTAL DEFINITIONS

Definition 1 A *square matrix* is a matrix in which the number of rows equals the number of columns. If in (7.2) $N = 3$, then $\mathbf{M}$ is a 3×3 square matrix.

Definition 2 The *transpose* of the $N \times q$ matrix $\mathbf{M}$, written $\mathbf{M}'$, is the $q \times N$ matrix

$$\mathbf{M}' = \begin{bmatrix} m_{11} & m_{21} & \cdots & m_{N1} \\ m_{12} & m_{22} & \cdots & m_{N2} \\ \vdots & \vdots & \ddots & \vdots \\ m_{1q} & m_{2q} & \cdots & m_{Nq} \end{bmatrix}$$

Definition 3 A *symmetric matrix* is equal to its transpose, that is, if $\mathbf{M} = \mathbf{M}'$ then $\mathbf{M}$ is symmetric (so is $\mathbf{M}'$).

Definition 4 A *diagonal matrix* is a square matrix whose nondiagonal elements are zero, that is, $m_{ij} = 0$ for $i \neq j$,

$$\mathbf{M} = \begin{bmatrix} m_{11} & 0 & 0 \\ 0 & m_{22} & 0 \\ 0 & 0 & m_{33} \end{bmatrix} = \text{diag}(m_{11}, m_{22}, m_{33})$$

Definition 5 A *scalar* is a single number (a 1×1 matrix).

Definition 6 Matrices that have the same dimension can be added together, element by element. They are said to be *conformable for addition*. For example,

$$\underset{2\times3}{\mathbf{M}} = \begin{bmatrix} m_{11} & m_{12} & m_{13} \\ m_{21} & m_{22} & m_{23} \end{bmatrix}, \qquad \underset{2\times3}{\mathbf{N}} = \begin{bmatrix} n_{11} & n_{12} & n_{13} \\ n_{21} & n_{22} & n_{23} \end{bmatrix}$$

and

$$\underset{2\times3}{\mathbf{M}+\mathbf{N}} = \begin{bmatrix} m_{11}+n_{11} & m_{12}+n_{12} & m_{13}+n_{13} \\ m_{21}+n_{21} & m_{22}+n_{22} & m_{23}+n_{23} \end{bmatrix}$$

Definition 7 The product of an $N \times q$ matrix $\mathbf{M}$ by a scalar c is the $N \times q$ matrix $c\mathbf{M} = [cm_{ij}]$.

Definition 8 Two matrices are said to be *conformable for multiplication* if the number of columns of the one on the left is equal to the number of rows of the matrix on the right, that is, if $\mathbf{M}$ is $N \times q$ and $\mathbf{T}$ is $q \times r$, then $\mathbf{MT}$ is an $N \times r$ matrix (but $\mathbf{TM}$ is not). For example

$$\underset{3\times3}{\begin{bmatrix} m_{11} & m_{12} & m_{13} \\ m_{21} & m_{22} & m_{23} \\ m_{31} & m_{32} & m_{33} \end{bmatrix}} \underset{3\times2}{\begin{bmatrix} t_{11} & t_{12} \\ t_{21} & t_{22} \\ t_{31} & t_{32} \end{bmatrix}} = \underset{3\times2}{\begin{bmatrix} m_{11}t_{11}+m_{12}t_{21}+m_{13}t_{31} & m_{11}t_{12}+m_{12}t_{22}+m_{13}t_{32} \\ m_{21}t_{11}+m_{22}t_{21}+m_{23}t_{31} & m_{21}t_{12}+m_{22}t_{22}+m_{23}t_{32} \\ m_{31}t_{11}+m_{32}t_{21}+m_{33}t_{31} & m_{31}t_{12}+m_{32}t_{22}+m_{33}t_{32} \end{bmatrix}}$$

Definition 9 The *rank* of a matrix is the maximum number of linearly independent rows or columns, whichever is less.

Definition 10 The *identity* matrix $\mathbf{I}$ is a diagonal matrix with ones on the diagonal.

Definition 11 The *inverse* of a square matrix $\mathbf{M}$ is denoted by $\mathbf{M}^{-1}$ and is such that $\mathbf{MM}^{-1} = \mathbf{M}^{-1}\mathbf{M} = \mathbf{I}$. A square matrix that has an inverse is said to be *nonsingular*. It is also said to be of *full rank*.

Definition 12 A square matrix $\mathbf{T}$ is said to be *orthogonal* if $\mathbf{T}' = \mathbf{T}^{-1}$, that is, the transpose of $\mathbf{T}$ is equal to its inverse. The transformation $\mathbf{w} = \mathbf{Tv}$ is called an *orthogonal* transformation.

Definition 13 A *quadratic form* in the q variables $x_1, x_2, \ldots, x_q$ is a scalar function of the form

$$Q = \sum_{i=1}^{q} \sum_{j=1}^{q} m_{ij} x_i x_j$$

where the m_{ij} are constants. In matrix notation, we may define the quadratic form by defining the vector $\mathbf{x} = (x_1, x_2, \ldots, x_q)'$ and the $q \times q$ matrix $\mathbf{M} = [m_{ij}]$ so that

$$Q = \mathbf{x}'\mathbf{M}\mathbf{x}$$

The symmetric matrix **M** is called the matrix of the quadratic form Q.

Definition 14 A quadratic form $Q = \mathbf{x}'\mathbf{M}\mathbf{x}$ which is positive for all values of $\mathbf{x}$ other than $\mathbf{x} = \mathbf{0}$ is said to be a *positive definite* quadratic form. The matrix **M** is called a *positive definite* matrix and the roots are all greater than zero. The rank r of a quadratic form is the rank of the matrix **M**.

Definition 15 If **M** is a symmetric positive definite matrix, the inequality $(\mathbf{x} - \mathbf{x}_0)'\mathbf{M}(\mathbf{x} - \mathbf{x}_0) \leq 1$ defines an ellipsoid with center at $\mathbf{x}_0$.

Definition 16 The *trace* of a square matrix **V** is the sum of the diagonal elements. The trace of an $N \times N$ matrix **V** is denoted by $\text{tr}(\mathbf{V}) = \sum_{i=1}^{N} v_{ii}$.

7.3. A REVIEW OF LEAST SQUARES

Let us assume provisionally that N observations of the response are expressible by means of the linear first-degree equation

$$y_u = \beta_1 x_{u1} + \beta_2 x_{u2} + \cdots + \beta_q x_{uq} + \epsilon_u \tag{7.3}$$

where y_u denotes the observed response for the uth trial, x_{ui} represents the proportion of component i at the uth trial, β_i represents an unknown parameter in the equation and ϵ_u represents the random error, $u = 1, 2, \ldots, N$. The *method of least squares* selects as the estimates b_i for the unknown parameters β_i, $i = 1, 2, \ldots, q$, those values which minimize the quantity

$$\sum_{u=1}^{N} (y_u - b_1 x_{u1} - b_2 x_{u2} - \cdots - b_q x_{uq})^2$$

In *matrix notation*, the first-degree model in Eq. (7.3), over the N observations, can be expressed as

$$\mathbf{y} = \mathbf{X}\boldsymbol{\beta} + \boldsymbol{\epsilon} \tag{7.4}$$

where

$$\underset{N \times 1}{\mathbf{y} = \begin{bmatrix} y_1 \\ y_2 \\ \vdots \\ y_N \end{bmatrix}} \quad \underset{N \times q}{\mathbf{X} = \begin{bmatrix} x_{11} & x_{12} & \cdots & x_{1q} \\ x_{21} & x_{22} & \cdots & x_{2q} \\ \vdots & \vdots & & \vdots \\ x_{N1} & x_{N2} & \cdots & x_{Nq} \end{bmatrix}} \quad \underset{q \times 1}{\boldsymbol{\beta} = \begin{bmatrix} \beta_1 \\ \beta_2 \\ \vdots \\ \beta_q \end{bmatrix}} \quad \underset{N \times 1}{\boldsymbol{\epsilon} = \begin{bmatrix} \epsilon_1 \\ \epsilon_2 \\ \vdots \\ \epsilon_N \end{bmatrix}} \tag{7.5}$$

Some Rules for Obtaining Means and Variances of Random Vectors

Let us assume the random variable y_u has expectation $E(y_u) = \mu_u$ where $\mu_u = \beta_1 x_{u1} + \beta_2 x_{u2} + \cdots + \beta_q x_{uq}$, $u = 1, 2, \ldots, N$. Then

$$E(\mathbf{y}) = E\begin{bmatrix} y_1 \\ y_2 \\ \vdots \\ y_N \end{bmatrix} = \begin{bmatrix} \mu_1 \\ \mu_2 \\ \vdots \\ \mu_N \end{bmatrix} = \boldsymbol{\mu}$$

that is, the expectation of a vector is the vector of expectations. If the variances and the covariances of the y_u are given by $\text{var}(y_u) = E(y_u - \mu_u)^2 = \sigma_u^2$, and $\text{cov}(y_u, y_{u'}) = E(y_u - \mu_u)(y_{u'} - \mu_{u'}) = \sigma_{uu'}$, $u \neq u'$, then the variance–covariance ($N \times N$ symmetric) matrix is denoted by

$$\mathbf{V} = E(\mathbf{y} - \boldsymbol{\mu})(\mathbf{y} - \boldsymbol{\mu})' = \begin{bmatrix} \sigma_1^2 & \sigma_{12} & \sigma_{13} & \cdots & \sigma_{1N} \\ \sigma_{21} & \sigma_2^2 & \sigma_{23} & \cdots & \sigma_{2N} \\ \vdots & \vdots & \vdots & & \vdots \\ \sigma_{N1} & \sigma_{N2} & \sigma_{N3} & \cdots & \sigma_N^2 \end{bmatrix}$$

If the variables y_u, $u = 1, 2, \ldots, N$, are jointly normally distributed with mean vector $\boldsymbol{\mu}$ and variance–covariance matrix $\mathbf{V}$, this is written as $\mathbf{y} \sim N(\boldsymbol{\mu}, \mathbf{V})$.

For the general case of the mathematical model in Eq. (7.4), $\mathbf{y}$ is an $N \times 1$ vector of observations, $\mathbf{X}$ is an $N \times p$ matrix whose elements are the mixture component proportions and functions of the component proportions, $\boldsymbol{\beta}$ is a $p \times 1$ vector of parameters, and $\boldsymbol{\epsilon}$ is an $N \times 1$ vector of random errors. When the model $\mathbf{y} = \mathbf{X}\boldsymbol{\beta} + \boldsymbol{\epsilon}$ is of the first-degree as in Eq. (7.3), then $p = q$. When the model is of the second-degree, then $p = q(q+1)/2$.

The *normal equations* which are set up to estimate the elements of the parameter vector $\boldsymbol{\beta}$ in Eq. (7.4) are

$$\mathbf{X}'\mathbf{X}\mathbf{b} = \mathbf{X}'\mathbf{y} \tag{7.6}$$

where the $p \times p$ square matrix $\mathbf{X}'\mathbf{X}$ consists of sums of squares and sums of cross products of the mixture proportions and the $p \times 1$ vector $\mathbf{X}'\mathbf{y}$ consists of sums of cross products of the x_{ui} and y_u. The least squares estimates of the elements of $\boldsymbol{\beta}$ are

$$\mathbf{b} = (\mathbf{X}'\mathbf{X})^{-1}\mathbf{X}'\mathbf{y} \tag{7.7}$$

where the $p \times p$ matrix $(\mathbf{X}'\mathbf{X})^{-1}$ is the inverse of $\mathbf{X}'\mathbf{X}$. Since $\mathbf{X}'\mathbf{X}$ is symmetric, so is $(\mathbf{X}'\mathbf{X})^{-1}$.

Properties of the Parameter Estimates

The statistical properties of the estimator $\mathbf{b}$ are easily verified once certain assumptions are made about the elements of $\boldsymbol{\epsilon}$. Writing the expectation of the elements of the random error vector in Eq. (7.4) as $E(\boldsymbol{\epsilon}) = \mathbf{0}$ and $\text{var}(\boldsymbol{\epsilon}) = E[\boldsymbol{\epsilon}\boldsymbol{\epsilon}'] = \sigma^2\mathbf{I}_N$, where $\mathbf{I}_N$ is the identity matrix of order N, then the expectation of $\mathbf{b}$ is

$$\begin{aligned} E[\mathbf{b}] &= E[(\mathbf{X}'\mathbf{X})^{-1}\mathbf{X}'\mathbf{y}] \\ &= E[(\mathbf{X}'\mathbf{X})^{-1}\mathbf{X}'(\mathbf{X}\boldsymbol{\beta} + \boldsymbol{\epsilon})] \\ &= \boldsymbol{\beta} + E(\mathbf{X}'\mathbf{X})^{-1}\mathbf{X}'\boldsymbol{\epsilon} \\ &= \boldsymbol{\beta} \end{aligned} \tag{7.8}$$

Thus if the model $\mathbf{y} = \mathbf{X}\boldsymbol{\beta} + \boldsymbol{\epsilon}$ is correct, $\mathbf{b}$ is an *unbiased estimator* of $\boldsymbol{\beta}$.

The *variance–covariance* matrix of the elements of $\mathbf{b}$ is expressed as

$$\begin{aligned} \text{var}(\mathbf{b}) &= \text{var}[(\mathbf{X}'\mathbf{X})^{-1}\mathbf{X}'\mathbf{y}] \\ &= (\mathbf{X}'\mathbf{X})^{-1}\mathbf{X}'\,\text{var}(\mathbf{y})\mathbf{X}(\mathbf{X}'\mathbf{X})^{-1} \\ &= (\mathbf{X}'\mathbf{X})^{-1}\mathbf{X}'\mathbf{V}\mathbf{X}(\mathbf{X}'\mathbf{X})^{-1} \end{aligned}$$

Since $\text{var}(\boldsymbol{\epsilon}) = \sigma^2\mathbf{I}_N$, then $\mathbf{V} = \sigma^2\mathbf{I}_N$ and

$$\text{var}(\mathbf{b}) = (\mathbf{X}'\mathbf{X})^{-1}\sigma^2 \tag{7.9}$$

Along the main diagonal of the $p \times p$ matrix $(\mathbf{X}'\mathbf{X})^{-1}\sigma^2$, the iith element is the variance of b_i, the ith element of $\mathbf{b}$. The ijth element of $(\mathbf{X}'\mathbf{X})^{-1}\sigma^2$ is the covariance between the elements b_i and b_j of $\mathbf{b}$. Furthermore, if the errors $\boldsymbol{\epsilon}$ are jointly normally distributed, then with the properties of $\mathbf{b}$ defined in Eqs. (7.8) and (7.9) the distribution of $\mathbf{b}$ is written as

$$\mathbf{b} \sim N(\boldsymbol{\beta}, (\mathbf{X}'\mathbf{X})^{-1}\sigma^2) \tag{7.10}$$

Predicted Response Values

Once the vector of estimates $\mathbf{b}$ is obtained using Eq. (7.7), a prediction of the value of the response at some point $\mathbf{x} = (x_1, x_2, \ldots, x_q)'$ in the experimental region can be made and is expressed in matrix notation as

$$\hat{y}(\mathbf{x}) = \mathbf{x}_p'\mathbf{b} \tag{7.11}$$

where $\mathbf{x}_p'$ is a $1 \times p$ vector whose elements correspond to the elements in a row of the matrix $\mathbf{X}$ in Eq. (7.4). Specifically, if the predicted value of the response corresponding to the uth observation is desired, then $\hat{y}(\mathbf{x}) = \mathbf{x}_u'\mathbf{b}$, where $\mathbf{x}_u'$ is the uth row of $\mathbf{X}$. [We use the notation $\hat{y}(\mathbf{x})$ to denote the predicted value of $\hat{y}$ at the point $\mathbf{x}$.]

A measure of the precision of the estimate $\hat{y}(\mathbf{x})$ is defined as the *variance* of $\hat{y}(\mathbf{x})$ and is expressed as

$$\begin{aligned} \text{var}[\hat{y}(\mathbf{x})] &= \text{var}[\mathbf{x}_p'\mathbf{b}] \\ &= \mathbf{x}_p' \, \text{var}(\mathbf{b})\mathbf{x}_p \\ &= \mathbf{x}_p'(\mathbf{X}'\mathbf{X})^{-1}\mathbf{x}_p\sigma^2 \end{aligned} \tag{7.12}$$

The inverse matrix $(\mathbf{X}'\mathbf{X})^{-1}$ used for obtaining $\mathbf{b}$ in Eq. (7.7) also determines the variances and covariances of the elements of $\mathbf{b}$ as well as the variance of $\hat{y}(\mathbf{x})$.

7.4. THE ANALYSIS OF VARIANCE

The results of the analysis of a set of data from a mixture experiment can be displayed in table form. The table is called an *analysis of variance* table. The entries in the table represent measures of information about the separate sources of variation in the data.

The total variation in a set of data is called the "total sum of squares" and is abbreviated as SST. The quantity SST is computed by summing the squares of the observed y_u's about their mean $\bar{y} = (y_1 + y_2 + \cdots + y_N)/N$,

$$\text{SST} = \sum_{u=1}^{N} (y_u - \bar{y})^2 \tag{7.13}$$

The quantity SST has associated with it $N-1$ degrees of freedom since there are only $N-1$ independent deviations $y_u - \bar{y}$ in the sum.

The sum of squares of the deviations of the observed y_u's from their predicted values is

$$\text{SSE} = \sum_{u=1}^{N} (y_u - \hat{y}_u)^2 \tag{7.14}$$

The difference between the sums of squares quantities is

$$\text{SSR} = \text{SST} - \text{SSE} = \sum_{u=1}^{N} (\hat{y}_u - \bar{y})^2$$

TABLE 7.1. Analysis of variance table

Source of Variation	Degrees of Freedom	Sum of Squares	Mean Square
Regression (Fitted Model)	$p-1$	$SSR = \sum_{u=1}^{N} (\hat{y}_u - \bar{y})^2$	$\frac{SSR}{(p-1)}$
Residual	$N-p$	$SSE = \sum_{u=1}^{N} (y_u - \hat{y}_u)^2$	$\frac{SSE}{(N-p)}$
Total	$N-1$	$SST = \sum_{u=1}^{N} (y_u - \bar{y})^2$	

and SSR represents the portion of SST attributable to the fitted regression equation. The quantity SSR is called the *sum of squares due to regression.* In matrix notation, short-cut formulas for SST, SSE, and SSR are, letting $\mathbf{1}'$ be a $1 \times N$ vector of 1's,

$$
\begin{aligned}
SST &= \mathbf{y}'\mathbf{y} - \frac{(\mathbf{1}'\mathbf{y})^2}{N} \\
SSE &= \mathbf{y}'\mathbf{y} - \mathbf{b}'\mathbf{X}'\mathbf{y} \\
SSR &= \mathbf{b}'\mathbf{X}'\mathbf{y} - \frac{(\mathbf{1}'\mathbf{y})^2}{N}
\end{aligned}
\qquad (7.15)
$$

The partitioning of the total sum of squares is summarized with the familiar analysis of variance Table 7.1 where we are assuming the fitted model in Eq. (7.11) contains p terms.

7.5. A NUMERICAL EXAMPLE: MODELING THE TEXTURE OF FISH PATTIES

Fish patties were formulated from three different types of saltwater fish. The fish were mullet (x_1), sheepshead (x_2), and croaker (x_3). The design chosen was a simplex-centroid design with the following fish percentages: mullet (100%), sheepshead (100%), croaker (100%), mullet:sheepshead (50%:50%), mullet:croaker (50%:50%), sheepshead:croaker (50%:50%) and mullet:sheepshead:croaker (33%:33%:33%). The data in Table 7.2 represent average texture readings in grams of force ($\times 10^{-3}$) for replicate patties where each average was computed from three readings taken on each patty. A second degree equation was fitted to the data.

TABLE 7.2. Average texture readings taken on replicate fish patties

Fish Percentages (%)			Component Proportions			Average Texture
Mullet	Sheepshead	Croaker	x_1	x_2	x_3	grams $\times 10^{-3}$
100	0	0	1	0	0	2.02, 2.08
0	100	0	0	1	0	1.47, 1.37
50	50	0	$\frac{1}{2}$	$\frac{1}{2}$	0	1.91, 2.00
0	0	100	0	0	1	1.93, 1.83
50	0	50	$\frac{1}{2}$	0	$\frac{1}{2}$	1.98, 2.13
0	50	50	0	$\frac{1}{2}$	$\frac{1}{2}$	1.80, 1.71
33	33	33	$\frac{1}{3}$	$\frac{1}{3}$	$\frac{1}{3}$	1.46, 1.50

In matrix notation, the second-degree model is written as $\mathbf{y} = \mathbf{X}\boldsymbol{\beta} + \boldsymbol{\epsilon}$ where

$$\mathbf{y} = \begin{bmatrix} 2.02 \\ 2.08 \\ 1.47 \\ 1.37 \\ 1.91 \\ 2.00 \\ 1.93 \\ 1.83 \\ 1.98 \\ 2.13 \\ 1.80 \\ 1.71 \\ 1.46 \\ 1.50 \end{bmatrix} \quad \mathbf{X} = \begin{array}{c} \begin{matrix} x_1 & x_2 & x_3 & x_1x_2 & x_1x_3 & x_2x_3 \end{matrix} \\ \begin{bmatrix} 1 & 0 & 0 & 0 & 0 & 0 \\ 1 & 0 & 0 & 0 & 0 & 0 \\ 0 & 1 & 0 & 0 & 0 & 0 \\ 0 & 1 & 0 & 0 & 0 & 0 \\ 0.5 & 0.5 & 0 & 0.25 & 0 & 0 \\ 0.5 & 0.5 & 0 & 0.25 & 0 & 0 \\ 0 & 0 & 1 & 0 & 0 & 0 \\ 0 & 0 & 1 & 0 & 0 & 0 \\ 0.5 & 0 & 0.5 & 0 & 0.25 & 0 \\ 0.5 & 0 & 0.5 & 0 & 0.25 & 0 \\ 0 & 0.5 & 0.5 & 0 & 0 & 0.25 \\ 0 & 0.5 & 0.5 & 0 & 0 & 0.25 \\ 0.33 & 0.33 & 0.33 & 0.11 & 0.11 & 0.11 \\ 0.33 & 0.33 & 0.33 & 0.11 & 0.11 & 0.11 \end{bmatrix} \end{array} \quad \boldsymbol{\beta} = \begin{bmatrix} \beta_1 \\ \beta_2 \\ \beta_3 \\ \beta_{12} \\ \beta_{13} \\ \beta_{23} \end{bmatrix}$$

The normal equations (7.6) are

$$\overset{\mathbf{X}'\mathbf{X}}{\begin{bmatrix} 3.222 & 0.722 & 0.722 & 0.324 & 0.324 & 0.074 \\ & 3.222 & 0.722 & 0.324 & 0.074 & 0.324 \\ & & 3.222 & 0.074 & 0.324 & 0.324 \\ & & & 0.150 & 0.025 & 0.025 \\ \text{(symmetric)} & & & & 0.150 & 0.025 \\ & & & & & 0.150 \end{bmatrix}} \overset{\mathbf{b}}{\begin{bmatrix} b_1 \\ b_2 \\ b_3 \\ b_{12} \\ b_{13} \\ b_{23} \end{bmatrix}} = \overset{\mathbf{X}'\mathbf{y}}{\begin{bmatrix} 9.10 \\ 7.54 \\ 8.56 \\ 1.31 \\ 1.36 \\ 1.21 \end{bmatrix}}$$

The solutions to the normal equations are

$$
\begin{array}{ccc}
\mathbf{b} & = & (\mathbf{X}'\mathbf{X})^{-1} \quad\quad \mathbf{X}'\mathbf{y}
\end{array}
$$

$$
\begin{bmatrix} 2.08 \\ 1.45 \\ 1.91 \\ 0.21 \\ -0.31 \\ -0.25 \end{bmatrix}
=
\begin{bmatrix}
0.496 & -0.004 & -0.004 & -0.924 & -0.924 & 0.076 \\
 & 0.496 & -0.004 & -0.924 & 0.076 & -0.924 \\
 & & 0.496 & 0.076 & -0.924 & -0.924 \\
 & & & 10.485 & 0.485 & 0.485 \\
\text{(symmetric)} & & & & 10.485 & 0.485 \\
 & & & & & 10.485
\end{bmatrix}
\begin{bmatrix} 9.10 \\ 7.54 \\ 8.56 \\ 1.31 \\ 1.36 \\ 1.21 \end{bmatrix}
$$

and the fitted model (7.11) is

$$
\begin{aligned}
\hat{y}(\mathbf{x}) &= \mathbf{x}_p'\mathbf{b} \\
&= \underset{(0.14)}{2.08}x_1 + \underset{(0.14)}{1.45}x_2 + \underset{(0.14)}{1.91}x_3 + \underset{(0.65)}{0.21}x_1x_2 - \underset{(0.65)}{0.31}x_1x_3 - \underset{(0.65)}{0.25}x_2x_3 \qquad (7.16)
\end{aligned}
$$

where the number in parentheses below each parameter estimate is the estimated standard error of the estimate.

The analysis of variance calculations are

$$\text{SST} = \mathbf{y}'\mathbf{y} - \frac{(\mathbf{1}'\mathbf{y})^2}{14} = 46.170 - 45.325 = 0.845$$

$$\text{SSE} = \mathbf{y}'\mathbf{y} - \mathbf{b}'\mathbf{X}'\mathbf{y} = 46.170 - 45.847 = 0.323$$

$$\text{SSR} = \text{SST} - \text{SSE} = 0.845 - 0.323 = 0.522$$

and the analysis of variance table is

Source of Variation	Degrees of Freedom	Sum of Squares	Mean Square
Regression	5	0.522	0.104
Residual	8	0.323	0.04
Total	13	0.845	

The estimated standard errors of the parameter estimates in Eq. (7.16) are the square roots of the diagonal elements of $(\mathbf{X}'\mathbf{X})^{-1}s^2$ where $s^2 = 0.04$. In other words,

$$\widehat{\text{s.e.}}(b_1) = \sqrt{0.496(0.04)} = 0.14$$

$$\widehat{\text{s.e.}}(b_{12}) = \sqrt{10.485(0.04)} = 0.65$$

7.6. THE ADJUSTED MULTIPLE CORRELATION COEFFICIENT

A measure of the goodness of fit of a standard regression equation is the multiple correlation coefficient denoted by R. The proportion of the total sum of squares which is explained by the fitted model is denoted by $R^2 = \text{SSR}/\text{SST}$.

With the canonical polynomials, although the sum of squares regression is corrected for the overall mean, the model does not contain a β_0 term. For this reason an adjusted R^2 statistic has been suggested for use with the canonical polynomials and is calculated as

$$R_A^2 = 1 - \frac{\text{SSE}/(N-p)}{\text{SST}/(N-1)} \tag{7.17}$$

where SSE and SST are defined in Eqs. (7.14) and (7.13) respectively. The R_A^2 statistic, unlike R^2, takes into account the number of parameters (p) in the fitted model in the degrees of freedom used to estimate σ^2 in the numerator $\text{SSE}/(N-p)$. For the fish patty data, the value of R_A^2 associated with the fitted second-degree model in Eq. (7.16) is $R_A^2 = 1 - [(0.323/8)/(0.845/13)] = 0.379$.

7.7. TESTING HYPOTHESES ABOUT THE FORM OF THE MODEL: TESTS OF SIGNIFICANCE

The properties of the parameter estimates shown in Eqs. (7.8) and (7.9) assumed the fitted model is correct (i.e., that $E(\boldsymbol{\epsilon}) = \mathbf{0}$, $E(\boldsymbol{\epsilon}\boldsymbol{\epsilon}') = \sigma^2\mathbf{I}_N$) and the additional assumption that the errors are jointly normally distributed was made which enabled us to say the distribution of the parameter estimates is normal. The normality assumption on the errors allows us additional flexibility, namely a significance test can be made of the goodness of fit of the fitted model using the entries from the analysis of variance table, Table 7.1.

Let us consider the general mixture model $\mathbf{y} = \mathbf{X}\boldsymbol{\beta} + \boldsymbol{\epsilon}$, where the matrix $\mathbf{X}$ is $N \times p$ and where the errors are normally distributed. Then SSE in Eq. (7.14) can be shown to be distributed as

$$\frac{\text{SSE}}{\sigma^2} \sim \chi^2_{(N-p)} \tag{7.18}$$

where the symbol $\sim$ means "is distributed as" and $\chi^2_{(N-p)}$ is a chi-square random variable with $N-p$ degrees of freedom. Under the null hypothesis that the mixture components do not affect the response, that is, for

$q=3$ the surface is a level plane above the triangle whose height is constant (equal to β) at all points,

$$H_0: \text{ all } \beta_i = \beta \text{ and all } \beta_{ij}, \beta_{ijk}, \text{ other than } \beta_i, = 0$$
$$i = 1, 2, \ldots, q \qquad i \neq j \neq k$$

then

$$\frac{\text{SSR}}{\sigma^2} \sim \chi^2_{(p-1)} \tag{7.19}$$

and SSR/σ^2 is independent of SSE/σ^2. Hence, the ratio

$$F = \frac{\text{SSR}/(p-1)}{\text{SSE}/(N-p)} \tag{7.20}$$

has an F-distribution with $p-1$ and $N-p$ degrees of freedom in the numerator and denominator, respectively. Computed values of the F-ratio in Eq. (7.20) can be compared with the tabled values of $F_{(p-1,N-p,\alpha)}$ to test the significance of the regression on $x_1, x_2, \ldots, x_q$ at the α level.

Occasionally we shall want to test an hypothesis of the form $\mathbf{C\beta} = \mathbf{m}$, where $\mathbf{C}$ is an $r \times p$ matrix and $\mathbf{m}$ is a $r \times 1$ vector of constants. The rows of $\mathbf{C\beta}$ are linearly independent estimable functions so that $\mathbf{C}$ has rank r $(r \leq p)$. For example, if the elements of the parameter vector are $\boldsymbol{\beta} = (\beta_1, \beta_2, \beta_3, \beta_{12}, \beta_{13}, \beta_{23})'$ and we wanted to test the hypothesis that $\beta_1 = \beta_2$ and $\beta_{12} = \beta_{23}$, we might have for the matrix $\mathbf{C}$ and the vector $\mathbf{m}$

$$\mathbf{C\beta} = \mathbf{m}$$

$$\begin{bmatrix} 1 & -1 & 0 & 0 & 0 & 0 \\ 0 & 0 & 0 & 1 & 0 & -1 \end{bmatrix} \boldsymbol{\beta} = \begin{bmatrix} 0 \\ 0 \end{bmatrix}$$

The test of the hypothesis $\mathbf{C\beta} = \mathbf{m}$ requires first defining the complete model as $\mathbf{y} = \mathbf{X\beta} + \boldsymbol{\epsilon}$ and a reduced model $\mathbf{y} = \tilde{\mathbf{X}}\tilde{\boldsymbol{\beta}} + \boldsymbol{\epsilon}$ where $\boldsymbol{\beta}$ is changed into $\tilde{\boldsymbol{\beta}}$ by the conditions of the hypothesis and $\tilde{\mathbf{X}}$ is the form of $\mathbf{X}$ corresponding to $\tilde{\boldsymbol{\beta}}$. Next the sum of squares due to fitting the complete model and that due to fitting the reduced model are obtained. These are

$$\begin{aligned} \text{SSR}_{\text{complete}} &= \mathbf{b}'\mathbf{X}'\mathbf{y} - \frac{(\mathbf{1}'\mathbf{y})^2}{N} && \text{with } p-1 \text{ degrees of freedom} \\ \text{SSR}_{\text{reduced}} &= \tilde{\mathbf{b}}'\tilde{\mathbf{X}}'\mathbf{y} - \frac{(\mathbf{1}'\mathbf{y})^2}{N} && \text{with } p-1-r \text{ degrees of freedom} \end{aligned} \tag{7.21}$$

where **b** and $\tilde{\mathbf{b}}$ are the estimators of $\boldsymbol{\beta}$ and $\tilde{\boldsymbol{\beta}}$, respectively. The test of $\mathbf{C}\boldsymbol{\beta} = \mathbf{m}$ involves the difference

$$F = \frac{[\text{SSR}_{\text{complete}} - \text{SSR}_{\text{reduced}}]/r}{\text{SSE}_{\text{complete}}/(N-p)} \tag{7.22}$$

or

$$F = \frac{[\text{SSE}_{\text{reduced}} - \text{SSE}_{\text{complete}}]/r}{\text{SSE}_{\text{complete}}/(N-p)} \tag{7.23}$$

where $\text{SSE}_{\text{complete}} = \mathbf{y}'\mathbf{y} - \mathbf{b}'\mathbf{X}'\mathbf{y}$ and $\text{SSE}_{\text{reduced}} = \mathbf{y}'\mathbf{y} - \tilde{\mathbf{b}}'\tilde{\mathbf{X}}'\mathbf{y}$. Comparing the computed F-ratio with tabulated values of $F_{(r,N-p,\alpha)}$ provides the test.

A more readily obtainable numerator for the F-ratio in (7.22) for testing $\mathbf{C}\boldsymbol{\beta} = \mathbf{m}$ is possible. A proof of a theorem in Searle (1966), Chapter 10, reveals that the F-ratio can be written as

$$F = \frac{(\mathbf{Cb} - \mathbf{m})'[\mathbf{C}(\mathbf{X}'\mathbf{X})^{-1}\mathbf{C}']^{-1}(\mathbf{Cb} - \mathbf{m})}{r\{\text{SSE}_{\text{complete}}/(N-p)\}} \tag{7.24}$$

where **Cb** is the estimator of $\mathbf{C}\boldsymbol{\beta}$. Hence, once **C** and **m** are specified, only the complete model estimates are necessary for the calculations of F in Eq. (7.24).

Let us test the hypothesis H_0: $\beta_{12} = \beta_{13} = \beta_{23} = 0$ versus H_A: *One or more equality signs is false*, using the fish patty data in Table 7.2. The matrix **C** and vector **m** are

$$\mathbf{C} = \begin{bmatrix} 0 & 0 & 0 & 1 & 0 & 0 \\ 0 & 0 & 0 & 0 & 1 & 0 \\ 0 & 0 & 0 & 0 & 0 & 1 \end{bmatrix}, \qquad \mathbf{m} = \begin{bmatrix} 0 \\ 0 \\ 0 \end{bmatrix}$$

The complete model is the second-degree equation (7.16) and $\text{SSE}_{\text{complete}} = 0.323$ with $N - p = 8$ degrees of freedom.

The numerator of the F-statistic in Eq. (7.24) is

$$(\mathbf{Cb} - \mathbf{m})'[\mathbf{C}(\mathbf{X}'\mathbf{X})^{-1}\mathbf{C}']^{-1}(\mathbf{Cb} - \mathbf{m})$$

$$= \begin{bmatrix} b_{12} \\ b_{13} \\ b_{23} \end{bmatrix}' [10\mathbf{I} + 0.485\mathbf{J}]^{-1} \begin{bmatrix} b_{12} \\ b_{13} \\ b_{23} \end{bmatrix}$$

$$= (0.21, -0.31, -0.25)\left[\frac{1}{10}\mathbf{I} - \frac{0.485}{114.55}\mathbf{J}\right]\begin{bmatrix} 0.21 \\ -0.31 \\ -0.25 \end{bmatrix} = 0.02$$

and therefore the value of the F-statistic is

$$F = \frac{0.02}{3(0.04)} = 0.16$$

Since the tabled value $F_{(3,8,\alpha=0.05)} = 4.07$, we do not reject H_0: $\beta_{12} = \beta_{13} = \beta_{23} = 0$. The use of the residual sums of squares quantities for the complete and reduced models where the reduced model is the 3-term first-degree model, according to Eq. (7.23) produces

$$F = \frac{[\text{SSE}_{\text{reduced}} - \text{SSE}_{\text{complete}}]/r}{\text{SSE}_{\text{complete}}/(N-p)}$$

$$= \frac{[0.3432 - 0.3240]/3}{0.3240/8} = 0.16$$

This value must be identical to the value of the previously calculated F.

7.8. REFERENCES AND RECOMMENDED READING

Draper, N. R. and H. Smith (1966). *Applied Regression Analysis*. Wiley, New York.

Graybill, F. A. (1961). An *Introduction to Linear Statistical Models*. Vol. 1, McGraw-Hill, New York.

Searle, S. R. (1966). *Matrix Algebra for the Biological Sciences*. Wiley, New York.

Answers to Selected Odd-Numbered Questions

CHAPTER 2

2.1. With a regular figure, all of its angles are congruent and all of its sides are congruent.

2.3. Synergistic in cases 1 and 5, antagonistic in cases 3 and 4, and, additive in case 2.

2.5.
$$\hat{\eta}(\mathbf{x}) = 11.7(0.40) + 9.4(0.30) + 16.4(0.30) + 19.0(0.12) + 11.4(0.12) - 9.6(0.09) = 15.20$$

$$\widehat{\text{var}}[\hat{\eta}(\mathbf{x})] = 0.73\left\{\frac{0.08^2 + 0.12^2 + 0.12^2}{2} + \frac{0.48^2 + 0.48^2 + 0.36^2}{3}\right\} = 0.1565$$

so that the 95% confidence interval is $[14.30 < \eta < 16.10]$.

2.7. 1. b 2. f 3. e 4. d 5. a 6. c.

2.9. At $x_i = x_j = \frac{1}{2}$, $b_{ij}x_ix_j = b_{ij}/4$ and $b_ix_i = b_i/2$. In order that $b_{ij}/4 = b_i/2$, then $b_{ij} = 2b_i$. At $x_1 = x_2 = x_3 = \frac{1}{3}$, $b_{123}x_1x_2x_3 = b_{123}/27$ and $b_{ij}x_ix_j = b_{ij}/9$ and $b_ix_i = b_i/3$. In order that $b_{123}/27 = b_i/3$ and $b_{123}/27 = b_{ij}x_ix_j/9$, then $b_{123}/b_i = 9$ and $b_{123}/b_{ij} = 3$, that is, the coefficient estimate b_{123} must be 9 times as large as the linear coefficient b_i and 3 times as large as the binary coefficient b_{ij} in order that the respective terms in the model contribute equally to $\hat{y}(\mathbf{x})$ at the centroid of the simplex.

2.11. **a.**
$$\underset{(0.11)}{\hat{y}(\mathbf{x}) = 4.85x_1} + \underset{(0.11)}{1.28x_2} + \underset{(0.11)}{2.74x_3} - \underset{(0.46)}{4.98x_1x_2} - \underset{(0.46)}{2.82x_1x_3} + \underset{(0.46)}{1.48x_2x_3}$$

b. $\hat{y}(7) = 2.25$, $\widehat{\text{var}}[\hat{y}(7)] = 0.022(0.216) = 0.0048$

$$t = \frac{y(7) - \hat{y}(7)}{\sqrt{s^2 + s^2(0.216)}} = \frac{0.50}{0.164} = 3.05$$

At the $\alpha = 0.05$ level of significance, $t_{(9,0.025)} = 2.262$. Reject the hypothesis that states no difference in the observed value and the value predicted by the model. Hence, refit the model using the observation $y = 2.75$ at the centroid.

c. $\hat{y}(8) = 3.08$, $t = -4.52$ Reject no difference

$\hat{y}(9) = 1.65$, $t = 1.29$

$\hat{y}(10) = 2.56$, $t = -5.81$ Reject no difference

Define the 4×1 vector of differences

$$\mathbf{d} = \begin{bmatrix} y(7) - \hat{y}(7) \\ y(8) - \hat{y}(8) \\ y(9) - \hat{y}(9) \\ y(10) - \hat{y}(10) \end{bmatrix} = \begin{bmatrix} y(7) \\ y(8) \\ y(9) \\ y(10) \end{bmatrix} - \mathbf{Pb} = \begin{bmatrix} 0.50 \\ -0.28 \\ 0.08 \\ -0.36 \end{bmatrix}$$

where

$$\mathbf{P} = \begin{bmatrix} \frac{1}{3} & \frac{1}{3} & \frac{1}{3} & \frac{1}{9} & \frac{1}{9} & \frac{1}{9} \\ \frac{2}{3} & \frac{1}{6} & \frac{1}{6} & \frac{1}{9} & \frac{1}{9} & \frac{1}{36} \\ \frac{1}{6} & \frac{2}{3} & \frac{1}{6} & \frac{1}{9} & \frac{1}{36} & \frac{1}{9} \\ \frac{1}{6} & \frac{1}{6} & \frac{2}{3} & \frac{1}{36} & \frac{1}{9} & \frac{1}{9} \end{bmatrix}$$

Then $\widehat{\text{var}}(\mathbf{d}) = \{\mathbf{I}_4 + \mathbf{P}(\mathbf{X}'\mathbf{X})^{-1}\mathbf{P}'\}s^2$ and the test statistic is the F-ratio, $F = \dfrac{\mathbf{d}'\{\mathbf{I}_4 + \mathbf{P}(\mathbf{X}'\mathbf{X})^{-1}\mathbf{P}'\}^{-1}\mathbf{d}}{4s^2} = 5.0$ which exceeds $F_{(4,9,0.05)} = 3.63$ hence reject the adequacy of the model.

d. Second-degree model seems as good as any other.

$$\hat{y}(\mathbf{x}) = 4.81x_1 + 1.29x_2 + 2.69x_3 - 4.83x_1x_2 - 2.71x_1x_3 + 1.62x_2x_3$$

(0.15) (0.15) (0.15) (0.65) (0.65) (0.65)

$$s^2 = 0.05$$

CHAPTER 3

3.1. **a.** Familiarity with standard independent variable design configurations and known design criteria.

b. Difficulty interpreting the coefficient estimates in the transformed variable model as they pertain to the original mixture system.

3.3. In the (w_1, w_2) system, the first-degree coefficient estimates -1.77 and -0.21 represent changes in the average response brought about by changing the level of the w_i from -0.707 to $+0.707$ when computed at the average level of z_1. That is $-1.77 = \sqrt{2}$ times the slope of the surface along the w_1-axis if $z_1 = 0$. The estimate 4.65 represents one-half the main effect of the process variable. The coefficient estimates -0.04 and -1.03 represent some measure of dependency between the effect of the process variable and the effects of the variables w_1 and w_2. To get positive values of $\hat{y}(\mathbf{w}, z)$, let w_1 and w_2 be at the low levels (negative values settings) and z_1 at the high level. This would correspond to a high setting for x_1 and just medium valued settings for x_2 and x_3 if $z_1 = +1$. $\hat{y}(0.707, 0, -1) = 13.45$ and the 95% confidence interval for η is

$$\Pr[13.45 - 2.54 < \eta < 13.45 + 2.54] = [10.91 < \eta < 15.99] = 0.95$$

3.5. Using a central design $\mathbf{D}_w$ of the type considered in (3.39) with $\mathbf{D}_3 = (0\ 0\ 0)$ where $c = \rho^*/\sqrt{3} = 1.04/\sqrt{3} = 0.6004$ and $g = \sqrt[4]{8}c = 1.01$, and from Appendix 3B if $\mathbf{T}_1$ is of the form

$$\mathbf{T}_1 = \begin{bmatrix} -0.3758 & -0.4474 & -0.4022 \\ 0.9268 & -0.1814 & -0.1630 \\ 0 & 0.8760 & -0.2391 \\ 0 & 0 & 0.8687 \end{bmatrix}$$

then $\mathbf{X} = \begin{bmatrix} 0.6422 & 0.0975 & 0.1359 & 0.1244 \\ 0.4752 & 0.2645 & 0.1359 & 0.1244 \\ 0.4434 & 0.0649 & 0.3673 & 0.1244 \\ 0.2765 & 0.2318 & 0.3673 & 0.1244 \\ 0.4635 & 0.0682 & 0.0727 & 0.3956 \\ 0.2966 & 0.2351 & 0.0727 & 0.3956 \\ 0.2647 & 0.0355 & 0.3041 & 0.3956 \\ 0.0978 & 0.2025 & 0.3041 & 0.3956 \\ 0.5104 & 0.0096 & 0.2200 & 0.2600 \\ 0.2296 & 0.2904 & 0.2200 & 0.2600 \\ 0.5372 & 0.1775 & 0.0253 & 0.2600 \\ 0.2028 & 0.1225 & 0.4146 & 0.2600 \\ 0.5203 & 0.1747 & 0.2731 & 0.0319 \\ 0.2197 & 0.1253 & 0.1669 & 0.4881 \\ 0.3700 & 0.1500 & 0.2200 & 0.2600 \end{bmatrix}$

3.7. $\hat{y}(\mathbf{w}) = \underset{(0.66)}{13.54} + \underset{(0.26)}{0.30}w_1 - \underset{(0.26)}{0.70}w_2$

The residual mean square is a composite measure of model lack of fit and the experimental error variance. If we feel $s^2 = 0.73$ is a valid estimate of the error variance, then lack of fit is significant. This is because the residual sum of squares can be partitioned into error sum of squares (due to replicates) plus lack of fit sum of squares and the analysis of variance table is

Analysis of variance table

Source	d.f.	S.S.	M.S.
Regression	2	57.63	28.82
Residual	12	77.23	6.44
Lack of fit	3	70.66	23.55
Error	9	6.57	0.73
Total	14	134.86	

$F = \frac{23.55}{0.73} = 32.3 > F_{(3,9,0.01)} = 6.99$

Reject the hypothesis that the first-degree model has zero lack of fit. Fit second-degree model.

Second-degree model:

$$\hat{y}(\mathbf{w}) = \underset{(0.40)}{14.81} + \underset{(0.11)}{0.96}w_1 - \underset{(0.11)}{0.53}w_2 - \underset{(0.04)}{0.33}w_1^2 + \underset{(0.04)}{0.13}w_2^2 + \underset{(0.05)}{0.12}w_1w_2$$

and at $x_1 = \frac{2}{3}$, $x_2 = \frac{1}{3}$, $x_3 = 0$ the values of w_1 and w_2 are $w_1 = \sqrt{6}$, $w_2 = \sqrt{2}$ so that $\hat{y}(\sqrt{6}, \sqrt{2}) = 15.2$ and the 95% confidence interval for η is $\Pr[14.2 < \eta < 16.2]$. The length of the interval computed from the model in w_1 and w_2 must be the same length as the interval computed from the model in x_1, x_2, and x_3.

3.9. **a.** Let x_A and x_B represent beef types A and B respectively and let c^1 and c^2 represent the linear and quadratic terms for the time-temperature conditions of the process variable. From replication 1,

$$\begin{aligned}\hat{y}(\mathbf{x}, c) = {} & 2.17x_A + 1.40x_B + 4.28x_Ax_B + 0.93x_Ax_B(x_A - x_B) \\ & + 0.75x_Ac^1 + 0.30x_Bc^1 - 2.36x_Ax_Bc^1 \\ & - 2.36x_Ax_B(x_A - x_B)c^1 + 0.08x_Ac^2 + 0.10x_Bc^2 \\ & - 1.01x_Ax_Bc^2 - 1.31x_Ax_B(x_A - x_B)c^2\end{aligned}$$

b. If from the restriction on the randomization we analyze the data as if the design was of the split-plot type where cooking levels are the main treatments and the mixture blends are the subtreatments, the analysis of variance table would be, with replications 1 and 2,

Source	d.f.	S.S.	M.S.	F
Reps.	1	0.16	0.16	
Cooking Temp.	2	0.61	0.30	$7.5 < F_{(2,2,0.05)} = 19.0$
E_A	2	0.08	0.04	
Components	3	8.59	2.86	$16.9 > F_{(3,9,0.01)} = 6.99$
Cook × Components	6	1.56	0.26	0.26
E_B	9	1.49	0.17	

According to the F test on the components source, the average textures of the beef patties made from the four blends are different. The average textures, across replications 1 and 2, are

$x_A : x_B$	$\bar{y}$
$1:0$	1.9
$\frac{2}{3}:\frac{1}{3}$	3.1
$\frac{1}{3}:\frac{2}{3}$	2.1
$0:1$	1.5

$$\begin{aligned}\hat{y}(\mathbf{x}, c) = {} & 2.09x_A + 1.35x_B + 4.13x_Ax_B + 0.54x_Ac^1 + 0.36x_Bc^1 \\ & - 2.31x_Ax_Bc^1\end{aligned}$$

From the coefficient estimates we make the following inferences; beef type A has a higher average texture than B or type A patties are firmer than B (since $2.09 > 1.35$) while blends of A and B have a higher average texture value than would be expected by simply averaging A and B (since $4.13 > 0$). From the value of $\hat{\gamma}^1_{AB} = -2.31 < 0$, the relative firmness of the $A:B$ blends to the average of A and B is greater at the low cooking temperature than at the high cooking temperature. In fact the negative synergism increases or the firmness of the texture of the blend decreases significantly as the cooking temperature is raised from the low level to the high level.

c. Using all 36 observations from a completely randomized design,

Source	d.f.	S.S.	M.S.	F
Regression (model below)	11	15.50		
Cooking Temp. (c)	2	1.39	0.69	$7.0 > F_{(2,24,0.01)} = 5.61$
Components (C)	3	11.11	3.70	$37.4 > F_{(3,24,0.01)} = 4.72$
Cook × Components	6	3.00	0.50	$5.0 > F_{(6,24,0.01)} = 3.67$
Residual	24	2.38	$0.099 = s_e^2$	
Reps	2	0.17		
Reps × c	4	0.23		
Reps × C	6	1.39		
Reps × C × c	12	0.59		
Total	35	17.88		

***Model*:**

$$\hat{y}(\mathbf{x}, c) = \underset{(0.11)}{2.02x_A} + \underset{(0.11)}{1.49x_B} + \underset{(0.47)}{3.68x_Ax_B} + \underset{(1.06)}{5.03x_Ax_B(x_A - x_B)}$$

$$+\underset{(0.13)}{0.63x_Ac^1} + \underset{(0.13)}{0.43x_Bc^1} - \underset{(0.58)}{2.74x_Ax_Bc^1}$$

$$+\underset{(1.29)}{0.11x_Ax_B(x_A - x_B)c^1} + \underset{(0.07)}{0.04x_Ac^2} + \underset{(0.07)}{0.01x_Bc^2}$$

$$-\underset{(0.33)}{0.64x_Ax_Bc^2} - \underset{(0.75)}{1.24x_Ax_B(x_A - x_B)c^2}$$

Over the three replicates, the beef type A patties appear to be firmer than the patties of beef B and blends of A and B are even firmer. The trend in the average texture values of the blends appears to be other than additive when averaged across the three cooking temperatures. These average textures are 2.02, 3.03, 2.11, 1.49 when $(x_A : x_B) = 1:0, \frac{2}{3}:\frac{1}{3}, \frac{1}{3}:\frac{2}{3}, 0:1$ which indicates that the average texture of the patties with a small proportion ($B = \frac{1}{3}$) of beef B with A is firmer than the pure A patties but as the proportion of beef B is increased, the firmness of the texture decreases.

Increasing the cooking temperature increases the firmness of the pure beef patties in a linear fashion (0.63, 0.43 are nonzero), whereas the increase in temperature does not affect blends of A and B linearly. The texture of the $(x_A : x_B) = (\frac{2}{3}:\frac{1}{3})$ blend behaves in a curvilinear fashion across the three temperature levels, being firmest at the c_2 level whereas the average texture of the $(x_A : x_B) = (\frac{1}{3}:\frac{2}{3})$ patties is unchanged as the level of temperature is changed.

CHAPTER 4

4.1. $$x'_1 = \frac{x_1 - a_1}{1 - (a_1 + a_2 + a_3)} = \frac{x_1 - 0.10}{0.75},$$

$$x'_2 = \frac{x_2 - a_2}{0.75} = \frac{x_2 - 0.15}{0.75}$$

$$x'_3 = \frac{x_3}{0.75}$$

Pseudocomponents: *Original components*:

x'_1	x'_2	x'_3	x_1	x_2	x_3
1	0	0	0.85	0.15	0
0	1	0	0.10	0.90	0
0	0	1	0.10	0.15	0.75
$\frac{1}{2}$	$\frac{1}{2}$	0	0.475	0.525	0
$\frac{1}{2}$	0	$\frac{1}{2}$	0.475	0.15	0.375
0	$\frac{1}{2}$	$\frac{1}{2}$	0.10	0.525	0.375
$\frac{1}{3}$	$\frac{1}{3}$	$\frac{1}{3}$	0.35	0.40	0.25

4.3.

z_1	z_2	z_3	:	x_1	x_2	x_3	y
$\frac{3}{4}$	$\frac{1}{8}$	$\frac{1}{8}$		0.70	0.25	0.05	22.4
$\frac{1}{8}$	$\frac{3}{4}$	$\frac{1}{8}$		0.20	0.75	0.05	9.2
$\frac{1}{8}$	$\frac{1}{8}$	$\frac{3}{4}$		0.20	0.25	0.55	20.4
$\frac{1}{2}$	$\frac{1}{2}$	0		0.50	0.50	0.00	16.5
$\frac{1}{2}$	0	$\frac{1}{2}$		0.50	0.15	0.35	18.6
0	$\frac{1}{2}$	$\frac{1}{2}$		0.10	0.55	0.35	10.7

$$\hat{y}(\mathbf{x}) = \underset{(3.28)}{27.68x_1} + \underset{(3.12)}{3.78x_2} + \underset{(4.09)}{20.49x_3} \quad s^2 = 6.46$$

4.5. **a.**

x_1	x_2	x_3	x_4	x_5	x_6
0	0.5	0	0.35	0	0.15
0	0.5	0	0.35	0.075	0.075
0	0.5	0	0.35	0.15	0
0	0.5	0.175	0.175	0	0.15
0	0.5	0.175	0.175	0.075	0.075
0	0.5	0.175	0.175	0.15	0
0	0.5	0.35	0	0	0.15
0	0.5	0.35	0	0.075	0.075
0	0.5	0.35	0	0.15	0

x_1	x_2	x_3	x_4	x_5	x_6
0.25	0.25	0	0.35	0	0.15
0.25	0.25	0	0.35	0.075	0.075
0.25	0.25	0	0.35	0.15	0
0.25	0.25	0.175	0.175	0	0.15
0.25	0.25	0.175	0.175	0.075	0.075
0.25	0.25	0.175	0.175	0.15	0
0.25	0.25	0.35	0	0	0.15
0.25	0.25	0.35	0	0.075	0.075
0.25	0.25	0.35	0	0.15	0

x_1	x_2	x_3	x_4	x_5	x_6
0.5	0	0	0.35	0	0.15
0.5	0	0	0.35	0.075	0.075
0.5	0	0	0.35	0.15	0
0.5	0	0.175	0.175	0	0.15
0.5	0	0.175	0.175	0.075	0.075
0.5	0	0.175	0.175	0.15	0
0.5	0	0.35	0	0	0.15
0.5	0	0.35	0	0.075	0.075
0.5	0	0.35	0	0.15	0

$$\mathbf{b.}\quad \mathbf{T}_1 = \begin{bmatrix} 0.7071 & -0.7071 & 0 & 0 & 0 & 0 \\ 0 & 0 & 0.7071 & -0.7071 & 0 & 0 \\ 0 & 0 & 0 & 0 & 0.7071 & -0.7071 \end{bmatrix}$$

$c = \rho^*/\sqrt{3} = 0.8165$

then

$$\mathbf{X} = \begin{bmatrix} 0.1224 & 0.3776 & 0.1234 & 0.2266 & 0.0343 & 0.1157 \\ 0.3776 & 0.1224 & 0.1234 & 0.2266 & 0.0343 & 0.1157 \\ 0.1224 & 0.3776 & 0.2766 & 0.0734 & 0.0343 & 0.1157 \\ 0.3776 & 0.1224 & 0.2766 & 0.0734 & 0.0343 & 0.1157 \\ 0.1224 & 0.3776 & 0.1234 & 0.2266 & 0.1057 & 0.0443 \\ 0.3776 & 0.1224 & 0.1234 & 0.2266 & 0.1057 & 0.0443 \\ 0.1224 & 0.3776 & 0.2766 & 0.0734 & 0.1057 & 0.0443 \\ 0.3776 & 0.1224 & 0.2766 & 0.0734 & 0.1057 & 0.0443 \\ 0.0354 & 0.4646 & 0.2000 & 0.1500 & 0.0700 & 0.0800 \\ 0.4646 & 0.0354 & 0.2000 & 0.1500 & 0.0700 & 0.0800 \\ 0.2500 & 0.2500 & 0.0712 & 0.2788 & 0.0700 & 0.0800 \\ 0.2500 & 0.2500 & 0.3288 & 0.0212 & 0.0700 & 0.0800 \\ 0.2500 & 0.2500 & 0.2000 & 0.1500 & 0.0099 & 0.1401 \\ 0.2500 & 0.2500 & 0.2000 & 0.1500 & 0.1301 & 0.0199 \\ 0.2500 & 0.2500 & 0.2000 & 0.1500 & 0.0700 & 0.0800 \end{bmatrix}$$

CHAPTER 5

5.1. The advantage to fitting the seven-term special cubic model is that the model provides us with information about the relative importance of the pure blends, of the binary blends where x_i and x_j each contribute $\frac{1}{2}$ to the mixture, and also gives us some measure of the importance of the complete blend where $x_1 = x_2 = x_3 = \frac{1}{3}$. However, the disadvantage to using the complete model is if there are no replicated observations, then we have saturated all of the data with the fitted model and no estimate of the observation variance is possible. With a reduced model form, an estimate of the observation variance is available. Also, with a reduced model, the task of predicting response values throughout the triangle (by hand calculations) is easier than with the complete model.

5.3. **a.** $\hat{y}(\mathbf{x}) = 10.60x_1 + 6.49x_2 + 6.77x_3$
$\qquad\qquad (1.83) \quad (1.83) \quad (1.83)$

Analysis of variance table

Source	d.f.	S.S.	M.S.
Regression	2	29.04	14.52
Residual	12	130.64	10.89

Test statistic: $F = \dfrac{14.52}{10.89} = 1.33 < F_{(2,12,0.10)} = 2.81$

Do not reject H_0: $\beta_1 = \beta_2 = \beta_3 =$ constant, at this stage.
Note: $R_A^2 = 0.0450$.

b. $\hat{y}(\mathbf{x}) = \underset{(0.74)}{6.50x_1} + \underset{(0.74)}{4.70x_2} + \underset{(0.74)}{3.20x_3} + \underset{(3.19)}{18.00x_1x_2} + \underset{(3.19)}{31.00x_1x_3}$

$+ \underset{(3.19)}{14.07x_2x_3}$

Increase in regression sum of squares $= 149.89 - 29.04 = 120.85$

Is it a significant increase? Yes, because

$$F = \frac{120.85/3}{1.09} = 36.96 > F_{(3,9,0.01)} = 6.99$$

$$C_3 = \frac{130.64}{1.09} - (15-6) \approx 111$$

$$C_6 = \frac{9.79}{1.09} - (15-12) \approx 6$$

Source	d.f.	S.S.	M.S.	
Regression	5	149.89	29.98	Note: $R_A^2 = 0.9044$
Residual	9	9.79	1.09	
Total	14	159.68		

5.5. If $\tilde{\mathbf{b}} = \mathbf{Ab}$, then $\text{var}(\tilde{\mathbf{b}}) = \mathbf{A}(\mathbf{X}'\mathbf{X})^{-1}\mathbf{A}'\sigma^2$. With the matrices $\mathbf{A}$ and $(\mathbf{X}'\mathbf{X})^{-1}$ defined as in Section 5.6,

$$\widehat{\text{var}}\begin{bmatrix} \tilde{b}_1 \\ \tilde{b}_2 \\ \tilde{b}_3 \\ \tilde{b}_{12} \\ \tilde{b}_{13} \\ \tilde{b}_{23} \end{bmatrix} = \begin{bmatrix} \frac{9}{20} & \frac{1}{20} & 0 & -1 & -\frac{1}{2} & -\frac{1}{2} \\ & \frac{9}{20} & 0 & -1 & -\frac{1}{2} & -\frac{1}{2} \\ & & \frac{1}{2} & 0 & -1 & -1 \\ & & & 12 & 2 & 2 \\ & \text{(symmetric)} & & & 7 & 7 \\ & & & & & 7 \end{bmatrix} (s^2 = 0.04)$$

and the standard errors of the restricted coefficient estimates are the positive square roots of the diagonal elements.

5.7. $\widehat{\text{slope}}(\text{at } x_1) = 5.69 - \frac{1}{2}\{5.82 + 4.46 + 3.44 - (-0.24 + 3.72)\}$
$+ \{1.72 - (-0.24) - 3.72\}x_1$

$= 0.57 - 1.76x_1 : \text{zero at } x_1 = 0.33$

$\widehat{\text{slope}}(\text{at } x_2) = 0.49 - 1.34x_2 : \text{zero at } x_2 = 0.37$

$\widehat{\text{slope}}(\text{at } x_3) = 2.41 - 7.28x_3 : \text{zero at } x_3 = 0.33$

Along each of the three axes, beginning at $x_i = 0$ and moving toward $x_i = 1$, the surface increases and reaches a maximum at $\widehat{\text{slope}} = 0$. Then the surface decreases as $x_i \to 1$.

Model: $\hat{y}(\mathbf{x}) = 5.74(x_1 + x_2) + 4.46x_3 + 3.62(x_1 + x_2)x_3$

$\widehat{\text{slope}}[\text{at}(x_1 + x_2)] = 4.90 - 7.24(x_1 + x_2) : \text{zero at } x_1 + x_2 = 0.68$

5.9. $\hat{y}(\mathbf{x}, \mathbf{z}) = \underset{(0.12)}{2.86x_1} + \underset{(0.12)}{1.06x_2} + \underset{(0.12)}{0.46x_1z_1} + \underset{(0.12)}{0.14x_2z_1} + \underset{(0.12)}{0.58x_1z_2} + \underset{(0.12)}{0.24x_2z_2}$

Source	d.f.	S.S.	M.S.
Regression	5	9.98	1.996
Residual	6	0.43	0.07
Total	11	10.41	

With this set of data, component 1 appears to produce patties with a firmer average texture than does component 2. Process variables 1 and 2 (represented by z_1 and z_2 respectively in the model) appear to affect in a positive manner the average firmness of patties made from component 1.

5.11. **a.** $\hat{y} = \underset{(0.21)}{9.26W} + \underset{(0.21)}{8.94P} + \underset{(0.21)}{5.78C} - \underset{(0.82)}{3.92WP} + \underset{(0.84)}{0.03WC} - \underset{(0.92)}{6.99PC}$

b. $\hat{y} = \underset{(0.19)}{9.26W} + \underset{(0.19)}{8.94P} + \underset{(0.19)}{5.78C} - \underset{(0.76)}{3.92WP} + \underset{(0.77)}{0.03WC} - \underset{(0.85)}{6.99PC}$
$+ \underset{(0.06)}{0.12\text{day}}$

	Model **a.**			Model **b.**		
Source	d.f.	S.S.	M.S.	d.f.	S.S.	M.S.
Regression	5	27.903	5.581	6	28.232	4.705
Residual	18	1.633	0.091	17	1.304	0.077
Total	23	29.536				

Test of H_0: Day effect = 0 vs H_A: Day effect $\neq 0$

$$F = \frac{(1.633 - 1.304)/1}{1.304/17} = 4.27 < F_{(1,17,0.05)} = 4.45.$$

Do not reject H_0.

c. All blends consisting of at least 60% wheat flour.
All binary blends consisting of wheat flour and peanut flour.

CHAPTER 6

6.1. Using the values of x_1, x_2 and x_3 from Table 6.5,

a. First-degree model:

$$\hat{y}(\mathbf{x}) = \underset{(4.01)}{79.35x_1} + \underset{(7.34)}{72.72x_2} + \underset{(6.01)}{106.53x_3} \qquad C_3 = 8$$

b. First-degree plus inverse terms:

$$\hat{y}(\mathbf{x}) = \underset{(25.96)}{146.68x_1} + \underset{(35.33)}{181.46x_2} + \underset{(14.65)}{109.19x_3} - \underset{(2.92)}{6.23x_1^{-1}} - \underset{(4.59)}{12.64x_3^{-1}} \qquad c_5 = 4$$

c. Second-degree model:

$$\hat{y}(\mathbf{x}) = \underset{(29.23)}{29.07x_1} - \underset{(99.61)}{2.12x_2} - \underset{(61.78)}{7.34x_3} + \underset{(112.81)}{100.73x_1x_2} + \underset{(157.71)}{308.54x_1x_3} + \underset{(291.01)}{321.49x_2x_3} \qquad c_6 = 8$$

Source	d.f.	Model **a** S.S.	d.f.	Model **b** S.S.	d.f.	Model **c** S.S.
Regression	2	50.69	4	70.87	5	66.85
Residual	6	28.54 ($s^2 = 4.76$)	4	8.36 ($s^2 = 2.09$)	3	12.38 ($s^2 = 4.13$)
Total	8	79.23				

6.4. Along each of the three rays $l = 1, 2, 3$, the model is

$$y_{lu} = \mu_l + \beta_l(1 - x_{3u}) + \gamma_l(1 - x_{3u})^2 + \epsilon_u$$

Ray 1 ($x_1 - x_3$ edge)

$$\mathbf{y} = \begin{bmatrix} 49.8 \\ 41.3 \\ 34.7 \\ 28.8 \\ 20.1 \end{bmatrix} \qquad \mathbf{X} = \left[\begin{array}{cc:c} 1 & 1.00 & 1 \\ 1 & 0.75 & 0.5625 \\ 1 & 0.50 & 0.2500 \\ 1 & 0.25 & 0.0625 \\ 1 & 0 & 0 \end{array}\right]$$

Ray 2 ($x_2 - x_3$ edge)

$$\mathbf{y} = \begin{bmatrix} 84.2 \\ 66.0 \\ 52.4 \\ 39.4 \\ 20.1 \end{bmatrix} \qquad \mathbf{X} = \left[\begin{array}{cc:c} 1 & 1 & 1 \\ 1 & 0.75 & 0.5625 \\ 1 & 0.50 & 0.2500 \\ 1 & 0.25 & 0.0625 \\ 1 & 0 & 0 \end{array}\right]$$

Ray 3 (x_3-axis)

$$\mathbf{y} = \begin{bmatrix} 35.8 \\ 32.7 \\ 29.6 \\ 27.9 \\ 26.1 \\ 23.3 \\ 20.1 \end{bmatrix} \qquad \mathbf{X} = \left[\begin{array}{cc:c} 1 & 1 & 1 \\ 1 & 0.8 & 0.64 \\ 1 & 0.6 & 0.36 \\ 1 & 0.5 & 0.25 \\ 1 & 0.4 & 0.16 \\ 1 & 0.2 & 0.04 \\ 1 & 0 & 0 \end{array}\right]$$

	First degree	SSE
Ray 1:	$\hat{y}(\mathbf{x}) = \underset{(0.68)}{20.56} + \underset{(1.11)}{28.76}(1 - x_3)$	2.27
Ray 2:	$\hat{y}(\mathbf{x}) = \underset{(1.56)}{21.46} + \underset{(2.55)}{61.92}(1 - x_3)$	12.19
Ray 3:	$\hat{y}(\mathbf{x}) = \underset{(0.09)}{20.06} + \underset{(0.16)}{15.74}(1 - x_3)$	0.08

	Second degree	SSE
Ray 1:	$\hat{y}(\mathbf{x}) = 20.60 + 28.42(1 - x_3) + 0.34(1 - x_3)^2$	2.28
	(1.01) (4.77) (4.57)	
Ray 2:	$\hat{y}(\mathbf{x}) = 21.23 + 63.75(1 - x_3) - 1.83(1 - x_3)^2$	12.00
	(2.31) (10.93) (10.47)	
Ray 3:	$\hat{y}(\mathbf{x}) = 20.10 + 15.43(1 - x_3) + 0.31(1 - x_3)^2$	0.08
	(0.13) (0.56) (0.53)	

On each ray the drop in SSE from fitting the quadratic model is not significant. Hence the quadratic term is not necessary, that is, do not reject H_0: $\gamma_1 = 0$. However, reject H_0: $\beta_1 = 0$.

6.5. $r_1 = x_1/x_2 \quad r_2 = x_2/x_3$ then $x_1 = \dfrac{r_1 r_2}{1 + r_2 + r_1 r_2} = \dfrac{2z_1 + z_1 z_2 + 2z_2 + 4}{2z_1 + z_1 z_2 + 4z_2 + 10}$

$$x_2 = \frac{r_2}{1 + r_2 + r_1 r_2} = \frac{2z_2 + 4}{2z_1 + z_1 z_2 + 4z_2 + 10}$$

$$x_3 = \frac{1}{1 + r_2 + r_1 r_2} = \frac{2}{2z_1 + z_1 z_2 + 4z_2 + 10}$$

Component settings

$z_1 =$	-1	1	-1	1	$-\sqrt{2}$	$\sqrt{2}$	0	0	0	0	0
$z_2 =$	-1	-1	1	1	0	0	$-\sqrt{2}$	$\sqrt{2}$	0	0	0
$x_1 =$	$\frac{1}{5}$	$\frac{3}{7}$	$\frac{3}{11}$	$\frac{9}{17}$	$\frac{1.17}{7.17}$	$\frac{6.83}{12.83}$	$\frac{1.17}{4.34}$	$\frac{6.83}{15.66}$	$\frac{4}{10}$	$\frac{4}{10}$	$\frac{4}{10}$
$x_2 =$	$\frac{2}{5}$	$\frac{2}{7}$	$\frac{6}{11}$	$\frac{6}{17}$	$\frac{4}{7.17}$	$\frac{4}{12.83}$	$\frac{1.17}{4.34}$	$\frac{6.83}{15.66}$	$\frac{4}{10}$	$\frac{4}{10}$	$\frac{4}{10}$
$x_3 =$	$\frac{2}{5}$	$\frac{2}{7}$	$\frac{2}{11}$	$\frac{2}{17}$	$\frac{2}{7.17}$	$\frac{2}{12.83}$	$\frac{2}{4.34}$	$\frac{2}{15.66}$	$\frac{2}{10}$	$\frac{2}{10}$	$\frac{2}{10}$

or

$x_1 =$	0.200	0.428	0.273	0.529	0.163	0.532	0.270	0.436	0.400	0.400	0.400
$x_2 =$	0.400	0.286	0.545	0.353	0.558	0.312	0.270	0.436	0.400	0.400	0.400
$x_3 =$	0.400	0.286	0.182	0.118	0.279	0.156	0.460	0.128	0.200	0.200	0.200

$$\hat{y}(\mathbf{z}) = \underset{(0.82)}{14.97} + \underset{(0.50)}{0.12z_1} + \underset{(0.50)}{1.06z_2} - \underset{(0.60)}{4.91z_1^2} - \underset{(0.60)}{3.61z_2^2} - \underset{(0.71)}{1.95z_1 z_2}$$

$$= -29.27 + 47.32r_1 + 19.40r_2 - 19.64r_1^2 - 3.61r_2^2 - 3.90r_1 r_2$$

6.7. **a.** $b_1 = \frac{1-s_1}{\Delta_1}[\bar{y}(\mathbf{x}) - \bar{y}(\mathbf{s})] = \frac{0.80}{0.60}[26.4 - 8.0] = 24.53$

$$\widehat{\text{var}}(b_1) = \frac{16}{9}\left[\frac{s^2}{2} + s^2\right] = 7.01 \text{ where } s^2 = 2.63$$

$$b_2 = \frac{0.80}{0.60}[14.6 - 8.0] = 8.80$$

$$b_3 = \frac{0.80}{0.60}[9.0 - 8.0] = 1.33, \qquad \widehat{\text{var}}(b_3) = \frac{16}{9}\left[\frac{s^2}{3} + s^2\right] = 6.23$$

$$b_4 = -4.00$$

$$b_5 = 21.33$$

tests: $t_1^{**} = \frac{b_1}{\sqrt{\widehat{\text{var}}(b_1)}} = \frac{24.53}{\sqrt{7.01}} = 9.26, \qquad t_2^{*} = 3.32$

$t_3 = 0.33, \qquad t_4 = -1.60, \qquad t_5^{**} = 8.55$

$t_{(8,0.005)} = 3.355, \qquad t_{(8,0.025)} = 2.306$

b. Including the response value 8.0 at **s**, the parameter estimates and the first degree model are

$$\hat{y}(\mathbf{x}) = \mathbf{x}'\mathbf{g} = \underset{(2.62)}{29.16x_1} + \underset{(2.62)}{13.43x_2} + \underset{(2.15)}{6.28x_3} + \underset{(2.15)}{0.95x_4} + \underset{(2.15)}{26.28x_5}$$

With these estimates, $g_i = 4.63 + b_i$, for $i = 1, 2$ and $g_i = 4.95 + b_i$, for $i = 3, 4, 5$

$$E_1 = 29.16 - \tfrac{1}{4}[13.43 + 6.28 + 0.95 + 26.28] = 17.43$$

$$E_2 = -2.24$$

$$E_3 = -11.18$$

$$E_4 = -17.84$$

$$E_5 = 13.83$$

These component effects sum to zero and therefore are meaningful only relative to one another, that is, $\Sigma_{i=1}^{5} E_i = 0$. If we compute each effect independently as in **a.**, then the E_i and the b_i are unrelated because the b_i are independent except for the observation value at **s**. If we estimate the six elements of $\mathbf{b}_1 = (b_0, b_1, b_2, b_3, b_4, b_5)'$ simultaneously as in Eq. (6.43) using the matrix $\mathbf{B}_1$ as defined in Eq. (6.44), then the estimates are

$$b_0 = (g_1 + g_2 + g_3 + g_4 + g_5)/5 = 76.1/5 = 15.22$$

$$b_1 = g_1 - b_0 = 29.16 - 15.22 = 13.94$$

$$b_2 = g_2 - b_0 = -1.79$$

$$b_3 = -8.94$$

$$b_4 = -14.27$$

$$b_5 = 11.06$$

and these estimates satisfy the same property as the E_i, that is, $\Sigma_{i=1}^{5} b_i = 0$. The b_i are estimated effects relative to the overall average of the g_i's whereas E_i compares g_i to the average of the other g_j's. In fact, $b_i = \dfrac{(q-1)}{q} E_i$.

Appendix

TABLE A. Percentage points of the t-distribution with ν degrees of freedom

	α										
ν	.45	.40	.35	.30	.25	.125	.05	.025	.0125	.005	.0025
1	0.158	0.325	0.510	0.727	1.000	2.414	6.314	12.71	25.45	63.66	127.3
2	0.142	0.289	0.445	0.617	0.817	1.604	2.920	4.303	6.205	9.925	14.09
3	0.137	0.277	0.424	0.584	0.765	1.423	2.353	3.183	4.177	5.841	7.453
4	0.134	0.271	0.414	0.569	0.741	1.344	2.132	2.776	3.495	4.604	5.598
5	0.132	0.267	0.408	0.559	0.727	1.301	2.015	2.571	3.163	4.032	4.773
6	0.131	0.265	0.404	0.553	0.718	1.273	1.943	2.447	2.969	3.707	4.317
7	0.130	0.263	0.402	0.549	0.711	1.254	1.895	2.365	2.841	3.500	4.029
8	0.130	0.262	0.399	0.546	0.706	1.240	1.860	2.306	2.752	3.355	3.833
9	0.129	0.261	0.398	0.543	0.703	1.230	1.833	2.262	2.685	3.250	3.690
10	0.129	0.260	0.397	0.542	0.700	1.221	1.813	2.228	2.634	3.169	3.581
11	0.129	0.260	0.396	0.540	0.697	1.215	1.796	2.201	2.593	3.106	3.500
12	0.128	0.259	0.395	0.539	0.695	1.209	1.782	2.179	2.560	3.055	3.428
13	0.128	0.259	0.394	0.538	0.694	1.204	1.771	2.160	2.533	3.012	3.373
14	0.128	0.258	0.393	0.537	0.692	1.200	1.761	2.145	2.510	2.977	3.326
15	0.128	0.258	0.392	0.536	0.691	1.197	1.753	2.132	2.490	2.947	3.286
20	0.127	0.257	0.391	0.533	0.687	1.185	1.725	2.086	2.423	2.845	3.153
25	0.127	0.256	0.390	0.531	0.684	1.178	1.708	2.060	2.385	2.787	3.078
30	0.127	0.256	0.389	0.530	0.683	1.173	1.697	2.042	2.360	2.750	3.030
40	0.126	0.255	0.388	0.529	0.681	1.167	1.684	2.021	2.329	2.705	2.971
60	0.126	0.254	0.387	0.527	0.679	1.162	1.671	2.000	2.299	2.660	2.915
120	0.126	0.254	0.386	0.526	0.677	1.156	1.658	1.980	2.270	2.617	2.860
∞	0.126	0.253	0.385	0.524	0.674	1.150	1.645	1.960	2.241	2.576	2.807

ν = degrees of freedom.

[a]Reproduced with permission from *Probability and Statistics in Engineering and Management Science* by W. W. Hines and D. C. Montgomery, The Ronald Press, New York, 1972.

TABLE B. Percentage points of the F-distribution

$F_{.25, \nu_1, \nu_2}$

Degrees of freedom for the numerator (ν_1); degrees of freedom for the denominator (ν_2)

ν_2 \ ν_1	1	2	3	4	5	6	7	8	9	10	12	15	20	24	30	40	60	120	∞
1	5.83	7.50	8.20	8.58	8.82	8.98	9.10	9.19	9.26	9.32	9.41	9.49	9.58	9.63	9.67	9.71	9.76	9.80	9.85
2	2.57	3.00	3.15	3.23	3.28	3.31	3.34	3.35	3.37	3.38	3.39	3.41	3.43	3.43	3.44	3.45	3.46	3.47	3.48
3	2.02	2.28	2.36	2.39	2.41	2.42	2.43	2.44	2.44	2.44	2.45	2.46	2.46	2.46	2.47	2.47	2.47	2.47	2.47
4	1.81	2.00	2.05	2.06	2.07	2.08	2.08	2.08	2.08	2.08	2.08	2.08	2.08	2.08	2.08	2.08	2.08	2.08	2.08
5	1.69	1.85	1.88	1.89	1.89	1.89	1.89	1.89	1.89	1.89	1.89	1.89	1.88	1.88	1.88	1.88	1.87	1.87	1.87
6	1.62	1.76	1.78	1.79	1.79	1.78	1.78	1.78	1.77	1.77	1.77	1.76	1.76	1.75	1.75	1.75	1.74	1.74	1.74
7	1.57	1.70	1.72	1.72	1.71	1.71	1.70	1.70	1.70	1.69	1.68	1.68	1.67	1.67	1.66	1.66	1.65	1.65	1.65
8	1.54	1.66	1.67	1.66	1.66	1.65	1.64	1.64	1.63	1.63	1.62	1.62	1.61	1.60	1.60	1.59	1.59	1.58	1.58
9	1.51	1.62	1.63	1.63	1.62	1.61	1.60	1.60	1.59	1.59	1.58	1.57	1.56	1.56	1.55	1.54	1.54	1.53	1.53
10	1.49	1.60	1.60	1.59	1.59	1.58	1.57	1.56	1.56	1.55	1.54	1.53	1.52	1.52	1.51	1.51	1.50	1.49	1.48
11	1.47	1.58	1.58	1.57	1.56	1.55	1.54	1.53	1.53	1.52	1.51	1.50	1.49	1.49	1.48	1.47	1.47	1.46	1.45
12	1.46	1.56	1.56	1.55	1.54	1.53	1.52	1.51	1.51	1.50	1.49	1.48	1.47	1.46	1.45	1.45	1.44	1.43	1.42
13	1.45	1.55	1.55	1.53	1.52	1.51	1.50	1.49	1.49	1.48	1.47	1.46	1.45	1.44	1.43	1.42	1.42	1.41	1.40
14	1.44	1.53	1.53	1.52	1.51	1.50	1.49	1.48	1.47	1.46	1.45	1.44	1.43	1.42	1.41	1.41	1.40	1.39	1.38
15	1.43	1.52	1.52	1.51	1.49	1.48	1.47	1.46	1.46	1.45	1.44	1.43	1.41	1.41	1.40	1.39	1.38	1.37	1.36
16	1.42	1.51	1.51	1.50	1.48	1.47	1.46	1.45	1.44	1.44	1.43	1.41	1.40	1.39	1.38	1.37	1.36	1.35	1.34
17	1.42	1.51	1.50	1.49	1.47	1.46	1.45	1.44	1.43	1.43	1.41	1.40	1.39	1.38	1.37	1.36	1.35	1.34	1.33
18	1.41	1.50	1.49	1.48	1.46	1.45	1.44	1.43	1.42	1.42	1.40	1.39	1.38	1.37	1.36	1.35	1.34	1.33	1.32
19	1.41	1.49	1.49	1.47	1.46	1.44	1.43	1.42	1.41	1.41	1.40	1.38	1.37	1.36	1.35	1.34	1.33	1.32	1.30
20	1.40	1.49	1.48	1.47	1.45	1.44	1.43	1.42	1.41	1.40	1.39	1.37	1.36	1.35	1.34	1.33	1.32	1.31	1.29
21	1.40	1.48	1.48	1.46	1.44	1.43	1.42	1.41	1.40	1.39	1.38	1.37	1.35	1.34	1.33	1.32	1.31	1.30	1.28
22	1.40	1.48	1.47	1.45	1.44	1.42	1.41	1.40	1.39	1.39	1.37	1.36	1.34	1.33	1.32	1.31	1.30	1.29	1.28
23	1.39	1.47	1.47	1.45	1.43	1.42	1.41	1.40	1.39	1.38	1.37	1.35	1.34	1.33	1.32	1.31	1.30	1.28	1.27
24	1.39	1.47	1.46	1.44	1.43	1.41	1.40	1.39	1.38	1.38	1.36	1.35	1.33	1.32	1.31	1.30	1.29	1.28	1.26
25	1.39	1.47	1.46	1.44	1.42	1.41	1.40	1.39	1.38	1.37	1.36	1.34	1.33	1.32	1.31	1.29	1.28	1.27	1.25
26	1.38	1.46	1.45	1.44	1.42	1.41	1.39	1.38	1.37	1.37	1.35	1.34	1.32	1.31	1.30	1.29	1.28	1.26	1.25
27	1.38	1.46	1.45	1.43	1.42	1.40	1.39	1.38	1.37	1.36	1.35	1.33	1.32	1.31	1.30	1.28	1.27	1.26	1.24
28	1.38	1.46	1.45	1.43	1.41	1.40	1.39	1.38	1.37	1.36	1.34	1.33	1.31	1.30	1.29	1.28	1.27	1.25	1.24
29	1.38	1.45	1.45	1.43	1.41	1.40	1.38	1.37	1.36	1.35	1.34	1.32	1.31	1.30	1.29	1.27	1.26	1.25	1.23
30	1.38	1.45	1.44	1.42	1.41	1.39	1.38	1.37	1.36	1.35	1.34	1.32	1.30	1.29	1.28	1.27	1.26	1.24	1.23
40	1.36	1.44	1.42	1.40	1.39	1.37	1.36	1.35	1.34	1.33	1.31	1.30	1.28	1.26	1.25	1.24	1.22	1.21	1.19
60	1.35	1.42	1.41	1.38	1.37	1.35	1.33	1.32	1.31	1.30	1.29	1.27	1.25	1.24	1.22	1.21	1.19	1.17	1.15
120	1.34	1.40	1.39	1.37	1.35	1.33	1.31	1.30	1.29	1.28	1.26	1.24	1.22	1.21	1.19	1.18	1.16	1.13	1.10
∞	1.32	1.39	1.37	1.35	1.33	1.31	1.29	1.28	1.27	1.25	1.24	1.22	1.19	1.18	1.16	1.14	1.12	1.08	1.00

[a]Adapted with permission from *Biometrika Tables For Statisticians*, Vol. 1, 3rd Edition, by E. S. Pearson and H. O. Hartley, Cambridge University Press, Cambridge, 1966.

$F_{.10,\, \nu_1,\, \nu_2}$

Degrees of freedom for the numerator (ν_1); Degrees of freedom for the denominator (ν_2)

ν_2 \ ν_1	1	2	3	4	5	6	7	8	9	10	12	15	20	24	30	40	60	120	∞
1	39.86	49.50	53.59	55.83	57.24	58.20	58.91	59.44	59.86	60.19	60.71	61.22	61.74	62.00	62.26	62.53	62.79	63.06	63.33
2	8.53	9.00	9.16	9.24	9.29	9.33	9.35	9.37	9.38	9.39	9.41	9.42	9.44	9.45	9.46	9.47	9.47	9.48	9.49
3	5.54	5.46	5.39	5.34	5.31	5.28	5.27	5.25	5.24	5.23	5.22	5.20	5.18	5.18	5.17	5.16	5.15	5.14	5.13
4	4.54	4.32	4.19	4.11	4.05	4.01	3.98	3.95	3.94	3.92	3.90	3.87	3.84	3.83	3.82	3.80	3.79	3.78	3.76
5	4.06	3.78	3.62	3.52	3.45	3.40	3.37	3.34	3.32	3.30	3.27	3.24	3.21	3.19	3.17	3.16	3.14	3.12	3.10
6	3.78	3.46	3.29	3.18	3.11	3.05	3.01	2.98	2.96	2.94	2.90	2.87	2.84	2.82	2.80	2.78	2.76	2.74	2.72
7	3.59	3.26	3.07	2.96	2.88	2.83	2.78	2.75	2.72	2.70	2.67	2.63	2.59	2.58	2.56	2.54	2.51	2.49	2.47
8	3.46	3.11	2.92	2.81	2.73	2.67	2.62	2.59	2.56	2.54	2.50	2.46	2.42	2.40	2.38	2.36	2.34	2.32	2.29
9	3.36	3.01	2.81	2.69	2.61	2.55	2.51	2.47	2.44	2.42	2.38	2.34	2.30	2.28	2.25	2.23	2.21	2.18	2.16
10	3.29	2.92	2.73	2.61	2.52	2.46	2.41	2.38	2.35	2.32	2.28	2.24	2.20	2.18	2.16	2.13	2.11	2.08	2.06
11	3.23	2.86	2.66	2.54	2.45	2.39	2.34	2.30	2.27	2.25	2.21	2.17	2.12	2.10	2.08	2.05	2.03	2.00	1.97
12	3.18	2.81	2.61	2.48	2.39	2.33	2.28	2.24	2.21	2.19	2.15	2.10	2.06	2.04	2.01	1.99	1.96	1.93	1.90
13	3.14	2.76	2.56	2.43	2.35	2.28	2.23	2.20	2.16	2.14	2.10	2.05	2.01	1.98	1.96	1.93	1.90	1.88	1.85
14	3.10	2.73	2.52	2.39	2.31	2.24	2.19	2.15	2.12	2.10	2.05	2.01	1.96	1.94	1.91	1.89	1.86	1.83	1.80
15	3.07	2.70	2.49	2.36	2.27	2.21	2.16	2.12	2.09	2.06	2.02	1.97	1.92	1.90	1.87	1.85	1.82	1.79	1.76
16	3.05	2.67	2.46	2.33	2.24	2.18	2.13	2.09	2.06	2.03	1.99	1.94	1.89	1.87	1.84	1.81	1.78	1.75	1.72
17	3.03	2.64	2.44	2.31	2.22	2.15	2.10	2.06	2.03	2.00	1.96	1.91	1.86	1.84	1.81	1.78	1.75	1.72	1.69
18	3.01	2.62	2.42	2.29	2.20	2.13	2.08	2.04	2.00	1.98	1.93	1.89	1.84	1.81	1.78	1.75	1.72	1.69	1.66
19	2.99	2.61	2.40	2.27	2.18	2.11	2.06	2.02	1.98	1.96	1.91	1.86	1.81	1.79	1.76	1.73	1.70	1.67	1.63
20	2.97	2.59	2.38	2.25	2.16	2.09	2.04	2.00	1.96	1.94	1.89	1.84	1.79	1.77	1.74	1.71	1.68	1.64	1.61
21	2.96	2.57	2.36	2.23	2.14	2.08	2.02	1.98	1.95	1.92	1.87	1.83	1.78	1.75	1.72	1.69	1.66	1.62	1.59
22	2.95	2.56	2.35	2.22	2.13	2.06	2.01	1.97	1.93	1.90	1.86	1.81	1.76	1.73	1.70	1.67	1.64	1.60	1.57
23	2.94	2.55	2.34	2.21	2.11	2.05	1.99	1.95	1.92	1.89	1.84	1.80	1.74	1.72	1.69	1.66	1.62	1.59	1.55
24	2.93	2.54	2.33	2.19	2.10	2.04	1.98	1.94	1.91	1.88	1.83	1.78	1.73	1.70	1.67	1.64	1.61	1.57	1.53
25	2.92	2.53	2.32	2.18	2.09	2.02	1.97	1.93	1.89	1.87	1.82	1.77	1.72	1.69	1.66	1.63	1.59	1.56	1.52
26	2.91	2.52	2.31	2.17	2.08	2.01	1.96	1.92	1.88	1.86	1.81	1.76	1.71	1.68	1.65	1.61	1.58	1.54	1.50
27	2.90	2.51	2.30	2.17	2.07	2.00	1.95	1.91	1.87	1.85	1.80	1.75	1.70	1.67	1.64	1.60	1.57	1.53	1.49
28	2.89	2.50	2.29	2.16	2.06	2.00	1.94	1.90	1.87	1.84	1.79	1.74	1.69	1.66	1.63	1.59	1.56	1.52	1.48
29	2.89	2.50	2.28	2.15	2.06	1.99	1.93	1.89	1.86	1.83	1.78	1.73	1.68	1.65	1.62	1.58	1.55	1.51	1.47
30	2.88	2.49	2.28	2.14	2.03	1.98	1.93	1.88	1.85	1.82	1.77	1.72	1.67	1.64	1.61	1.57	1.54	1.50	1.46
40	2.84	2.44	2.23	2.09	2.00	1.93	1.87	1.83	1.79	1.76	1.71	1.66	1.61	1.57	1.54	1.51	1.47	1.42	1.38
60	2.79	2.39	2.18	2.04	1.95	1.87	1.82	1.77	1.74	1.71	1.66	1.60	1.54	1.51	1.48	1.44	1.40	1.35	1.29
120	2.75	2.35	2.13	1.99	1.90	1.82	1.77	1.72	1.68	1.65	1.60	1.55	1.48	1.45	1.41	1.37	1.32	1.26	1.19
∞	2.71	2.30	2.08	1.94	1.85	1.77	1.72	1.67	1.63	1.60	1.55	1.49	1.42	1.38	1.34	1.30	1.24	1.17	1.00

Table B. (Continued)

$F_{.05, \nu_1, \nu_2}$

Degrees of freedom for the numerator (ν_1); rows: degrees of freedom for the denominator (ν_2)

ν_2 \ ν_1	1	2	3	4	5	6	7	8	9	10	12	15	20	24	30	40	60	120	∞
1	161.4	199.5	215.7	224.6	230.2	234.0	236.8	238.9	240.5	241.9	243.9	245.9	248.0	249.1	250.1	251.1	252.2	253.3	254.3
2	18.51	19.00	19.16	19.25	19.30	19.33	19.35	19.37	19.38	19.40	19.41	19.43	19.45	19.45	19.46	19.47	19.48	19.49	19.50
3	10.13	9.55	9.28	9.12	9.01	8.94	8.89	8.85	8.81	8.79	8.74	8.70	8.66	8.64	8.62	8.59	8.57	8.55	8.53
4	7.71	6.94	6.59	6.39	6.26	6.16	6.09	6.04	6.00	5.96	5.91	5.86	5.80	5.77	5.75	5.72	5.69	5.66	5.63
5	6.61	5.79	5.41	5.19	5.05	4.95	4.88	4.82	4.77	4.74	4.68	4.62	4.56	4.53	4.50	4.46	4.43	4.40	4.36
6	5.99	5.14	4.76	4.53	4.39	4.28	4.21	4.15	4.10	4.06	4.00	3.94	3.87	3.84	3.81	3.77	3.74	3.70	3.67
7	5.59	4.74	4.35	4.12	3.97	3.87	3.79	3.73	3.68	3.64	3.57	3.51	3.44	3.41	3.38	3.34	3.30	3.27	3.23
8	5.32	4.46	4.07	3.84	3.69	3.58	3.50	3.44	3.39	3.35	3.28	3.22	3.15	3.12	3.08	3.04	3.01	2.97	2.93
9	5.12	4.26	3.86	3.63	3.48	3.37	3.29	3.23	3.18	3.14	3.07	3.01	2.94	2.90	2.86	2.83	2.79	2.75	2.71
10	4.96	4.10	3.71	3.48	3.33	3.22	3.14	3.07	3.02	2.98	2.91	2.85	2.77	2.74	2.70	2.66	2.62	2.58	2.54
11	4.84	3.98	3.59	3.36	3.20	3.09	3.01	2.95	2.90	2.85	2.79	2.72	2.65	2.61	2.57	2.53	2.49	2.45	2.40
12	4.75	3.89	3.49	3.26	3.11	3.00	2.91	2.85	2.80	2.75	2.69	2.62	2.54	2.51	2.47	2.43	2.38	2.34	2.30
13	4.67	3.81	3.41	3.18	3.03	2.92	2.83	2.77	2.71	2.67	2.60	2.53	2.46	2.42	2.38	2.34	2.30	2.25	2.21
14	4.60	3.74	3.34	3.11	2.96	2.85	2.76	2.70	2.65	2.60	2.53	2.46	2.39	2.35	2.31	2.27	2.22	2.18	2.13
15	4.54	3.68	3.29	3.06	2.90	2.79	2.71	2.64	2.59	2.54	2.48	2.40	2.33	2.29	2.25	2.20	2.16	2.11	2.07
16	4.49	3.63	3.24	3.01	2.85	2.74	2.66	2.59	2.54	2.49	2.42	2.35	2.28	2.24	2.19	2.15	2.11	2.06	2.01
17	4.45	3.59	3.20	2.96	2.81	2.70	2.61	2.55	2.49	2.45	2.38	2.31	2.23	2.19	2.15	2.10	2.06	2.01	1.96
18	4.41	3.55	3.16	2.93	2.77	2.66	2.58	2.51	2.46	2.41	2.34	2.27	2.19	2.15	2.11	2.06	2.02	1.97	1.92
19	4.38	3.52	3.13	2.90	2.74	2.63	2.54	2.48	2.42	2.38	2.31	2.23	2.16	2.11	2.07	2.03	1.98	1.93	1.88
20	4.35	3.49	3.10	2.87	2.71	2.60	2.51	2.45	2.39	2.35	2.28	2.20	2.12	2.08	2.04	1.99	1.95	1.90	1.84
21	4.32	3.47	3.07	2.84	2.68	2.57	2.49	2.42	2.37	2.32	2.25	2.18	2.10	2.05	2.01	1.96	1.92	1.87	1.81
22	4.30	3.44	3.05	2.82	2.66	2.55	2.46	2.40	2.34	2.30	2.23	2.15	2.07	2.03	1.98	1.94	1.89	1.84	1.78
23	4.28	3.42	3.03	2.80	2.64	2.53	2.44	2.37	2.32	2.27	2.20	2.13	2.05	2.01	1.96	1.91	1.86	1.81	1.76
24	4.26	3.40	3.01	2.78	2.62	2.51	2.42	2.36	2.30	2.25	2.18	2.11	2.03	1.98	1.94	1.89	1.84	1.79	1.73
25	4.24	3.39	2.99	2.76	2.60	2.49	2.40	2.34	2.28	2.24	2.16	2.09	2.01	1.96	1.92	1.87	1.82	1.77	1.71
26	4.23	3.37	2.98	2.74	2.59	2.47	2.39	2.32	2.27	2.22	2.15	2.07	1.99	1.95	1.90	1.85	1.80	1.75	1.69
27	4.21	3.35	2.96	2.73	2.57	2.46	2.37	2.31	2.25	2.20	2.13	2.06	1.97	1.93	1.88	1.84	1.79	1.73	1.67
28	4.20	3.34	2.95	2.71	2.56	2.45	2.36	2.29	2.24	2.19	2.12	2.04	1.96	1.91	1.87	1.82	1.77	1.71	1.65
29	4.18	3.33	2.93	2.70	2.55	2.43	2.35	2.28	2.22	2.18	2.10	2.03	1.94	1.90	1.85	1.81	1.75	1.70	1.64
30	4.17	3.32	2.92	2.69	2.53	2.42	2.33	2.27	2.21	2.16	2.09	2.01	1.93	1.89	1.84	1.79	1.74	1.68	1.62
40	4.08	3.23	2.84	2.61	2.45	2.34	2.25	2.18	2.12	2.08	2.00	1.92	1.84	1.79	1.74	1.69	1.64	1.58	1.51
60	4.00	3.15	2.76	2.53	2.37	2.25	2.17	2.10	2.04	1.99	1.92	1.84	1.75	1.70	1.65	1.59	1.53	1.47	1.39
120	3.92	3.07	2.68	2.45	2.29	2.17	2.09	2.02	1.96	1.91	1.83	1.75	1.66	1.61	1.55	1.55	1.43	1.35	1.25
∞	3.84	3.00	2.60	2.37	2.21	2.10	2.01	1.94	1.88	1.83	1.75	1.67	1.57	1.52	1.46	1.39	1.32	1.22	1.00

$F_{.025,\, \nu_1,\, \nu_2}$

ν_2 \ ν_1	Degrees of freedom for the numerator (ν_1) 1	2	3	4	5	6	7	8	9	10	12	15	20	24	30	40	60	120	∞
1	647.8	799.5	864.2	899.6	921.8	937.1	948.2	956.7	963.3	968.6	976.7	984.9	993.1	997.2	1001	1006	1010	1014	1018
2	38.51	39.00	39.17	39.25	39.30	39.33	39.36	39.37	39.39	39.40	39.41	39.43	39.45	39.46	39.46	39.47	39.48	39.49	39.50
3	17.44	16.04	15.44	15.10	14.88	14.73	14.62	14.54	14.47	14.42	14.34	14.25	14.17	14.12	14.08	14.04	13.99	13.95	13.90
4	12.22	10.65	9.98	9.60	9.36	9.20	9.07	8.98	8.90	8.84	8.75	8.66	8.56	8.51	8.46	8.41	8.36	8.31	8.26
5	10.01	8.43	7.76	7.39	7.15	6.98	6.85	6.76	6.68	6.62	6.52	6.43	6.33	6.28	6.23	6.18	6.12	6.07	6.02
6	8.81	7.26	6.60	6.23	5.99	5.82	5.70	5.60	5.52	5.46	5.37	5.27	5.17	5.12	5.07	5.01	4.96	4.90	4.85
7	8.07	6.54	5.89	5.52	5.29	5.12	4.99	4.90	4.82	4.76	4.67	4.57	4.47	4.42	4.36	4.31	4.25	4.20	4.14
8	7.57	6.06	5.42	5.05	4.82	4.65	4.53	4.43	4.36	4.30	4.20	4.10	4.00	3.95	3.89	3.84	3.78	3.73	3.67
9	7.21	5.71	5.08	4.72	4.48	4.32	4.20	4.10	4.03	3.96	3.87	3.77	3.67	3.61	3.56	3.51	3.45	3.39	3.33
10	6.94	5.46	4.83	4.47	4.24	4.07	3.95	3.85	3.78	3.72	3.62	3.52	3.42	3.37	3.31	3.26	3.20	3.14	3.08
11	6.72	5.26	4.63	4.28	4.04	3.88	3.76	3.66	3.59	3.53	3.43	3.33	3.23	3.17	3.12	3.06	3.00	2.94	2.88
12	6.55	5.10	4.47	4.12	3.89	3.73	3.61	3.51	3.44	3.37	3.28	3.18	3.07	3.02	2.96	2.91	2.85	2.79	2.72
13	6.41	4.97	4.35	4.00	3.77	3.60	3.48	3.39	3.31	3.25	3.15	3.05	2.95	2.89	2.84	2.78	2.72	2.66	2.60
14	6.30	4.86	4.24	3.89	3.66	3.50	3.38	3.29	3.21	3.15	3.05	2.95	2.84	2.79	2.73	2.67	2.61	2.55	2.49
15	6.20	4.77	4.15	3.80	3.58	3.41	3.29	3.20	3.12	3.06	2.96	2.86	2.76	2.70	2.64	2.59	2.52	2.46	2.40
16	6.12	4.69	4.08	3.73	3.50	3.34	3.22	3.12	3.05	2.99	2.89	2.79	2.68	2.63	2.57	2.51	2.45	2.38	2.32
17	6.04	4.62	4.01	3.66	3.44	3.28	3.16	3.06	2.98	2.92	2.82	2.72	2.62	2.56	2.50	2.44	2.38	2.32	2.25
18	5.98	4.56	3.95	3.61	3.38	3.22	3.10	3.01	2.93	2.87	2.77	2.67	2.56	2.50	2.44	2.38	2.32	2.26	2.19
19	5.92	4.51	3.90	3.56	3.33	3.17	3.05	2.96	2.88	2.82	2.72	2.62	2.51	2.45	2.39	2.33	2.27	2.20	2.13
20	5.87	4.46	3.86	3.51	3.29	3.13	3.01	2.91	2.84	2.77	2.68	2.57	2.46	2.41	2.35	2.29	2.22	2.16	2.09
21	5.83	4.42	3.82	3.48	3.25	3.09	2.97	2.87	2.80	2.73	2.64	2.53	2.42	2.37	2.31	2.25	2.18	2.11	2.04
22	5.79	4.38	3.78	3.44	3.22	3.05	2.93	2.84	2.76	2.70	2.60	2.50	2.39	2.33	2.27	2.21	2.14	2.08	2.00
23	5.75	4.35	3.75	3.41	3.18	3.02	2.90	2.81	2.73	2.67	2.57	2.47	2.36	2.30	2.24	2.18	2.11	2.04	1.97
24	5.72	4.32	3.72	3.38	3.15	2.99	2.87	2.78	2.70	2.64	2.54	2.44	2.33	2.27	2.21	2.15	2.08	2.01	1.94
25	5.69	4.29	3.69	3.35	3.13	2.97	2.85	2.75	2.68	2.61	2.51	2.41	2.30	2.24	2.18	2.12	2.05	1.98	1.91
26	5.66	4.27	3.67	3.33	3.10	2.94	2.82	2.73	2.65	2.59	2.49	2.39	2.28	2.22	2.16	2.09	2.03	1.95	1.88
27	5.63	4.24	3.65	3.31	3.08	2.92	2.80	2.71	2.63	2.57	2.47	2.36	2.25	2.19	2.13	2.07	2.00	1.93	1.85
28	5.61	4.22	3.63	3.29	3.06	2.90	2.78	2.69	2.61	2.55	2.45	2.34	2.23	2.17	2.11	2.05	1.98	1.91	1.83
29	5.59	4.20	1.61	3.27	3.04	2.88	2.76	2.67	2.59	2.53	2.43	2.32	2.21	2.15	2.09	2.03	1.96	1.89	1.81
30	5.57	4.18	3.59	3.25	3.03	2.87	2.75	2.65	2.57	2.51	2.41	2.31	2.20	2.14	2.07	2.01	1.94	1.87	1.79
40	5.42	4.05	3.46	3.13	2.90	2.74	2.62	2.53	2.45	2.39	2.29	2.18	2.07	2.01	1.94	1.88	1.80	1.72	1.64
60	5.29	3.93	3.34	3.01	2.79	2.63	2.51	2.41	2.33	2.27	2.17	2.06	1.94	1.88	1.82	1.74	1.67	1.58	1.48
120	5.15	3.80	3.23	2.89	2.67	2.52	2.39	2.30	2.22	2.16	2.05	1.94	1.82	1.76	1.69	1.61	1.53	1.43	1.31
∞	5.02	3.69	3.12	2.79	2.57	2.41	2.29	2.19	2.11	2.05	1.94	1.83	1.71	1.64	1.57	1.48	1.39	1.27	1.00

Degrees of freedom for the denominator (ν_2)

Table B. (Continued)

$F_{.01, \nu_1, \nu_2}$

ν_2 \ ν_1 (Degrees of freedom for the denominator (ν_2) / Degrees of freedom for the numerator (ν_1))	1	2	3	4	5	6	7	8	9	10	12	15	20	24	30	40	60	120	∞
1	4052	4999.5	5403	5625	5764	5859	5928	5982	6022	6056	6106	6157	6209	6235	6261	6287	6313	6339	6366
2	98.50	99.00	99.17	99.25	99.30	99.33	99.36	99.37	99.39	99.40	99.42	99.43	99.45	99.46	99.47	99.47	99.48	99.49	99.50
3	34.12	30.82	29.46	28.71	28.24	27.91	27.67	27.49	27.35	27.23	27.05	26.87	26.69	26.00	26.50	26.41	26.32	26.22	26.13
4	21.20	18.00	16.69	15.98	15.52	15.21	14.98	14.80	14.66	14.55	14.37	14.20	14.02	13.93	13.84	13.75	13.65	13.56	13.46
5	16.26	13.27	12.06	11.39	10.97	10.67	10.46	10.29	10.16	10.05	9.89	9.72	9.55	9.47	9.38	9.29	9.20	9.11	9.02
6	13.75	10.92	9.78	9.15	8.75	8.47	8.26	8.10	7.98	7.87	7.72	7.56	7.40	7.31	7.23	7.14	7.06	6.97	6.88
7	12.25	9.55	8.45	7.85	7.46	7.19	6.99	6.84	6.72	6.62	6.47	6.31	6.16	6.07	5.99	5.91	5.82	5.74	5.65
8	11.26	8.65	7.59	7.01	6.63	6.37	6.18	6.03	5.91	5.81	5.67	5.52	5.36	5.28	5.20	5.12	5.03	4.95	4.86
9	10.56	8.02	6.99	6.42	6.06	5.80	5.61	5.47	5.35	5.26	5.11	4.96	4.81	4.73	4.65	4.57	4.48	4.40	4.31
10	10.04	7.56	6.55	5.99	5.64	5.39	5.20	5.06	4.94	4.85	4.71	4.56	4.41	4.33	4.25	4.17	4.08	4.00	3.91
11	9.65	7.21	6.22	5.67	5.32	5.07	4.89	4.74	4.63	4.54	4.40	4.25	4.10	4.02	3.94	3.86	3.78	3.69	3.60
12	9.33	6.93	5.95	5.41	5.06	4.82	4.64	4.50	4.39	4.30	4.16	4.01	3.86	3.78	3.70	3.62	3.54	3.45	3.36
13	9.07	6.70	5.74	5.21	4.86	4.62	4.44	4.30	4.19	4.10	3.96	3.82	3.66	3.59	3.51	3.43	3.34	3.25	3.17
14	8.86	6.51	5.56	5.04	4.69	4.46	4.28	4.14	4.03	3.94	3.80	3.66	3.51	3.43	3.35	3.27	3.18	3.09	3.00
15	8.68	6.36	5.42	4.89	4.36	4.32	4.14	4.00	3.89	3.80	3.67	3.52	3.37	3.29	3.21	3.13	3.05	2.96	2.87
16	8.53	6.23	5.29	4.77	4.44	4.20	4.03	3.89	3.78	3.69	3.55	3.41	3.26	3.18	3.10	3.02	2.93	2.84	2.75
17	8.40	6.11	5.18	4.67	4.34	4.10	3.93	3.79	3.68	3.59	3.46	3.31	3.16	3.08	3.00	2.92	2.83	2.75	2.65
18	8.29	6.01	5.09	4.58	4.25	4.01	3.84	3.71	3.60	3.51	3.37	3.23	3.08	3.00	2.92	2.84	2.75	2.66	2.57
19	8.18	5.93	5.01	4.50	4.17	3.94	3.77	3.63	3.52	3.43	3.30	3.15	3.00	2.92	2.84	2.76	2.67	2.58	2.49
20	8.10	5.85	4.94	4.43	4.10	3.87	3.70	3.56	3.46	3.37	3.23	3.09	2.94	2.86	2.78	2.69	2.61	2.52	2.42
21	8.02	5.78	4.87	4.37	4.04	3.81	3.64	3.51	3.40	3.31	3.17	3.03	2.88	2.80	2.72	2.64	2.55	2.46	2.36
22	7.95	5.72	4.82	4.31	3.99	3.76	3.59	3.45	3.35	3.26	3.12	2.98	2.83	2.75	2.67	2.58	2.50	2.40	2.31
23	7.88	5.66	4.76	4.26	3.94	3.71	3.54	3.41	3.30	3.21	3.07	2.93	2.78	2.70	2.62	2.54	2.45	2.35	2.26
24	7.82	5.61	4.72	4.22	3.90	3.67	3.50	3.36	3.26	3.17	3.03	2.89	2.74	2.66	2.58	2.49	2.40	2.31	2.21
25	7.77	5.57	4.68	4.18	3.85	3.63	3.46	3.32	3.22	3.13	2.99	2.85	2.70	2.62	2.54	2.45	2.36	2.27	2.17
26	7.72	5.53	4.64	4.14	3.82	3.59	3.42	3.29	3.18	3.09	2.96	2.81	2.66	2.58	2.50	2.42	2.33	2.23	2.13
27	7.68	5.49	4.60	4.11	3.78	3.56	3.39	3.26	3.15	3.06	2.93	2.78	2.63	2.55	2.47	2.38	2.29	2.20	2.10
28	7.64	5.45	4.57	4.07	3.75	3.53	3.36	3.23	3.12	3.03	2.90	2.75	2.60	2.52	2.44	2.35	2.26	2.17	2.06
29	7.60	5.42	4.54	4.04	3.73	3.50	3.33	3.20	3.09	3.00	2.87	2.73	2.57	2.49	2.41	2.33	2.23	2.14	2.03
30	7.56	5.39	4.51	4.02	3.70	3.47	3.30	3.17	3.07	2.98	2.84	2.70	2.55	2.47	2.39	2.30	2.21	2.11	2.01
40	7.31	5.18	4.31	3.83	3.51	3.29	3.12	2.99	2.89	2.80	2.66	2.52	2.37	2.29	2.20	2.11	2.02	1.92	1.80
60	7.08	4.98	4.13	3.65	3.34	3.12	2.95	2.82	2.72	2.63	2.50	2.35	2.20	2.12	2.03	1.94	1.84	1.73	1.60
120	6.85	4.79	3.95	3.48	3.17	2.96	2.79	2.66	2.56	2.47	2.34	2.19	2.03	1.95	1.86	1.76	1.66	1.53	1.38
∞	6.63	4.61	3.78	3.32	3.02	2.80	2.64	2.51	2.41	2.32	2.18	2.04	1.88	1.79	1.70	1.59	1.47	1.32	1.00

Source: Adapted from E. S. Pearson and H. O. Hartley, *Biometrika Tables for Statisticians*, Vol. II, with permission of the Biometrika Trustees.

Bibliography and Index of Authors

The boldface numbers at the end of some of the entries are page numbers in this work.

Becker, N. G. (1968). Models for the response of a mixture. *Journal of the Royal Statistical Society*, B, Vol. 30, pp. 349-358; **227, 228.**

Becker, N. G. (1969). Regression problems when the predictor variables are proportions. *Journal of the Royal Statistical Society*, B, Vol. 31, pp. 107-112; **14, 75.**

Becker, N. G. (1970). Mixture designs for a model linear in the proportions. *Biometrika*, Vol. 57, pp. 329-338.

Becker, N. G. (1978). Models and designs for experiments with mixtures. *Australian Journal of Statistics*, Vol. 20, pp. 195-208.

Bownds, J. M., I. S. Kurotori, and D. R. Cruise (1965). *Notes on simplex models in the study of multi-component mixtures.* U.S. Naval Ordinance Test Station, NAVWEPS Report 8670, NOTS TP 3719, Copy 197, China Lake, California, January, pp. 1-40.

Box, G. E. P. (1952). Multi-Factor Designs of First Order. *Biometrika*, Vol. 39, pp. 49-57; **84.**

Box, G. E. P. and N. R. Draper (1959). A basis for the selectio of a response surface design. *Journal of the American Statistical Association*, Vol. 54, pp. 622-654; **85.**

Box, G. E. P. and C. J. Gardiner (1966). *Constrained designs - part 1, First order designs.* Technical Report No. 89, Department of Statistics, University of Wisconsin, Madison, Wisconsin, pp. 1-14.

Box, G. E. P. and J. S. Hunter (1957). Multifactor experimental designs for exploring response surfaces. *Annals of Mathematical Statistics*, Vol. 28, pp. 195-242.

Box, G. E. P., W. G. Hunter and J. S. Hunter (1978). *Statistics for Experimenters: An Introduction to Design, Data Analysis, and Model Building.* Wiley, New York; **40.**

Box, G. E. P. and K. G. Wilson (1951). On the Experimental Attainment of Optimum Conditions. *Journal of the Royal Statistical Society*, B, Vol. 13, pp. 1-45; **86.**

Claringbold, P. J. (1955). Use of the simplex design in the study of the joint action of related hormones. *Biometrics*, Vol. 11, pp. 174-185; **12, 14.**

Cornell, J. A. (1971). Process variables in the mixture problem for categorized components. *Journal of the American Statistical Association*, Vol. 66, pp. 42-48.

Cornell, J. A. (1973). Experiments with mixtures: A review. *Technometrics*, Vol. 15, pp. 437-455; **14.**

Cornell, J. A. (1975). Some comments on designs for Cox's mixture polynomial. *Technometrics*, Vol 17, pp. 25-35; **51, 239, 260.**

Cornell, J. A. (1977). Weighted versus unweighted estimates using Scheffé's mixture model for symmetrical error variances patterns. *Technometrics*, Vol. 19, pp. 237-247.

Cornell, J. A. (1979). Experiments with mixtures: An update and bibliography. *Technometrics,* Vol. 21, pp. 95-106.

Cornell, J. A. and I. J. Good (1970). The mixture problem for categorized components. *Journal of the American Statistical Association,* Vol. 65, pp. 339-355; **146.**

Cornell, J. A. and L. Ott (1975). The use of gradients to aid in the interpretation of mixture response surfaces. *Technometrics,* Vol. 17, pp. 409-424; **194.**

Cornell, J. A. and J. W. Gorman (1978). On the detection of an additive blending component in multicomponent mixtures. *Biometrics,* Vol. 34, pp. 251-263; **221.**

Cornell, J. A. and A. I. Khuri (1979). Obtaining constant prediction variance on concentric triangles for ternary mixture systems. *Technometrics,* Vol. 21, pp. 147-157.

Cox, D. R. (1971). A note on polynomial response functions for mixtures. *Biometrika,* Vol. 58, pp. 155-159; **54, 237.**

Cruise, D. R. (1966). Plotting the composition of mixtures on simplex coordinates. *Journal of Chemical Education,* Vol. 43, pp. 30-33.

Daniel, C. and F. S. Wood (1971). *Fitting Equations to Data.* Computer Analysis of Multifactor Data for Scientists and Engineers. Wiley, New York; **40, 41.**

Diamond, W. J. (1967). Three dimensional models of extreme vertices designs for four component mixtures. *Technometrics,* Vol. 9, pp. 472-475.

Donelson, D. H. and J. T. Wilson (1960). Effect of the relative quantity of flour fractions on cake quality. *Cereal Chemistry,* Vol. 37, pp. 241-262.

Draper, N. R. and W. E. Lawrence (1965). Mixture designs for three factors. *Journal of the Royal Statistical Society,* B, Vol. 27, pp. 450-465; **14, 75.**

Draper, N. R. and W. E. Lawrence (1965). 473-478; Mixture designs for four factors. *Journal of the Royal Statistical Society,* B, Vol. 27, pp. 473-478; **14, 75.**

Draper, N. R. and R. C. St. John (1977). A mixtures model with inverse terms. *Technometrics,* Vol. 19, pp. 37-46; **212, 214.**

Draper, N. R. and R. C. St. John (1977). Designs in three and four components for mixtures models with inverse terms. *Technometrics,* Vol. 19, pp. 177-130.

Draper, N. R. and H. Smith (1966). *Applied Regression Analysis.* Wiley, New York, **69.**

Drew, B. A. (1967). Experiments with mixtures. *The Minnesota Chemist,* Vol. 19, pp. 4-9.

Farrand, E. A. (1969). Starch damage and alpha-amylase as bases for mathematical models relating to flour-water absorption. *Cereal Chemistry,* Vol. 46, pp. 103-116.

Galil, Z. and J. Kiefer (1977). Comparison of Box-Draper and D-optimum designs for experiments with mixtures. *Technometrics,* Vol. 19, pp. 441-444.

Galil, Z. and J. Kiefer (1977). Comparison of simplex designs for quadratic mixture models. *Technometrics,* Vol. 19, pp. 445-453.

Gorman, J. W. (1966). Fitting equations to mixture data with restraints on compositions. Research and Development Department, American Oil Company, Whiting, Indiana, pp. 1-20. (1970). *Journal of Quality Technology,* Vol. 2, pp. 186-194. **151.**

Gorman, J. W. (1966). Discussion of extreme vertices designs of mixture experiments by R. A. McLean and V. L. Anderson, *Technometrics,* Vol. 8, pp. 455-456.

Gorman, J. W. and J. E. Hinman (1962). Simplex-lattice designs for multicomponent systems. *Technometrics,* Vol. 4, pp. 463-487; **14, 28, 63.**

Graybill, F. A. (1961). *An Introduction to Linear Statistical Models.* Vol. 1, McGraw-Hill, New York.

Hackler, W. C., W. W. Kriegel, and R. J. Hader (1956). Effect of raw-material ratios on absorption of whiteware compositions. *Journal of the American Ceramic Society,* Vol. 39, pp. 20-25.

Hahn, G. J., W. Q. Meeker, and P. I. Feder (1976). The evaluation and comparison of experimental designs for fitting regression relationships. *Journal of Quality Technology,* Vol. 8, pp. 140-157.

Hare, L. B. (1974). Mixture designs applied to food formulation. *Food Technology,* Vol. 28, pp. 50-62.

Hare, L. B. and P. L. Brown (1977). Plotting response surface contours on a three-component mixture space. *Journal of Quality Technology,* Vol. 9, pp. 193-197.

Hare, L. B. (1979). Designs for mixture experiments involving process variables. *Technometrics,* Vol. 21, pp. 159-173.

Henika, R. G. and M. R. Henselman (1969). Stable brew concentrate for whey cysteine breadmaking and methods. *Cereal Science Today,* Vol. 14, pp. 248-253.

Hewlett, P. S. (1969). Measurement of the potencies of drug mixtures. *Biometrics,* Vol. 25, pp. 477-487.

Hewlett, P. S. and R. L. Plackett (1961). Models for quantal responses to mixtures of two drugs. *Quantitative Methods in Pharmacology,* pp. 328-336, North Holland, Amsterdam.

Hines, W. W. and D. C. Montgomery (1972). *Probability and Statistics in Engineering and Management Science.* Ronald, New York.

Healy, W. C. Jr., C. W. Maassen, and R. T. Peterson (1959). *A New Approach to Blending Octanes.* Paper presented at Midyear Meeting of American Petroleum Institute. Division of Refining, New York, NY.

Huor, S. S., E. M. Ahmed, P. V. Rao, and J. A. Cornell (1980). Formulation and sensory evaluation of a fruit punch containing watermelon juice. *Journal of Food Science,* Vol. 45, pp. 809-813.

Kenworthy, O. O. (1963). Factorial experiments with mixtures using ratios. *Industrial Quality Control,* Vol. 19, pp. 24-26.

Kissell, L. T. (1959). A lean-formula cake method for varietal evaluation and research. *Cereal Chemistry,* Vol. 36, pp. 168-175.

Kissell, L. T. (1967), Optimization of white layer cake formulations by a multiple-factor experimental design. *Cereal Chemistry,* Vol. 44, pp. 253-268.

Kissell, L. T. and B. D. Marshall (1962). Multi-factor responses of cake quality to basic ingredient ratios. *Cereal Chemistry,* Vol. 39, pp. 16-30.

Krol, L. H. (1966). Butadiene and isoprene rubber in giant tire treads. *Rubber Chemistry and Technology,* Vol. 39, pp. 452-459.

Kurotori, I. S. (1966). Experiments with mixtures of components having lower bounds. *Industrial Quality Control,* Vol. 22, pp. 592-596; **14**.

Laake, P. (1975). On the optimal allocation of observations in experiments with mixtures. *Scandinavian Journal of Statistics,* Vol. 2, pp. 153-157.

Lambrakis, D. P. (1968). Estimated regression function of the $\{q,m\}$ simplex-lattice design. *Bulletin of Hellenic Mathematical Society,* Athens, Greece, Vol. 9, pp. 13-19.

Lambrakis, D. P. (1968). Experiments with mixtures: A generalization of the simplex-lattice design. *Journal of the Royal Statistical Society,* B, Vol. 30, pp. 123-136; **142**.

Lambrakis, D. P. (1969). Experiments with *p*-component mixtures. *Journal of the Royal Statistical Society,* B, Vol. 30, pp. 137-144.

Lambrakis, D. P. (1969). Experiments with mixtures: An alternative to the simplex-lattice design. *Journal of the Royal Statistical Society,* B, Vol. 31, pp. 234-245.

Lambrakis, D. P. (1969). Experiments with mixtures: Estimated regression function of the multiple-lattice design. *Journal of the Royal Statistical Society,* B, Vol. 31, pp. 276-284.

Lee, H. H. and J. C. Warner (1935). The system biphenyl-bibenzyl-Naphthalene. Nearly ideal binary and ternary systems. *Journal of the American Chemical Society*, **Vol. 37, pp. 318-321.**

Leonpacker, R. M. (1978). *The Ethyl Technique of Octane Prediction.* Ethyl Corporation Petroleum Chemicals Division Report No. RTM-400, Ethyl Corporation, Baton Rouge, LA.

Li, J. C. (1971). Design of experiments with mixtures and independent variable factors. Ph.D. dissertation, Department of Statistics, Rutgers University, New Brunswick, NJ.

MacDonald, I. A. and D. A. Bly (1966). Determination of optimal levels of several emulsifiers in cake mix shortenings. *Cereal Chemistry,* Vol. 43, pp. 571-584.

Mallows, C. L. (1973). Some comments on C_p. **Technometrics, Vol. 15, pp. 661-675; 71.**

Marquardt, D. W. and R. D. Snee (1974). Test statistics for mixture models. *Technometrics,* Vol. 16, pp. 533-537; **151, 159.**

McLean, R. A. and V. L. Anderson (1966). Extreme vertices design of mixture experiments. *Technometrics,* Vol. 8, pp. 447-454; **14, 124, 127.**

Mendieta, E. J., H. N. Linssen, and R. Doornbos (1975). Optimal designs for linear mixture models. *Statistical Neerlandica,* Vol. 29, pp. 145-150.

Morris, W. E. (1975). *The Interaction Approach to Gasoline Blending.* E. I. duPont de Nemours and Co., Inc. Petroleum Chemicals Division Report AM-75-30; **249.**

Morris, W. E. and R. D. Snee (1979). Blending relationships among gasoline component mixtures. *Automotive Technologist,* Vol. 3, pp. 56-61.

Murty, J. S. and M. N. Das (1968). Design and analysis of experiments with mixtures. *Annals of Mathematical Statistics,* Vol. 39, pp. 1517-1539.

Myers, R. H. (1964). Methods for estimating the composition of a three component liquid mixture. *Technometrics,* Vol. 6, pp. 343-356.

Myers, R. H. (1971). *Response Surface Methodology.* Allyn and Bacon, Inc., Boston; **74, 83.**

Narcy, J. P. and J. Renaud (1972). Use of simplex experimental designs in detergent formulation. *Journal of the American Oil Chemists' Society,* Vol. 49, pp. 598-608.

Nigam, A. K. (1970). Block designs for mixture experiments. *Annals of Mathematical Statistics,* Vol. 41, pp. 1861-1869.

Nigam, A. K. (1973). Multifactor mixture experiments. *Journal of the Royal Statistical Society,* B, Vol. 35, pp. 51-66.

Nigam, A. K. (1974). Some designs and models for mixture experiments for the sequential exploration of response surfaces. *Journal of Indian Society of Agricultural Statistics,* Vol. 26, pp. 120-124.

Paku, G. A., A. R. Manson, and L. A. Nelson (1971). Minimum bias estimation in the mixture problem. North Carolina State University Institute of Statistics Mimeo. Series No. 757, Raleigh, NC.

Park, S. H. (1978). Selecting contrasts among parameters in Scheffé's mixture models: Screening components and model reduction. *Technometrics,* Vol. 20, pp. 273-279; **171, 174.**

Pearson, E. S. and H. O. Hartley (1972). *Biometrika Tables for Statisticians.* Vol. 2, Cambridge University Press, Cambridge.

Quenouille, M. H. (1953). *The Design and Analysis of Experiments.* Charles Griffin and Company, London; **12.**

Quenouille, M. H. (1959). Experiments with mixtures. *Journal of the Royal Statistical Society,* B, Vol. 21, pp. 201-202.

Saxena, S. K. and A. K. Nigam (1973). Symmetric-simplex block designs for mixtures. *Journal of the Royal Statistical Society,* B, Vol. 35, pp. 466-472.

Saxena, S. K. and A. K. Nigam (1977). Restricted exploration of mixtures by symmetric-simplex designs. *Technometrics,* Vol. 19, pp. 47-52; **133.**

Scheffé, H. (1958). Experiments with mixtures. *Journal of the Royal Statistical Society,* B, Vol. 20, pp. 344-360; **12, 14, 16, 56.**

Scheffé, H. (1961). Reply to Mr. Quenouille's comments about my paper on mixtures.

Journal of the Royal Statistical Society, B, Vol. 23, pp. 171-172.

Scheffé, H. (1963). The simplex-centroid design for experiments with mixtures. *Journal of the Royal Statistical Society,* B, Vol. 25, pp. 235-263; **14**.

Snee, R. D. (1971). Design and analysis of mixture experiments. *Journal of Quality Technology,* Vol. 3, pp. 159-169.

Snee, R. D. (1973). Techniques for the analysis of mixture data. *Technometrics,* Vol. 15, pp. 517-528; **14, 227**.

Snee, R. D. (1975). Experimental designs for quadratic models in constrained mixture spaces. *Technometrics,* Vol. 17, pp. 149-159; **127, 133**.

Snee, R. D. (1975). Discussion of the use of gradients to aid in the interpretation of mixture response surfaces. *Technometrics,* Vol. 17, pp. 425-430.

Snee, R. D. (1979). Experimental designs for mixture systems with multicomponent constraints. *Communications in Statistics, Theory and Methods,* Vol. A8, No. 4, pp. 303-326.

Snee, R. D. (1979). Experimenting with mixtures. *CHEMTECH,* Vol. 9, pp. 702-710.

Snee, R. D. and D. W. Marquardt (1974). Extreme vertices designs for linear mixture models. *Technometrics,* Vol. 16, pp. 399-408; **127, 130**.

Snee, R. D. and D. W. Marquardt (1976). Screening concepts and designs for experiments with mixtures. *Technometrics,* Vol. 18, pp. 19-29; **178, 187**.

Sommerville, D. M. Y. (1958). *An Introduction to the Geometry of N Dimensions.* Methue, London, 1929; Dover, New York, 1958.

Thompson, W. O. and R. H. Myers (1968). Response surface designs for experiments with mixtures. *Technometrics,* Vol. 10, pp. 739-756; **14**.

Uranisi, H. (1964). Optimum design for the special cubic regression on the q-simplex. *Mathematical Reports,* Vol. 1, pp. 7-12. General Education Department, Kyushu University.

Van Schalkwyk, D. J. (1971). On the design of mixture experiments. Ph.D. dissertation, University of London, London, England.

Vuchkov, I. N., H. A. Yonchev, D. L. Damgaliev, V. K. Tsochev and T. D. Dikova (1978). *Catalogue of Sequentially Generated Designs.* Department of Automation, Higher Institute of Chemical Technology, Sofia, Bulgaria.

Wagner, T. O. and J. W. Gorman (1962). The lattice method for design of experiments with fuels and lubricants. *Application of Statistics and Computers to Fuel and Lubricant Research Problems*, pp. 123-145. Proceedings of a symposium sponsored by the Office of theChief of Ordinance, Department of the Army, San Antonio. TX.

Wagner, T. O. and J. W. Gorman (1963). Fuels, lubricants, engines, and experimental design. *Transactions of the Society of Automotive Engineers,* Vol. 196, pp. 684-701.

Watson, G. S. (1969). Linear regression on proportions. *Biometrics,* Vol. 25, pp. 585-588.

Subject Index